VOLCANISM ON IO

A Comparison with Earth

The most powerful volcanoes in the Solar System are found not on Earth but on Io, a tiny moon of Jupiter. Earth and Io are the only bodies in the Solar System with active, high-temperature volcanoes, but volcanoes on Io are larger and hotter – as well as more violent.

This book, the first dedicated to volcanism on Io, contains the latest results from the *Galileo* mission. In addition to investigating the different styles and scales of volcanic activity on Io, it compares those volcanoes to their contemporaries on Earth. The book also provides background on how volcanoes form and how they erupt, and explains quantitatively how remote-sensing data from spacecraft and telescopes are analyzed to reveal underlying volcanic processes.

This richly illustrated book will be a fascinating reference for advanced undergraduate and graduate students, as well as researchers in planetary sciences, volcanology, remote sensing, and geology.

ASHLEY DAVIES is a volcanologist at the Jet Propulsion Laboratory – California Institute of Technology in Pasadena, California. He was a member of the *Galileo* Near-Infrared Mapping Spectrometer Team, is Principal Investigator on several studies investigating volcanic activity on Io and Earth, and was a recipient of the 2005 NASA Software of the Year Award for his work on spacecraft autonomy.

VOLCANISM ON IO

A Comparison with Earth

ASHLEY GERARD DAVIES

Jet Propulsion Laboratory – California Institute of Technology

CAMBRIDGE
UNIVERSITY PRESS

CAMBRIDGE
UNIVERSITY PRESS

University Printing House, Cambridge CB2 8BS, United Kingdom

Published in the United States of America by Cambridge University Press, New York

Cambridge University Press is part of the University of Cambridge.

It furthers the University's mission by disseminating knowledge in the pursuit of education, learning and research at the highest international levels of excellence.

www.cambridge.org
Information on this title: www.cambridge.org/9781107665408

© A. G. Davies 2007

First published 2007
First paperback edition (with corrections) 2014

A catalogue record for this publication is available from the British Library

Library of Congress Cataloguing in Publication data
Davies, Ashley Gerard, 1961–
Volcanism on Io : a comparison with Earth / by Ashley Gerard Davies.
p. cm.
Includes bibliographical references and index.
ISBN-13: 978-0-521-85003-2
ISBN-10: 0-521-85003-7
1. Io (Satellite) – Volcanism. 2. Io (Satellite) – Volcanoes. 3. Planetary volcanism – Remote sensing.
4. Volcanism. I. Title.
QB404.D38 2007
551.21099925 – dc22 2006037426

ISBN 978-0-521-85003-2 Hardback
ISBN 978-1-107-66540-8 Paperback

Contents

Preface

I have always been fascinated by volcanoes, and especially by Io, a tiny moon that beyond any expectation turned out to be the most volcanically active body in the Solar System. Now that the NASA *Galileo* mission is over and initial data analyses have been completed, this is an appropriate time to assess the "state of the satellite" and review what has been learned about Io over the past few decades.

A fascination with volcanoes is understandable, but I am also inspired to understand, through modeling of volcanic processes, how volcanoes *work*. Such motivation was instilled in me as a post-graduate student by Lionel Wilson and Harry Pinkerton at Lancaster University in the UK.

In this book, therefore, I have endeavored not only to describe what *Galileo* saw, but also to provide the necessary background for understanding the physical, volcanological processes taking place on Io, and to demonstrate how remote-sensing data of volcanic activity can be used to peel back the layers of a planet to reveal interior processes and structure. To put the majestic scale of volcanism on Io into proper context, comparison is made wherever possible with volcanic activity on Earth.

It has taken nearly two years to write this book. Along the way, I have had a great deal of help from friends, family, and colleagues. I thank Simon Mitton of Cambridge University Press, who originally suggested that I write this book and helped me prepare the proposal for Cambridge University Press; Diana Blaney, Nathan Bridges, Julie Castillo-Rogez, Torrence Johnson, Dennis Matson, Dave Pieri, and Glenn Veeder at the Jet Propulsion Laboratory, California Institute of Technology (JPL); Giovanni Leone and Lionel Wilson (Lancaster University); Laszlo Keszthelyi (U.S. Geological Survey [USGS] Astrogeology Branch), Jani Radebaugh (Brigham Young University), and Alison Canning Davies, all of whom reviewed chapters (multiple chapters in some cases); Tammy Becker (USGS) for supplying me with *Voyager* imagery; Paul Geissler (USGS) for a cylindrical projection of the magnificent Io global mosaic he helped create; Ju Zhang for high-resolution images from his volcanic plume modeling; Bonnie Buratti (JPL), who gave me encouragement and sound advice over the years; my editors at Cambridge University Press, Jacqueline Garget and Susan Francis; and assistant editor Helen Morris. I wish also to thank Eleanor Umali of Aptara, Inc., and Dianne Scent for their invaluable help during the copyediting and indexing processes, and Mary Eleanor Johnson for her meticulous proofreading. My deepest thanks I reserve for my wife, Alison, who had to put up with me as I became more and more fixated on volcanoes and Io (well, more fixated than usual), and who was of immeasurable help, carefully reading, editing, and correcting text as it was written. This work I dedicate to her.

Abbreviations

AIDA	Adaptive Image Deconvolution Algorithm
ALI	Advanced Land Imager (*EO-1*)
AO	Adaptive optics
ASE	Autonomous Sciencecraft Experiment (*EO-1*)
ASTER	Advanced Spaceborne Thermal Emission and Reflection Radiometer (*Terra*)
CAI	Calcium-aluminum-rich inclusion
CRB	Columbia River Flood Basalts
CRISM	Compact Reconnaissance Imaging Spectrometers for Mars (*MRO*)
ELT	Extremely Large Telescope
EOS	Earth Observing System
EO-1	*Earth Observing 1*
ESA	European Space Agency
ESO	European Southern Observatory
FLIR	Forward-looking infrared camera, built by FLIR Systems, Inc.
GEM	*Galileo* Europa Mission
GMM	*Galileo* Millennium Mission
GOES	*Geostationary Operational Environmental Satellite*
HGA	High Gain Antenna (*Galileo*)
HiRise	High-Resolution Imaging Science Experiment (*MRO*)
HRIR	High-Resolution Infrared Radiometer (*Nimbus 1*)
HST	Hubble Space Telescope
IRIS	Infrared Radiometer Interferometer and Spectrometer (*Voyager*)
IRTF	Infrared Telescope Facility
ISS	Imaging Sub-System (*Cassini*)
ISS	Imaging Sub-System (*Voyager*)
JPL	Jet Propulsion Laboratory, California Institute of Technology
LGA	Low Gain Antenna (*Galileo*)

LLRI	Long-lived radioisotope
MODIS	Moderate-Resolution Imaging Spectroradiometer (*Aqua* and *Terra*)
MRO	*Mars Reconnaissance Orbiter*
NASA	National Aeronautics and Space Administration
NGAO	Next Generation Adaptive Optics
NIMS	Near-Infrared Mapping Spectrometer (*Galileo*)
OWL	OverWhelmingly Large telescope
PPR	Photo-Polarimeter Radiometer (*Galileo*)
PPS	Photo-Polarimeter (*Voyager*)
SLRI	Short-lived radioisotope
SSI	Solid State Imaging experiment (*Galileo*)
SWIR	Short-wavelength infrared
TM	Thematic Mapper (*Landsat*)
TMT	Thirty-Meter Telescope
UVS	Ultra-Violet Spectrometer (*Voyager*)
UVS-EUVS	Ultra-Violet/Extreme Ultra-violet Spectrometer (*Galileo*)
VLT	Very Large Telescope
WFPC2	Wide-Field Planetary Camera 2 (Hubble)

Note: Spacecraft names are in italic.

Reproduction permissions

Figures 1.9a and 1.9b are from "Io" by Douglas B. Nash, Michael H. Carr, Jonathan Gradie, Donald M. Hunten, and Charles F. Yoder in *Satellites* by Joseph A. Burns and Mildred Shapley Matthews, editors. © 1986 The Arizona Board of Regents. Reprinted by permission of the University of Arizona Press.

Figure 5.4 is from "Dynamics and thermodynamics of volcanic eruptions: implications for the plumes on Io" by Susan Werner Kieffer in *Satellites of Jupiter* by David Morrison, editor. © 1982 The Arizona Board of Regents. Reprinted by permission of the University of Arizona Press.

The following material is reprinted with permission from Elsevier:

Plate 9e from *Icarus*, 169, Radebaugh, J. *et al.*, Observations and temperatures of Io's Pele Patera from *Cassini* and *Galileo* spacecraft images, 65–79, © 2004;

Plate 7b from *Icarus*, 169, Lopes, R. *et al.*, Lava lakes on Io: Observations of Io's volcanic activity from *Galileo* NIMS during the 2001 fly-bys, 140–174, © 2004;

Figures 3.2, 7.9, 7.10, and Plate 7a from *Icarus*, 148, Davies, A. G. *et al.*, Silicate Cooling Model Fits to *Galileo* NIMS Data of Volcanism on Io, 211–225, © 2000;

Figures 2.5, 7.5, 7.13, 7.14, and 8.5 from *Icarus*, 124, Davies, A. G., Io's Volcanism: Thermo-Physical Models of Silicate Lava Compared with Observations of Thermal Emission, 45–61, © 1996;

Figures 7.15a, 7.15b, 7.16, and Plates 6d and 13a from *Icarus*, 184, Davies, A. G. *et al.*, The pulse of the volcano: the discovery of episodic volcanism at Prometheus on Io, 460–477, © 2006;

Figures 15.2a to 15.2c based on *Icarus*, 169, Keszthelyi, L. P. *et al.*, A post-*Galileo* view of Io's interior, 271–286, © 2004;

Figure 15.4a from *Icarus*, 169, Schenk, P. *et al.*, Shield volcano topography and the rheology of lava flows on Io, 98–110, © 2004;

Introduction

Volcanism: *the manifestation at the surface of a planet or satellite of internal thermal processes through the emission at the surface of solid, liquid, or gaseous products.*
Peter Francis (1993), *Volcanoes: A Planetary Perspective*

Few geological phenomena inspire as much awe as a volcanic eruption. Eruptions are, quite frankly, extremely exciting to watch and experience. Could there then be any more exciting place to a volcanologist than the jovian moon Io (Plate 1), which has more active volcanoes per square kilometer than anywhere else in the Solar System? Io is the only body in the Solar System other than the Earth where current volcanic activity can be witnessed on such a wide scale. As a result of this high level of volcanic activity, Io has the most striking appearance of any planetary satellite.

The detection of an umbrella-shaped plume extending high above the surface of Io was the most spectacular discovery made by NASA's *Voyager* spacecraft during their encounters at Jupiter; in fact, it was one of the most important results from NASA's planetary exploration program. The discovery of active extraterrestrial volcanism meant that Earth was no longer the only planetary body where the surface was being reworked by volcanoes. With this exciting discovery a revolution in planetary sciences began, leaving behind the perception of planetary satellites as geologically dead worlds, where any dynamic process had been damped down into extinction over geologic time (billions of years).

The two *Voyager* spacecraft would continue through the Solar System on a grand tour of the gas-giant planets, passing Saturn, Uranus, and Neptune. More volcanic activity, this time cryovolcanic in nature, was discovered by *Voyager* on the neptunian moon Triton (Smith *et al.*, 1989; Kirk *et al.*, 1995). The *Cassini* spacecraft, which at the time of this writing was orbiting Saturn, has detected anomalies in the atmosphere of Titan that may indicate ongoing volcanic activity (Sotin *et al.*, 2005) and active volcanism on Enceladus (Porco *et al.*, 2006). These planetary satellites are active, dynamic worlds. This realization began at Io.

This book is divided into six sections. Section 1 (Chapters 1 to 3) deals chronologically with Io observations and discoveries, which began at the very dawn of telescope-based astronomy in the seventeenth century; covers the discoveries made by technological masterpieces – the two *Voyager* spacecraft; and examines other studies of Io after *Voyager* and up to the arrival of the *Galileo* spacecraft at Jupiter at the end of 1995.

Sections 2 and 3 provide essential background for understanding observations of volcanic activity. The dynamic nature of Io having been established, just why Io and Earth are so volcanically active today is examined in detail in Section 2. Chapter 4 summarizes current theories of formation and evolution of Earth and Io and the root causes of volcanism. Chapter 5 then describes the genesis, properties, and behavior of magmas and the effects of dissolved volatiles, common on Io, on magma behavior.

Section 3 covers how volcanoes on Io and Earth are studied using remote-sensing techniques and how mathematical models yield a quantitative understanding of volcanic activity. As noted by Peter Cattermole in his excellent book *Planetary Volcanism* (Cattermole, 1996), the exploration of volcanoes throughout the Solar System follows principles that are well known and fundamental to geology. For the specific case of understanding observations of volcanic activity from remote-sensing platforms, such investigations can be broken down into three stages.

The first stage consists of the acquisition of data, ideally at as high temporal, spatial, and spectral resolutions as possible. Visible-wavelength image data reveal volcanic plumes, lava flows, and surface morphology. Infrared data yield surface temperatures. Reflectance spectra are analyzed to reveal, or at least constrain, composition. Observations at shorter wavelengths, in the ultraviolet, detect molecular gas transitions. Accordingly, Chapter 6 describes the remote-sensing techniques available for studying Io's volcanism.

The second stage of investigation involves the formation of a theory, based on all available data, of the physical processes involved that might explain the observations. From theory, a mathematical expression of the physical process is often developed. Chapter 7 describes those models with a particular focus on understanding volcanism on Io.

The third stage of investigation consists of applying the models to data. To be an accurate representation of the process taking place, any model should be able to predict subsequent physical conditions. Once the accuracy of a model is established, it becomes a valuable tool for data interpretation. Chapter 8 describes how different volcanic eruption styles can be identified from measurements of thermal emission.

Section 4 (Chapters 9 to 14) examines *Galileo*'s discoveries and looks at volcanic activity at individual locations: a tour of the volcano bestiary. *Galileo* revealed many different styles of volcanic activity taking place on Io, both effusive and explosive in nature. As on Earth, differences in magma composition, gas content, and tectonic setting play a role in the surface expression of volcanic activity. Where possible, Section 4 looks at individual volcanic centers, examines what was seen by *Galileo*, and then quantifies the volcanic processes using the techniques in Section 3. Volcanism on Io produces features that are familiar to terrestrial volcanologists: lava flows, lava lakes and ponds, pyroclastic deposits, and interactions between hot lava and surface volatiles. Where possible, ionian eruptions are quantitatively compared with similar styles of activity on Earth.

On Io, we observe volcanic activity on a scale that would be quite catastrophic on Earth, as well as processes that may have been extinct on Earth for millions and, in some cases, billions of years. By watching how eruptions in the extreme ionian environment evolve, we gain insight into similiar terrestrial eruptions both today and in Earth's distant past. Resulting hypotheses and mathematical models, developed to explain extinct processes, can then be tested against new data of the process in action.

Every chapter in Section 4 highlights a different facet of Io's volcanism. Chapter 9 assesses the major discoveries made by *Galileo* and how the satellite had changed since *Voyager*. The next four chapters deal with very different expressions of volcanism at five locations:

- At Pele, an active, persistent lava lake;
- At Pillan and Tvashtar Paterae, rapidly emplaced lava fountains and voluminous flows;
- At Prometheus, an extensive insulated flow field emplaced over 16 years; and
- At Loki Patera, a quiescent, periodically overturning lava sea covering over $20\,000$ km^2.

These are mostly familiar scenarios to a volcanologist, seen in many locations on Earth, but on Io they are on a vastly different scale. In the case of Loki Patera, there is no terrestrial analogue of this size, and there may never have been.

Different analyses performed at each location highlight a different technique of interpreting data of volcanic processes. In all cases, spectral signature constrains, to some extent, the lava emplacement or exposure mechanism. At Pele, the steadiness of wavelength of peak thermal emission is an important clue to the nature of activity. At Pillan, the modeling of variability in discharge rate yields clues to the emplacement of large, voluminous flows on Earth and Mars. At Prometheus, modeling the variability of effusion rate yields a picture of crustal structure and magma supply from a magma chamber. The temperature distribution at Loki Patera, derived from fitting data with thermal emission models, reveals the mechanism by which the patera is being resurfaced.

Chapter 14 looks at other ionian volcanoes and eruptions that exhibit other interesting facets of volcanic activity.

Section 5 summarizes, in a global context, Io's geomorphology (Chapter 15), volcanic plumes (Chapter 16), and hot spots (Chapter 17), including assessment of the role played by volcanism as a medium for transporting heat on Earth and Io.

Section 6 takes a broad look at Io after the *Galileo* mission, assesses what has been learned, and identifies the important questions raised after *Galileo* (Chapter 18). The prospects for future observations of Io that may answer these questions are covered in Chapter 19.

This book includes two appendices. Appendix 1 contains the locations of volcanic hot spots identified on Io. Appendix 2 contains maps of Io showing the locations of all named features.

Section 1

Io, 1610 to 1995: Galileo to *Galileo*

1

Io, 1610–1979

This chapter reviews the history of Io observations up to and including the *Voyager* encounters. The material in this chapter is drawn primarily from *Satellites of Jupiter* (Morrison, 1982) and *Satellites* (Burns and Matthews, 1986), both published by the University of Arizona Press, and *Time-Variable Phenomena in the Jovian System*, NASA Special Publication 494 (Belton *et al.*, 1989).

1.1 Io before *Voyager*

The study of Io dates from the very beginning of telescope-based astronomy. In his observation notes for January 7, 1610, Galileo Galilei wrote: "when I was viewing the heavenly bodies with a spyglass, Jupiter presented itself to me; and because I had prepared a very excellent instrument for myself I perceived that beside the planet there were three little stars, small indeed, but very bright." Subsequent observations revealed a fourth "little star."

Thus were discovered the Galilean satellites, named by Simon Marius (a contemporary of Galileo) Io, Europa, Ganymede, and Callisto. Subsequently, little attention was paid to the satellite system except as a means of measuring the speed of light (Roemer's method). Not until the nineteenth century did physical observations become important. In 1805, Laplace used the orbital resonant properties to estimate satellite masses. New refracting telescopes at Lick and Yerkes measured satellite sizes, and bulk densities were obtained.

The development of photographic and photoelectric techniques in the early twentieth century led to the determination of satellite light curves, proved that all four Galilean satellites were in Jupiter-synchronous rotation (Stebbins, 1927; Stebbins and Jacobson, 1928), and led to the discovery of prograde and retrograde groups of smaller outer jovian satellites. Since the latter half of the twentieth century, larger telescopes and modern techniques of photometry, spectro-photometry, and polarimetry have been used to determine colors, albedoes, and more accurate values of sizes and densities. Io was found to have the reddest surface in the Solar System (this work is summarized by Harris [1961]).

The occultation of a star by Io in 1971 yielded the first high-precision radius, 1818 km (Taylor *et al.*, 1971). Spectroscopy soon showed that water was a major component of the surface material of Ganymede and Europa (Pilcher *et al.*, 1972; Fink *et al.*, 1973) but was absent (down to the 1% level) on Io, already marked as being anomalous in the jovian system by its high albedo (0.6), red color, and post-eclipse brightening (first noticed by Binder and Cruikshank [1964]). In 1971, Sodium D line emission from Io was discovered (Brown, 1974) and a cloud of neutral sodium in the vicinity of Io was mapped (Trafton *et al.*, 1974; Trafton, 1975a), which led to the discovery of a potassium cloud (Trafton, 1975b). These features appeared to have their genesis on Io. The sputtering of material from the surface of Io by bombardment of charged particles trapped in the intense jovian magnetic field was proposed as the removal mechanism (Matson *et al.*, 1974).

The first spacecraft observations of the jovian system were made by *Pioneer 10* in 1973. The primary task of *Pioneer 10*, apart from proving the feasibility of deep-space missions, was the measurement of fields and particles in the spacecraft's environment. Particular attention was focused on the jovian magnetosphere, the most powerful planetary magnetosphere in the Solar System. Although few satellite measurements were made, improved values for satellite masses were calculated from analysis of *Pioneer*'s trajectory through the jovian system (Anderson *et al.*, 1974). With a more accurate radius determination of 1815 km (Davies, 1982), the bulk density of Io was calculated to be 3540 kg m^{-3}, very similar to that of the Moon and indicating a composition dominated by silicates. A radio occultation of the spacecraft revealed that Io had an ionosphere or possibly an extended atmosphere (Kliore *et al.*, 1974, 1975). *Pioneer 10* also detected an extended hydrogen cloud near the orbit of Io (Judge and Carlson, 1974) as well as a high-energy flux tube linking Io to Jupiter. This flux tube was later found to be a control on Jupiter's decametric radio emission (Dessler and Hill, 1979; Desch, 1980). In December, 1974, *Pioneer 11* obtained a low-spatial-resolution image of Io from high above the north pole, showing that the northern hemisphere had some low-albedo areas.

In the wake of *Pioneer*, considerable attention was focused on the spectrum of Io. Wamsteker *et al.* (1974) were among those who noted a strong similarity between the spectrum of Io and that of sulphur, and it was proposed that sulphur was abundant on the surface of Io. Ionized sulphur emission from Io was discovered by Kupo *et al.* (1976). Spectral observations and laboratory work refined the Io spectrum and revealed a strong ultraviolet absorption feature, although the cause of this feature remained unknown until *Voyager* and laboratory work identified sulphur dioxide on the surface of Io.

Pre-*Voyager* spectral work is reviewed by Johnson and Pilcher (1977), and the reader is also directed to Cruikshank *et al.* (1977), Nash and Fanale (1977), Cruikshank *et al.* (1978), Pollack *et al.* (1978), and Fink *et al.* (1978).

The strong absorption feature in Io's spectrum at 4.1 microns (μm), observed by Cruikshank *et al.* (1978) and Pollack *et al.* (1978), was identified from laboratory studies as sulphur dioxide in the form of a frost or adsorbate (Fanale *et al.*, 1979; Smythe *et al.*, 1979). Again, no water absorption bands were seen in telescope observations, demonstrating that Io's surface was water-ice-free and therefore very different from the water-rich surfaces of the other Galilean satellites.

1.2 Prediction of volcanic activity

Even prior to *Voyager*, it was evident from ground-based instruments that Io had unusual far-infrared photometry and radiometry, with higher brightness temperatures at 10 μm than at 20 μm (Morrison *et al.*, 1972) and unusual thermal inertia as Io emerged from eclipse (e.g., Hansen, 1973; Morrison and Cruikshank, 1973). These observations were difficult to interpret in the context of Io's being a dead, inactive world.

Just before the *Voyager 1* encounter with Io in March, 1979, a notable discovery was made. Witteborn *et al.* (1979) announced that an intense, temporary brightening at 2 to 5 μm in the infrared had been observed, which they explained as an isolated surface area at a temperature of $\approx$600 K (on a planet where the peak daytime temperature is $\approx$130 K). Another hint of Io's dynamic nature came from Nelson and Hapke (1978), who suggested fumarolic activity as a possible mechanism for producing short-chain sulphur allotropes on Io's surface to explain features in Io's spectrum.

Only a few days before *Voyager 1* observed Io (and demonstrating exquisite timing), Peale *et al.* (1979) published an epochal paper on the heating of Io by tidal forces, in which they predicted "widespread and recurrent volcanism."

Voyager was to prove their theory spectacularly correct.

1.3 *Voyager* to Jupiter

Before the spinning *Pioneer 10* spacecraft had even reached Jupiter, development had already started on its successor. *Voyager* was based on the more sophisticated three-axis-stabilized *Mariner* spacecraft developed by NASA's Jet Propulsion Laboratory (JPL) in Pasadena, California. At launch, *Voyager* was eight times the weight of *Pioneer 10*, requiring NASA's most powerful available booster, the Titan IIIE, plus a Centaur upper stage to send it on its way to Jupiter. The *Voyager* project was developed as a multi-planet investigation of at least ten years' duration, for visiting

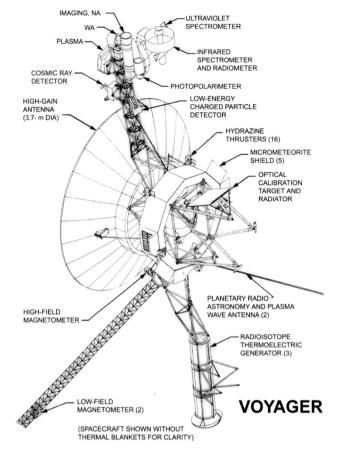

Figure 1.1 The fully deployed *Voyager* spacecraft. Courtesy of NASA.

Jupiter and Saturn and, hopefully, Uranus and Neptune as well: the long cherished "Grand Tour."

Each *Voyager* spacecraft (Figure 1.1) weighed 815 kg in full deployment and carried three classes of instruments designed to collect information about the physical and chemical natures and radiation environments of 15 or more planetary bodies. The first group of instruments was mounted on a stabilized scan platform, a feature absent from the *Pioneer* design, which was essential to compensate for the speed of the spacecraft as it passed close to planets and their satellites. These instruments were two television cameras (imaging sub-system [ISS], not to be confused with the instrument with the same acronym on *Cassini*), one wide- and the other, narrow-angle; an infrared radiometer interferometer and spectrometer (IRIS); an ultra-violet spectrometer (UVS) to assess gas composition; and a photo-polarimeter (PPS) to measure molecular hydrogen, methane, and ammonia in planetary atmospheres.

Table 1.1 Voyager *Galilean satellite encounters*

Satellite	Closest approach (km)	Best resolution (km/line pair)
Voyager 1		
Io	20 570	1[a]
Europa	733 800	33
Ganymede	114 700	2
Callisto	126 400	2
Voyager 2		
Io	1 129 900	20
Europa	205 700	4
Ganymede	62 100	1
Callisto	214 900	4

From Stone and Lane (1979a, 1979b).
[a] Best Io resolution limited by image smear.

Another set of instruments measured the strength of magnetic fields and the energies of charged particles: a plasma detector, a low-energy particle detector, two solid-state cosmic ray telescopes, and four magnetometers.

Finally, two 10-m antennas were used for radio astronomy experiments. The radio communications system was also used to sound planetary atmospheres and ionospheres by transmitting through them. All data were transmitted back to Earth at 23 W (watts) of power at rates up to 115.2 kB/s, to be received by stations of the Deep Space Network. Data were assembled, collated, and transmitted to JPL.

Voyager 1 was launched on September 5, 1977, 16 days after *Voyager 2* (which was on a slower trajectory), and began substantive monitoring of the jovian system on January 6, 1979. Its closest approach to Jupiter was at a distance of 348 890 km on March 5, 1979. *Voyager 2* arrived at Jupiter 18 weeks later and encountered Jupiter at a perijove of 721 670 km on July 9, 1979 (Stone and Lane, 1979a, 1979b).

At Jupiter, the two *Voyagers* were on complementary trajectories. *Voyager 1* made close fly-bys of Io, Ganymede, and Callisto after perijove, making a south polar passage by Io (which would confirm the existence of the flux tube linking Io to Jupiter). *Voyager 2* made close approaches to Europa, Ganymede, and Callisto, but not Io (see Table 1.1), before perijove on a trajectory that would swing the spacecraft onto a course not only toward Saturn, but Uranus and Neptune as well.

Voyager 1 passed by Io at a periapsis of 20 570 km (Table 1.1) and obtained images of moderate to high resolution (0.5–5 km/line pair) covering ≈40% of the surface (see Strom *et al.* [1981]). Table 1.2 shows the number of images taken as a function of resolution. The highest resolution images taken by *Voyager 1* were of the southern hemisphere of Io. *Voyager 2* took only low-spatial-resolution images of Io.

Table 1.2 *Total number of* Voyager *images of Io as function of image resolution*

Resolution/(km/line pair)	<0.5	0.5–2	2–5	5–20	>20
Voyager 1	33	96	75	61	278
Voyager 2	–	–	–	200	54

From Schaber (1982).

1.4 Discovery of active volcanism

The images that *Voyager* returned of Io (Plate 2) revealed a world the likes of which had never been seen before. There were no impact craters, a ubiquitous feature of every other solid planetary and satellite surface in the Solar System that had been observed at resolutions where such features could be resolved. Io was multi-colored, with black, white, yellow, and red units. Broad plains were dotted with vents and calderas, and features of volcanic origin were identified, including long dark flows, broad dark flows, light flows (Smith *et al.*, 1979c), and reddish and white ring deposits. Isolated mountains reared up from the plains. This new, unexpected Io was full of surprises, the first of which was the discovery of active volcanism.

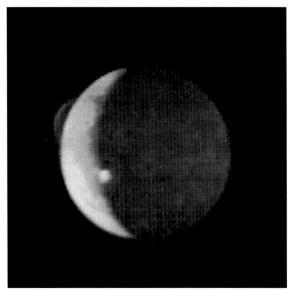

Figure 1.2 Discovery of volcanism on Io as seen by *Voyager 1* in an image obtained on March 8, 1979 (Morabito *et al.*, 1979). Two plumes are visible: the 260-km-high Pele plume on the limb and the plume at Loki illuminated on the terminator. (North is down.) Courtesy of NASA.

Table 1.3 Voyager 1 *and* Voyager 2 *selected plume observations*

Plume/site	Filter	Frame no.	Coordinates		Height[a] (km)	Width[a] (km)
			Long. W	Lat.		
1 Pele	C	16368.28[c]	256.8°	19.4°S	298	1 214
1 Pele	UV	16368.50	256.8°	19.4°S	312	1 164
2 Loki	UV	20513.48[d]	305.3°	19.0°N	382	–
2 Loki	C	20621.33	305.3°	19.0°N	148	445
3 Prometheus	C	16377.48	153.0°	2.9°S	77	272
4 Volund	V	16382.17	177.0°	21.5°N	98	–
4 Volund	UV	16382.23	177.0°	21.5°N	96	–
5 Amirani	C	16372.36	118.7°	27.2°N	114	184
5 Amirani	UV	16372.50	118.7°	27.2°N	137	387
6 Maui	C	16375.28	122.4°	18.9°N	68	343
6 Maui	C	20621.33	122.4°	18.9°N	76	185[b]
7 Marduk	B	16389.21	209.7°	27.9°S	90	200[b]
7 Marduk	UV	20608.01	209.7°	27.9°S	55	230[b]
8 Masubi	C	20641.52	52.7°	45.2°S	64	177
9 Loki	C	16375.34	300.6°	16.9°N	16	–
9 Loki	C	20621.33	300.6°	16.9°N	35	–

From Strom *et al.* (1981).
Filters: C, clear; UV, ultra-violet; V, violet; B, blue, with wide-angle camera.
[a] Corrected for distance from limb.
[b] Uncorrected for limb position.
[c] Image numbers beginning with 1 taken by *Voyager 1*.
[d] Image numbers beginning with 2 taken by *Voyager 2*.

The first two volcanic eruption plumes were discovered by Morabito *et al.* (1979) in an image of Io (Figure 1.2) taken at 13:28 GMT on March 8, 1979, at a range of 4.5 million km. As part of the program to determine both the trajectory of the spacecraft and the ephemerides of the five innermost jovian satellites, the image was enhanced to determine the position of two faint stars. In doing so, a dim cloud of less than 10% of Io's brightness was revealed above the satellite's eastern limb, and another plume was seen catching the light on the terminator. When compared with the position of landforms observed on the satellite's surface, the former cloud was found to lie over, or nearly over, a 1400-km-diameter, heart-shaped feature centered at ≈250°W longitude and 30°S latitude that had already independently been identified as being of volcanic origin. This was the plume from the volcano Pele. In total, the two *Voyager* spacecraft discovered nine active plumes, which were named after deities and mythological characters associated with volcanoes, fire, and mayhem in general: Pele, Loki (two plumes, one of which is the terminator plume in Figure 1.2), Prometheus, Volund, Amirani, Maui, Marduk, and Masubi (see Table 1.3).

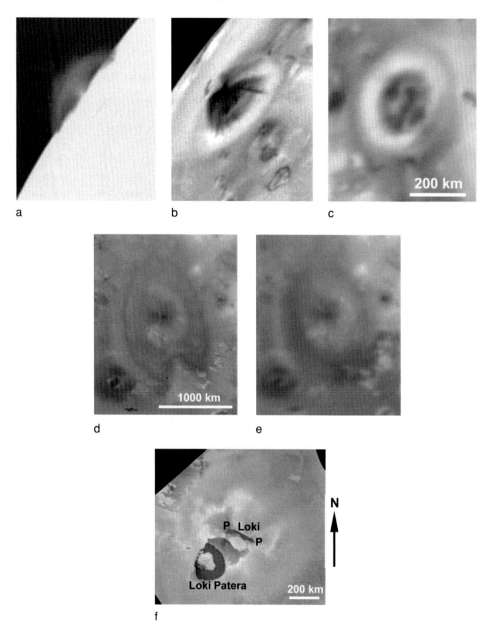

a   b   c

d   e

f

Figure 1.3 Plume and deposit classes as seen by *Voyager*: (a) and (b) The Prometheus plume is 75 km high, and structure in the plume implies the presence of some optically thick material. (c) The resulting bright annular surface deposits are approximately 300 km across. (d) The Pele plume laid down a deposit over 1200 km along the north–south axis and was active during the *Voyager 1* encounter. The unusual shape may have been caused by a vent obstruction impeding flow to the south. (e) By *Voyager 2*, the plume deposit had changed shape and the plume was not seen. It may be that the plume was mostly gas during this encounter, making direct observation difficult. (f) Loki, just north of the 200-km-diameter Loki Patera, was the site of two diffuse plumes (P), erupting from the ends of a possible fissure, that laid down irregular deposits. Courtesy of NASA.

Figure 1.4 Between *Voyager 1* and *2* encounters, eruptions took place at Surt and Aten Paterae resulting in deposits similar to those at Pele. The absence of plumes at these sites led to the classification of Io's plumes into two main classes (McEwen and Soderblom, 1983). Pele-type plumes were larger, were from short-duration eruptions, were from hotter sources, and were dominated by sulphur, hence the darker deposits. Pele proved to have greater longevity than proposed by this model. Prometheus-type plumes were persistent, relatively small, and dominated by SO_2. Courtesy of NASA.

When *Voyager 2* arrived on the scene four months after *Voyager 1*, some obvious changes had taken place (Smith *et al.*, 1979b). The largest plume, Pele, had ceased erupting or activity had reached a level so meager as to remain undetected. Nevertheless, the pattern of Pele's ejecta blanket had been significantly altered (Figure 1.3). Six of the plumes *Voyager 1* had discovered were still erupting in very much the same way, and two (both in the vicinity of the Loki hot spot) had increased in size. There was also evidence to suggest that other eruptions on the scale of Pele had taken place at Surt and Aten Paterae (Figure 1.4) (McEwen and Soderblom, 1983), where giant Pele-like deposits had been laid down.

The volcanic plumes were best viewed against the background of space, where they showed up bright in contrast, but some were also seen against the planetary surface, where they appeared quite dark when compared with the average surface brightness. *Voyager 1* observed about 45% of the surface of Io in the limb regions, at a resolution of 10 km/line pair or better, and discovered eight plumes (Smith *et al.*, 1979c). In an attempt to place limits on the extent and duration of this type

Table 1.4 *Io hot spots identified from Voyager 1 IRIS data*

Feature	Long. (W)	Lat.	Area[a] (km²)	Temp[a] (K)	Total emitted power (×10¹² W) = % Io total[b]	Power emitted at 5 μm (W)	Power emitted at 12 μm (W)	Power emitted at 20 μm (W)	% of total hot-spot output
Loki Patera	309°	12°N	1 385	450	3.2	2.77×10^{11}	1.56×10^{11}	7.73×10^{10}	8.3
"			45 996	245	9.4	4.37×10^{10}	5.22×10^{11}	1.82×10^{12}	24.1
Babbar Patera[c]	272°	40°S	88	322	0.1	1.39×10^{9}	3.27×10^{9}	1.06×10^{7}	0.1
"			16 286	175	0.9	1.41×10^{8}	2.59×10^{10}	4.53×10^{11}	2.2
Ulgen Patera[c]	289°	40°S	616	355	0.6	2.23×10^{10}	3.27×10^{10}	1.87×10^{10}	1.4
"			19 113	191	1.4	6.55×10^{8}	5.41×10^{10}	5.82×10^{11}	3.7
Svarog Patera[c]	268°	48°S	2 827	221	0.4	7.49×10^{8}	1.88×10^{10}	8.92×10^{10}	1.0
Pele	256°	18°S	113	654	1.2	1.68×10^{11}	3.23×10^{10}	1.32×10^{8}	3.0
"			20 106	175	1.1	1.74×10^{8}	3.20×10^{10}	5.62×10^{11}	2.7
Amaterasu Patera	307°	38°N	5 000	283	1.8	2.30×10^{10}	1.10×10^{11}	2.14×10^{11}	4.7
Amirani/Maui	120°	22°N	531	395	0.7	4.36×10^{10}	4.03×10^{10}	1.61×10^{10}	1.9
"			7 543	200	0.7	5.10×10^{8}	2.83×10^{10}	2.34×10^{11}	1.8
nr. NW Colchis	208°	31°N	254	385	0.3	1.73×10^{10}	1.78×10^{10}	2.82×10^{9}	0.8
"			9 503	165	0.4	3.03×10^{7}	9.99×10^{9}	2.45×10^{11}	1.0
Creidne Patera	344°	53°S	1 257	231	0.2	5.85×10^{8}	1.06×10^{10}	3.50×10^{10}	0.5
nr. Nemea Planum	330°	81°S	908	225	0.1	3.03×10^{8}	6.65×10^{9}	2.18×10^{10}	0.3
Mazda Paterae[d]	313°	9°S	60	420	0.1	7.61×10^{9}	5.51×10^{9}	3.05×10^{5}	0.3
"			65 000	160	1.4	1.20×10^{8}	5.44×10^{10}	1.68×10^{12}	3.5

Mbali Patera[d]	7°	31°S	2 000	159	0.04	3.31×10^6	1.60×10^9	4.29×10^{10}	0.1
"			2	574	0.01	1.60×10^9	4.25×10^8	<1	0.03
Viracocha Patera[d]	281°	61°S	6 600	195	0.4	3.08×10^8	2.12×10^{10}	1.98×10^{11}	1.1
Daedalus Patera[d]	275°	19°N	35 000	182	1.6	5.70×10^8	7.26×10^{10}	1.02×10^{12}	4.1
Aten Patera[d]	310°	48°S	4 400	191	0.3	1.51×10^8	1.25×10^{10}	1.26×10^{11}	0.7
Mihr/Gibil/Kibero Paterae[d]	294°–306°	15°–17°N	70 000	238	11.6	4.70×10^{10}	6.87×10^{11}	2.69×10^{12}	29.7
Nusku Patera/ELT[d,e]	5°	65°S	2 500	170	0.1	1.33×10^7	3.25×10^9	5.96×10^{10}	0.2
Creidne/Nusku Paterae[d]	5°–344°	53°–66°S	370	354	0.3	1.31×10^{10}	1.95×10^{10}	6.91×10^9	0.8
"			11 000	155	0.2	1.14×10^7	7.23×10^9	2.68×10^{11}	0.5
Unnamed active center[d]	≈334°	≈12°N	7 000	186	0.4	1.60×10^8	1.67×10^{10}	2.01×10^{22}	0.9
ELT[d]	280°	75°S	2 800	175	0.1	2.42×10^7	4.46×10^9	6.99×10^{10}	0.3
ELT[d]	50°	75°S	10	417	0.02	1.21×10^9	8.99×10^8	<1	0.0
"			86 000	63	0.1	<1	7.02×10^5	8.77×10^{11}	0.2
Total					**39.0**				**100**

[a] Temperatures and areas from Pearl and Sinton (1982), except for Amaterasu (Pearl, 1985) and hot spots discovered by McEwen et al. (1996): see note d.

[b] Total Io thermal output $= 10^{14}$ W (Veeder et al., 1994).

[c] Ulgen, Babbar, and Svarog were combined into one hot spot due to IRIS footprint location and size by McEwen et al. (1989). The shading of the % of total hot-spot output column denotes this combination.

[d] Hot spots identified in IRIS data by McEwen et al. (1992; 1996). Power output per unit area determined using $\sigma(T^4 - 130^4)$ for areas where $T > 130$ K.

[e] ELT = eroded layered terrain (also called eroded layered plain).

of volcanic activity, the *Voyager 2* imaging system observed Io intermittently over a five-day period, including a continuous eight-hour "volcano watch." *Voyager 2* observed 80% of the surface on the limb, which with the *Voyager 1* observations brought the total coverage to nearly 100%. All plumes greater than 100 km in height that were active during the encounters were probably recorded in this way; most were seen several times (Smith *et al.*, 1979c). No plume activity was observed at high latitudes, nor were any polar ice caps found. Indeed, the poles of Io were dark and reddish (Dollfus, 1975; Murray, 1975; Graham and Hapke, 1986).

1.5 IRIS and volcanic thermal emission

The *Voyager* IRIS observed approximately 30% of the surface of Io at resolutions between 70 km and 700 km, with the best resolution at high southern latitudes (Pearl and Sinton, 1982). *Voyager 2* was too far from Io to take any usable IRIS readings.

IRIS obtained useful spectra from wavelengths of 4 μm to 55 μm (Hanel *et al.*, 1979). The fields of view appeared on the Io surface as circular footprints ranging from 250 km to 1000 km in diameter. Uncertainties in the pointing of the *Voyager* scan platform resulted in possible errors in footprint locations on the surface of about half a footprint diameter when concurrent imaging was available, and larger possible errors otherwise (McEwen *et al.*, 1985). Probably the most valuable contribution IRIS made during the Jupiter system encounter was the identification of individual hot spots associated with areas of intense volcanism on Io (Smith *et al.*, 1979c). Major hot spots identified in IRIS data are shown in Table 1.4.

Co-registering IRIS and imaging data showed that some of these hot spots were relatively dark-floored calderas (albedo <0.3) with temperatures less than 400 K, although a small area >600 K at the base of the Pele plume was identified. Detailed modeling of these data showed that Pele IRIS data could be best fitted with a three-component fit. The highest temperature was at least 654 K, with a very small area (Figure 1.5) (Pearl and Sinton, 1982). This was the highest temperature derived from IRIS data and hinted at the possible presence of silicate volcanism.

IRIS also detected sulphur dioxide in the Loki plume (Pearl *et al.*, 1979a), determining the driving volatile for at least some of the plumes.

Hot spots and plumes were soon assigned names and associated, where possible, with a surface feature. The low spatial resolution of IRIS meant that in some cases determining the location of the thermal emission was difficult, with multiple sources in the IRIS field of view. Loki was a case in point. The Loki plumes issued from the ends of what appeared to be a fissure, and this was logically the source of the thermal emission; however, just to the south of the fissure and within the IRIS field of view was Loki Patera, which had the appearance of a huge, 200-km-diameter lava lake and another likely source of the thermal emission detected by IRIS.

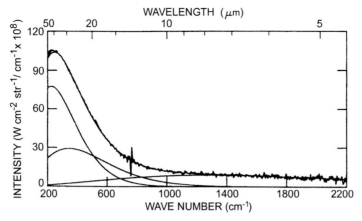

Figure 1.5 *Voyager 1* IRIS spectrum of the Pele region. The smooth curve fitted to the data is the sum of three temperatures and fractions of the IRIS field of view at 114 K (0.899), 175 K (0.100), and 654 K (5.77×10^{-4}) by Pearl and Sinton (1982). IRIS was not sensitive to smaller areas at hotter temperatures. A temperature of 654 K is too hot for liquid sulphur to be stable on the surface. The feature at 12 μm is an artifact.

1.6 Io: the view after *Voyager*

1.6.1 Surface features

The *Voyager* spacecraft revealed an Io that was unlike any other planetary surface known at the time, with a morphology and composition dominated by volcanism. This was a decade before the *Magellan* mission would reveal the volcanism-dominated surface of Venus. Not a single impact crater was seen even in the highest resolution images (in which a crater 1 km to 2 km in diameter could have been identified). Instead, over 100 caldera-like depressions up to 200 km in diameter were identified (Smith *et al.*, 1979a). These features were named "paterae" (see Figures 1.6 and 1.7). Some paterae and escarpments were 1–3 km deep (Arthur, 1981; Schaber, 1982; Davies and Wilson, 1987) and some had complex radiating flows. Smith *et al.* (1979c) concluded that Io was being resurfaced by volcanic processes at a prodigious burial rate, later estimated to be 0.1 cm year^{-1} to 10 cm year^{-1} (Johnson *et al.*, 1979; Johnson and Soderblom, 1982). This explained the absence of impact craters, which are being rapidly buried.

A geologic map covering about a third of Io's surface was created from high- and moderate-resolution *Voyager 1* images (Masursky *et al.*, 1979; Schaber, 1980, 1982). The wide array of landforms and albedo features was divided into three main broad categories: mountains, plains, and vent regions (Carr *et al.*, 1979; Smith *et al.*, 1979c; Schaber, 1982).

Voyager saw mountains as high as 9 km and more than 200 km across, such as Haemus Montes (Figure 1.7). Such high relief pointed to a composition

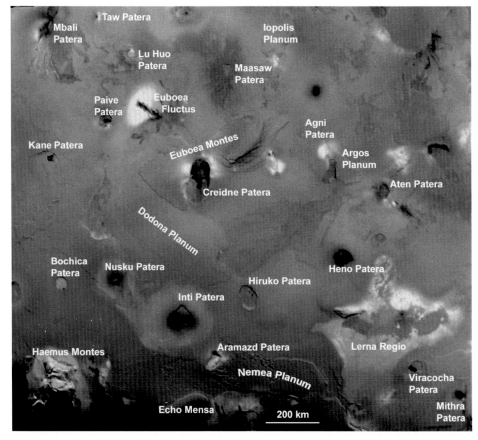

Figure 1.6 One of the best images of Io obtained by *Voyager 1* of the south polar region from longitude 270°W to 60°W and latitude 30°S to 85°S. The image contains a plethora of surface features and different morphologies including shield-like volcanoes with nested craters and radiating flows; other deep, steep-walled paterae set into flat plains; eroded plains; and jagged mountains, indicative of a predominantly silicate lithosphere. *Voyager* Wide Angle camera image FDS 16392.39. Courtesy of NASA.

predominantly of silicate material (Clow and Carr, 1980). The surface of the mountains appeared to be tectonically disrupted, with the mountain blocks exhibiting massive fracturing. The mountains north of Creidne Patera had lobate scarps that resembled large landslides, which may have been fault-controlled. Some mountains were possibly volcanic in origin, surrounded in some cases by bright aureoles (Haemus and Boösaule Montes, for example). Others were interpreted in the 1980s as exposures of silicate lithosphere that elsewhere were covered by sulphur deposits (Carr, 1986).

Plains were divided into three classes: inter-vent plains, layered plains, and eroded layered plains (also called eroded layered terrains). Multi-colored units (in exaggerated color images) ranged in hue from yellow and orange to red-brown.

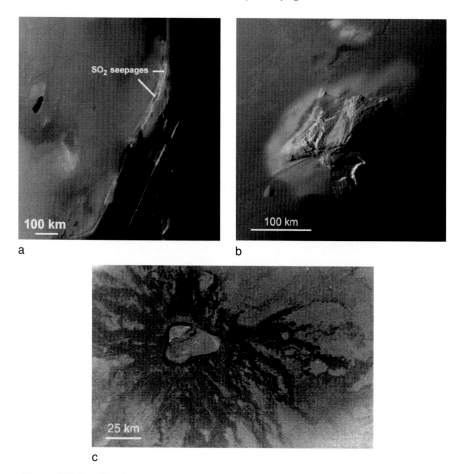

Figure 1.7 Details of selected features on the surface of Io as imaged by *Voyager 1*: (a) south polar region scarps in layered plains with seepages of SO$_2$; (b) Haemus Montes, a block of mountains ≈9 km high; (c) Maasaw Patera, ≈25 km × 35 km in size and ≈2 km deep, has a nested caldera similar to some terrestrial and martian volcanoes. Courtesy of NASA.

Inter-vent plains made up ≈40% of the mapped area and had relatively smooth surfaces, with an albedo that was regionally consistent. In the highest-resolution images, the plains were seen to contain abundant low relief scarps, both straight and sinuous (Figure 1.7). The inter-vent plains unit was thought to be composed of stratified materials of volcanic plume fallout, smaller-scale fumarole deposits, and material emplaced effusively from volcanic vents.

Layered plains appeared as extensive, flat plains with boundary scarps ranging in height from 500 m to 1700 m. This terrain was most markedly developed near the south pole. Layered plain boundaries were thought to be temporary erosion boundaries with the main erosion agent being sulphur dioxide (McCauley *et al.*, 1979). Sulphur dioxide supply was possibly controlled by local tectonic forces.

Also in the southern polar region, but covering only a small fraction (<1%) of the mapped area, were the eroded layered plains. These formed isolated, smooth, flat-topped mesas and were thought to be the remnants of layered plains. One postulated mechanism for their formation was through re-volatilization of sulphur dioxide, which led to the collapse of the plains unit. This area was found to be a high heat-flow area, which provided a driving mechanism for the erosion theory. An alternative formation mechanism was the removal of material by flows, possibly through massive outpouring of liquid sulphur dioxide or sulphur eroding through thick layers rich in sulphur dioxide frost. The debris from such an event would be rapidly buried by plume deposits.

1.6.2 Volcanic vents

Paterae

Volcanic vents included paterae, sources of radial lava flows, centers of bright halos, and dark, roughly circular markings that were poorly illuminated or resolved. None of the observed paterae on Io appeared to be of impact origin: because of the high resurfacing rate, such craters are rapidly buried. Some paterae have multiple-level floor and wall structures (e.g., Maasaw Patera; see Figure 1.7), indicating multiple eruption episodes and looking very similar to the nested summit calderas of Olympus Mons on Mars and the Hawaiian volcanoes Mauna Loa and Kilauea (Figure 15.1). The morphology of paterae, with steep walls more than 3 km high (e.g., Schaber, 1982), indicated that the supporting material was considerably stronger than pure sulphur – which would be molten at depths of paterae floors if Io had a very high thermal gradient (Clow and Carr, 1980; Carr, 1986). This pointed to a predominantly silicate crustal composition, perhaps interbedded silicates and silicate/sulphur-rich mixtures. Many paterae floors were dark, which led to much speculation about the nature of the floors: could they be lakes of dark liquid sulphur (e.g., Nelson *et al.*, 1983)?

Lava flows

Lava flows were classified into two main classes. Pit crater flows appeared to be massive flow fields. Some were traced for more than 700 km, indicating either very high eruption rates (Wilson and Head, 1983) or very fluid lava – or both. More sinuous "shield crater flows" emanated from paterae. The most striking of these were the Ra Patera flows (Figure 1.8). Erupting from a caldera on the summit of a 2000-m-high volcano, these narrow, sinuous flows, more than 200 km long, changed

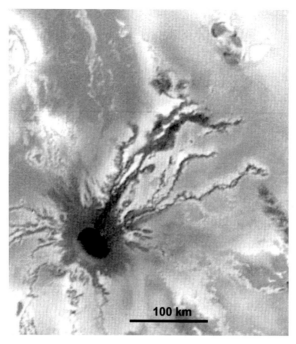

Figure 1.8 Ra Patera (325° W, 8° S) imaged by *Voyager 1*. The field of lava flows at Ra Patera (width, 250 km; area, 250 000 km^2) showed color changes along the flows that suggested the presence of sulphur-rich lava. The Ra Patera shield is low, $\approx$1 km high, with slopes of only $\approx$0.3°, shallower than terrestrial shield volcanoes (typically $\approx$6°), but similar to those found on lunar mare (Schenk *et al.*, 1997). Courtesy of NASA.

color along their length. This color change was interpreted as being the result of sulphur allotropes of different colors forming as the temperature of the liquid sulphur dropped (Sagan, 1979; Pieri *et al.*, 1984). Sulphur flows appeared to be common on Io (e.g., Carr, 1985). The color interpretation was strongly challenged (Young, 1984) on the basis of both sulphur chemistry and the response of the *Voyager* imaging instruments. Reprocessing of *Voyager* images by Alfred McEwen of Arizona University shifted Io's bright colors toward the ultraviolet (to shorter wavelengths): to the eye, Io became predominantly pastel yellow (see Plate 1 for a *Galileo* image of Io). However, Smythe and Nelson (1985) noted that if sulphur allotropes were quenched, then the color–temperature relationship could be preserved. Bright auras around these and other flows were interpreted as frosts mobilized by the heat of the lava flow.

The bright halos, which had the appearance of condensed vapor deposits, sometimes emanated from fissures. Others had no discernible source. Dark areas were often poorly illuminated or were observed at low resolution and were also thought to be volcanic vents.

1.6.3 Plumes

From *Voyager* observations, Io's plumes were categorized into two main classes (McEwen and Soderblom, 1983), named after Pele and Prometheus, the sites that typified the different styles of activity. Pele-type plumes were thought to be eruptions of short duration, laying down reddish deposits in a broad annulus. A high eruption velocity was necessary to create the ≈300-km-high plumes. The temperature of the Pele hot spot, ≈650 K, was the temperature of low-viscosity black liquid sulphur. The dark deposits laid down were then rich in sulphur. These plumes were modeled as sulphur-rich material erupting from a reservoir in the crust where sulphur was heated by contact with silicate magma (Kieffer, 1982).

Prometheus-type plumes (Figures 1.3a–c) were smaller, about 60 km to 100 km in height (implying an eruption velocity of ≈0.5 km s^{-1}), with deposits up to 200 km in diameter. Because all the Prometheus-type plumes seen during the first *Voyager* encounter were active during the second encounter, a minimum lifetime of two years was ascribed to this type (McEwen and Soderblom, 1983). The more persistent Prometheus-type plumes were modeled as resulting from sulphur dioxide in a reservoir being heated by contact with liquid, low-viscosity sulphur (Kieffer, 1982).

The placement of plumes into "Prometheus" or "Pele" classes was not always straightforward. A third type of plume, named after Loki, appeared to be a hybrid of the Prometheus and Pele classes. It was proposed that Loki may have started as a Pele-type eruption and then transitioned to a Prometheus-type eruption.

1.7 Summary

The view of Io after *Voyager* was of a highly volcanic world, with potentially hundreds of active or recently active centers. This high level of volcanism was the result of tidal heating. The surface was very young: Io was being resurfaced at a prodigious rate. Most of the heat flow came from volcanic hot spots. The most dramatic manifestation of volcanism was the volcanic plumes, of which there were two main classes. The crust was determined to be predominantly silicate in composition, based on strength arguments, and it was realized that if Io had been active at this level for a long period of geologic time, then potentially the entire crust may have been recycled multiple times (Carr, 1986).

The tidal heating model proposed by Peale *et al.* (1979) postulated the dissipation of heat in a liquid layer (the aesthenosphere) below the lithosphere, from which buoyant silicate magma would rise until the upper lithosphere – rich in sulphur and sulphur dioxide – was reached. The heat from the silicates then drove the sulphur and sulphur dioxide volcanism seen on the surface.

In the aftermath of *Voyager*, major questions about Io's volcanism remained.

Composition

What was the main driver of surface volcanism – sulphur or silicates? For individual hot spots, what were the temperature and area distributions, which were poorly constrained by IRIS? The answers to these questions would constrain the structure of the upper kilometers of the lithosphere. If sulphur was the dominant lava, then the surface of Io would consist of landforms dominated by sulphur and sulphur dioxide in thick layers above solid silicates (see Nash *et al.*, 1986). If dominated by silicate lavas, then only relatively thin layers of sulphur and sulphur dioxide, interbedded with silicates, would be present. In both cases, the sulphur-rich materials would be rapidly recycled. What therefore were the compositions, styles, and emplacement modes of effusive volcanism? And globally, how were plumes and hot spots distributed, and what did this tell us about Io's interior?

Volcanic volatiles

How stable were sulphur allotropes in an ionian environment and what would their colors be? Did sulphur exist in its pure form, as metastable allotropes, or in compounds? What other materials were present? What was the form of sulphur dioxide on the surface – frost or adsorbate? What were the main drivers of the volcanic plumes – sulphur or sulphur dioxide, or a mixture of these or other materials? Was the source of the sodium in the Io torus volcanic?

Temporal nature of volcanic activity

How did volcanic activity change with time on local and global scales? Variability was poorly constrained by data from the two *Voyager* encounters.

Heating

What was the non-volcanic surface heat flow? How was heat partitioned between volcanic and non-volcanic sources? Globally, was heat volcanically vented constantly or cyclically? Regarding the internal heating of Io, questions remained about the linkage between tidal heating and global heat loss.

Internal structure

What was Io's moment of inertia, which would constrain the internal distribution of mass? Did Io have a magnetic field, indicative of a large, conducting, partially molten core? Where was tidal energy being dissipated within Io, and how would this influence volcanic activity? The internal structure of Io and the makeup of the lithosphere as envisioned at the end of the *Voyager* epoch are shown in Figure 1.9.

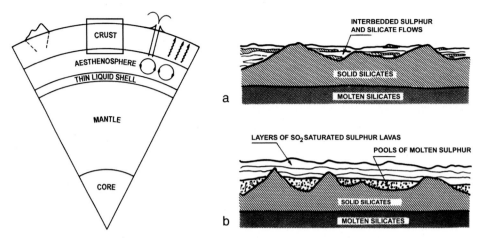

Figure 1.9 *Voyager*-epoch internal and near-surface structure of Io (not to scale). The major global structural units include an outer shell consisting of crust and aesthenosphere, a liquid layer where tidal energy is dissipated, a lower mantle, and an iron-rich core. Convection carries heat to the base of the lithosphere (crust) (Nash *et al.*, 1986). A predominantly silicate-composition lithosphere is needed to support large mountain blocks. Also shown are possible structures of the upper lithosphere (again, not to scale) as proposed by Kieffer (1982). (a) Sulphur compounds are interbedded with silicate lavas, the result of a mixture of silicate and sulphur volcanism. (b) Sulphur and sulphur compounds overlie hot silicates, resulting in a mostly sulphur-compound-rich surface. The need to support steep caldera walls pointed to a model more in tune with the top (silicate-rich) model (Clow and Carr, 1980). © 1986 the Arizona Board of Regents. Reproduced by permission of the University of Arizona Press.

The *Voyager* spacecraft continued through the rest of the Solar System, past Saturn, Uranus, and Neptune and into interstellar space. Back on Earth, the follow-up mission to Jupiter was already being planned. In deference to the great Italian astronomer, it was named *Galileo*.

2

Between *Voyager* and *Galileo*: 1979–1995

After the *Voyager* encounters, analysis of the data continued. More hot spots were identified in IRIS data (see McEwen *et al.*, 1996; also see Appendix 1). Early post-*Voyager* models of Io incorporated a sulphur-rich crust with an ocean of sulphur beneath (e.g., Sagan, 1979; Smith *et al.*, 1979b, 1979c). The sulphur was kept liquid and mobile through interaction with hot or molten silicates within the crust (e.g., Lunine and Stevenson, 1985). This model was supported by the abundance of sulphur compounds on Io's surface and sulphur in the Io plasma torus and neutral clouds, and by the fact that silicates had not been detected on the surface. The occasional eruption of silicate lava was still possible, but Io's volcanism was dominated by sulphur lavas.

2.1 Silicate volcanism on Io?

The case for ubiquitous silicate activity was made by Carr (1986), who argued on the basis of crustal strength that the "sulphur ocean" postulated to lie close to the surface was unlikely, based on the strength of the crust needed to support the steep walls of many paterae, where temperatures in excess of 650 K would literally melt the caldera walls. Heat transport from Io's interior to the surface was via silicate volcanism. To support the observed mountains, Io required a predominantly silicate lithosphere at least as thick as the mountains were high (>15 km in some cases).

The high temperatures associated with active silicate volcanism (in excess of 1000 K) were far from the minds of the scientists and engineers who designed the Infrared Radiometer Interferometer and Spectrometer (IRIS). Although the highest reliable temperature derived from IRIS data was 650 K at Pele (Pearl and Sinton, 1982) (Figure 1.5), it was from data obtained in the lowest signal-to-noise range of the instrument (i.e., the range with the highest uncertainties).

To test the silicate hypothesis, Carr developed a mathematical model of thermal emission from an active silicate lava flow and, by fitting model output to *Voyager*

IRIS data, showed that Loki's thermal emission was not inconsistent with a long-lived eruption of silicate lava with a liquidus temperature of ≈1400 K (typical of basalt). Carr realized that if Io's eruptions were indeed silicate in nature then, to produce the large observed thermal fluxes, the volumes of erupted material at individual eruption sites on Io were greater than contemporary eruptions on Earth. Most paterae appeared to be simply inserted into flat plains. Conversely, volcanic calderas on Earth and Mars are found at the summit of volcanic edifices created by the piling up of lava flows and pyroclastic deposits to form Mauna Loa- or Kilauea-like shield volcanoes. The general absence of such edifices, with the exception of Io's enigmatic shield-volcano-like mountains (*tholi*), suggests that ionian silicate lavas are of relatively low viscosity, allowing fluid emplacement. Eruptions additionally have to be of sufficient volumetric flux to allow resulting flows to spread over great distances before cooling effects halt them.

Carr's lava cooling model, as noted by Howell (1997), did not fit IRIS data of thermal emission from Loki in the range 5 to 10 μm as well as it fitted longer-wavelength data. Conversely, the model more closely fitted the IRIS Pele data at short wavelengths. An analytical model of silicate lava cooling developed by Howell produced fits to IRIS data that also implied silicate eruptions, showing that the IRIS hot-spot spectra were very unlikely to be produced by sulphur flows (Howell, 1997).

Once a lava composition was decided on, Howell's model had the advantage of inherently few input variables and produced the evolving thermal emission spectrum for a cooling lava flow emplaced at a steady areal coverage rate. This model produced good fits to IRIS data from Loki and Amaterasu Patera, indicating the presence of cooling silicates, but had less success with fitting thermal emission from Pele. Howell realized, as did Carr, that a very vigorous style of activity was taking place at Pele, where both models predicted rapid emplacement of lava and a relatively young surface age. Howell's model returned a surface age of ≈1 day. This analysis enabled constraints to be imposed on the lava emplacement mechanism at Pele: either (a) very thin flows (≈0.5 m thick) were being emplaced, which would rapidly solidify and cool; or (b) the apparent age of the surface was a resurfacing time representing the overturn time for a lava lake of area ≈140 km^2 (a circle ≈13 km in diameter; see Howell, 2006); although (c) it was also possible that *Voyager* just happened upon the early stages of a very large eruption (Howell, 1997). From a single IRIS observation, no further constraint on eruption style was possible, as there was no temporal information.

In another modeling paper, the case was made for Loki Patera's being a massive lake of liquid sulphur above a liquid silicate magma chamber (Lunine and Stevenson, 1985). Heat transfer between the silicates and sulphur produced a convecting, stable sulphur lake at a temperature of ≈500 K, which represented the high-temperature component in a two-temperature fit to the IRIS spectrum. The cool component was taken to be the cool (non-volcanic) background that took up most of the IRIS field of view.

The IRIS dataset, with no temporal evolution of thermal emission to further constrain activity and with low sensitivity at the most important wavelengths for observing active silicate volcanism thermal emission, could not therefore conclusively show the presence of silicate volcanism at the few hot spots identified, although the collective weight of arguments was certainly leaning that way.

2.2 Ground-based observations

Ground-based observations continued with increasing temporal resolution after the "snapshot" spacecraft encounters. Observations were initially hampered by Io's small apparent size (≈ 1 arcsecond). With the available spatial resolutions of telescopes in the 1980s and most of the 1990s, observations were limited to the integrated thermal emission from the hemisphere of Io currently in view. In the aftermath of *Voyager*, ground-based observations utilized several techniques to study Io's hot spots.

2.2.1 Light-curve photometry

As thermal sources on Io rotate into and out of the field of view of the observing instrument, the detected thermal emission intensity changes. Multiple-wavelength observations of Io allowed color temperature fits to the data (see Johnson *et al.*, 1984; Veeder *et al.*, 1994). Observations could be in sunlight or in eclipse.

2.2.2 Photometry of occultations

During each 1.7-day orbit, Io is occulted by Jupiter. When Io is close to Jupiter, observations can often be made of Io in Jupiter's shadow, where the absence of sunlight makes the detection of relatively faint hot spots possible. In 1980, it was discovered that during occultations, Io did not disappear when in eclipse at 4.8, 3.8, or even 2.2 μm (Sinton, 1980a; Sinton *et al.*, 1980), an indication of the magnitude of endogenic thermal emission. Io then passes behind Jupiter, and the masking of emission from hot spots as they disappear causes a decrease in observed integrated thermal emission, and an increase as they re-appear (Figures 2.1 and 2.2). Knowing the precise timing of the event allowed longitudinal constraints to be imposed on hot-spot location, although longitudinal positioning is degraded by atmospheric effects (see Spencer *et al.*, 1990; Deschamps *et al.*, 1992; Spencer *et al.*, 1994).

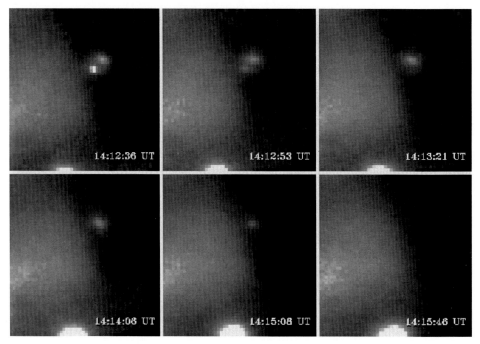

Figure 2.1 Sequence of images showing the disappearance of Io behind Jupiter. Images were obtained at 3.8 μm with the NASA-Caltech Infrared Telescope Facility (IRTF), Mauna Kea, Hawai'i, on October 12, 1990. Two hot spots can be seen: the spot on the left corresponds to Loki Patera, and the spot on the right to Kanehekili. The bright object at the bottom of each frame is Ganymede. Courtesy of John Spencer, Lowell Observatory.

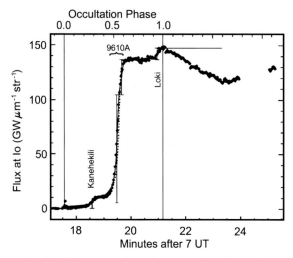

Figure 2.2 Example of the light curve of Io as it emerges from behind Jupiter on October 6, 1996. Io begins appearing at 07:17:37 (Phase = 0) and has fully emerged less than 3 minutes later at 07:21:08 (Phase = 1). The reappearance of the hot spots Loki and Kanehekili are indicated by the labels. From Stansberry *et al.* (1997).

Io is also occulted by other Galilean satellites (e.g., Goguen *et al.*, 1988; Spencer *et al.*, 1994). Every 6 years (twice every Jupiter orbit), the Earth passes through Jupiter's equatorial plane. The jovian satellites eclipse and occult each other in a series of mutual occultation events. The sharp edge presented by the occulting satellite allows very precise hot-spot location; however, in daylight, faint hot spots are difficult to see. If a hot spot can be observed entering into eclipse and exiting, positioning can be determined within 100 km.

2.2.3 Infrared polarimetry

Io's thermal emission is linearly polarized, and infrared measurements of the degree and position angle of polarization have been used to determine the position of a small number of thermal sources on Io (Goguen and Sinton, 1985; Sinton *et al.*, 1988). Interestingly, the value determined for the refraction index of the emitting surfaces was similar to that of terrestrial basalts and was somewhat less than that of sulphur. Loki Patera was again identified as a major heat source (Sinton and Kaminski, 1988).

2.2.4 Speckle interferometric imaging

Speckle interferometric imaging (Knox and Thompson, 1974; Lawrence *et al.*, 1992) and, more recently, adaptive optics (AO) allow ground-based telescopes to correct for blurring caused by atmospheric turbulence. These techniques result in telescopes operating at their theoretical limit of resolution, ≈0.04 arcsecond for a 10-m telescope, or ≈30 pixels across Io (a resolution of ≈120 km). Using speckle interferometry, Howell and McGinn (1985) identified a hot spot in the vicinity of Loki, with a temperature of ≈400 K and an area of 11 400 km^2; and McLeod *et al.* (1991) detected three hot spots, including Pele, seen for the first time since the *Voyager* encounters. Macintosh *et al.* (2003) observed Io in 1998 at 2.2 μm and detected 17 distinct thermal sources. AO observations of Io allowed multi-spectral monitoring of individual volcanic centers (Figure 2.3; see Marchis *et al.*, 2002; de Pater *et al.*, 2004; Marchis *et al.*, 2005).

2.2.5 Multi-wavelength observations

A new understanding of Io's volcanism resulted from observations yielding contemporaneous multi-wavelength observations (e.g., Veeder *et al.*, 1994; Blaney *et al.*, 1995). Multi-wavelength data allowed color temperatures to be derived and, with a series of observations obtained over a few hours, the temporal variability of activity could also be studied. From a decade of infrared monitoring of Io, Veeder *et al.*

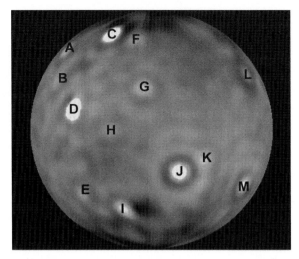

Figure 2.3 Detections of hot spots using the Keck telescope and adaptive optics. Image made from a hot-spot distribution map by Marchis *et al.* (2005), with the center of the globe at 270°W, 0°N. Hot spots were observed at different wavelengths, allowing derivation of color temperatures. A = Surt, B = Fuchi, C = Dazhbog, D = Loki Patera, E = Sengen, F = unnamed, G = Daedalus, H = Hephaestus, I = Ulgen, J = Pele, K = Pillan, L = Isum, M = Marduk. Base image courtesy of Franck Marchis.

(1994) reported that the global heat flow from Io exceeded 10^{14} W, or ≈ 2.5 W m^{-2}. The latter is an extraordinary figure, more than an order of magnitude greater than Earth's average heat flow (0.07 W m^{-2}). Large, warm (≤ 200 K) anomalies apparently dominated global heat flow, with smaller, hotter (≥ 300 K) areas (hot spots) contributing little to the global heat flow. High-temperature (>600 K) eruptions were seen on about 4% of the nights when observations were made.

Loki was the most persistent and brightest volcano, always seen to be active, although its intensity varied with time. Tellingly, three different temperature components were needed to match the Loki data – an indication of the complexity of activity at this site where, if silicate volcanism were the source of the thermal emission, a mixture of flow surface temperatures was present.

Pele was also identified as a persistent thermal source and was modeled using two components. Io's thermal anomalies were modeled with a total of ten source components at five locations defined by longitude (Veeder *et al.*, 1994).

On Christmas Eve, 1989, new hot spots were discovered in eclipse observations obtained from the Mauna Kea Observatory, Hawai'i, using a new infrared camera that yielded 0.5-arcsecond resolution, thus resolving Io (Spencer *et al.*, 1990). Coupled with occultation observations as Io disappeared behind Jupiter, at least three hot spots were identified in addition to Loki. Using a 3.8-μm observation, Loki was modeled as a hot spot of at least 370 K, which yielded an area of radius 120 km. The calculated area was close to the entire surface area of Loki Patera imaged by *Voyager*. Also noted was a possible order-of-magnitude drop in the 3.8-μm flux from Loki between December, 1989 and March, 1990.

2.3 Observations of Io from Earth orbit

In the 1990s, NASA's Hubble Space Telescope (HST) obtained disk-resolved observations at ultraviolet, visible, and infrared wavelengths. In 1992, observations in the ultraviolet using the HST Faint Object Camera, yielding a resolution of 210 km, showed no large-scale changes on Io's surface in the 13 years since the *Voyager* encounters (Sartoretti *et al.*, 1994). Io was exhibiting contradictory behavior. Large changes were seen in the four months between *Voyager* encounters, yet over periods of time of decadal length and longer, surface features appeared to be relatively stable, a conclusion reached in 1979 by a review of historical data dating back to 1927 (Morrison *et al.*, 1979).

After the HST repair mission in December, 1993 and the installation of the Wide-Field Planetary Camera 2 (WFPC2), spatial resolution of Io observations improved again. In March, 1994, disk-resolved HST images of Io were obtained at five wavelengths between 0.34 and 1.02 μm with a resolution of 160 km (Spencer *et al.*, 1997a). These were the best images of Io since *Voyager* and again showed little large-scale change on Io's surface. Two noticeable changes were detected, however: a darkening south of Colchis Regio and a darkening near 130°W, 30°N. Additionally, 1-μm observations of Io's Jupiter-facing hemisphere were made in March, 1994 to search for high-temperature thermal anomalies. None were detected, although activity may have been present below the detection limit of the observation.

Then dramatically, in July, 1995, WFPC2 observed a new 320-km-diameter yellowish-white feature in the location of Ra Patera, the site of the magnificent flows seen by *Voyager*. This was a larger surface change in 16 months than had been seen in the previous 15 years. The cause of this albedo change (Figure 2.4) was proposed to be a large eruption depositing pyroclastic material, lava, and volatile deposits (Spencer *et al.*, 1997a).

2.4 The Pele plume

HST observations of Io led to the discovery of an unusual spectral feature, a deep infrared absorption between 0.55 and 0.7 μm in the ejecta blanket around Pele, emplaced by the gigantic volcanic plume that was observed by *Voyager*. This absorption was attributed to the presence of the short-chain sulphur allotropes S_3 and S_4 (Spencer *et al.*, 1997a). These allotropes are not stable at Io background temperatures (≈ 100 K) and can be formed only by high-temperature processes, such as decomposition of S_2O, itself formed from the high-temperature dissociation of SO_2

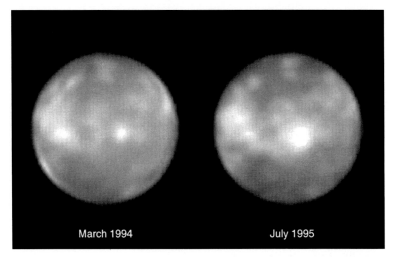

Figure 2.4 The NASA Hubble Space Telescope observed Io in March, 1994 and July, 1995 and detected a large, 320-km-diameter yellowish-white feature in the location of Ra Patera (center of right image). Image courtesy of John Spencer, Lowell Observatory, NASA (STScI-PRC1995–37: NASA PIA02160).

(Spencer *et al.*, 1997a) or the heating of stable S_8 (a process discussed in Chapter 5). Short-chain sulphur allotropes would certainly explain the red Pele plume deposits. The presence of this absorption feature elsewhere on Io indicated that high temperatures may be present in other locations. Silicate volcanism could have mobilized existing sulphur deposits or the sulphur could have degassed from the silicate magma.

2.5 Outburst eruptions

2.5.1 Outburst definition

The ground-based Io observation program identified a class of eruptions on Io that were called outburst eruptions, the largest eruptions seen on Io as measured by emitted power. The first of these events was discovered in 1978 by Witteborn *et al.* (1979) and a team led by Urey Fink, although these latter data were not published (see Spencer and Schneider, 1996). Outbursts are marked by dramatic increases in thermal emission. Changes take place rapidly and events are short-lived, ranging from hours to a few days. Outbursts were originally characterized by Blaney *et al.* (1995) as events yielding large increases in Io's 4.8-µm thermal emission. In 1980, an anomalous 4.8-µm flux was observed by William Sinton on one night between the *Voyager* Io encounters. This event correlated with a change in albedo at Surt, the site of a between-encounter eruption forming a Pele-like plume deposit. It

Table 2.1 *Observed outburst eruptions on Io**

UT Date	Position	Temperature (K)	Diameter (km)	Wavelength (μm)	Notes
1978/01/26	293°–113°W	?	?	2–4	*a*
1978/02/20	341°–138°W	600	54	3.5–5.4	*b*
1979/06/11	313°–125°W	(600)	70	5.0	*c*
1986/08/07	358°–140°W	1550	15	4.8–8.7	*d*
1990/01/09	258°–55°W	1225–1600	11	4.8–8.7	*e*
1995/03/02	80°–110°W	(600)	25	3.5–4.8	*f*
1996/08/28	16°±3°W, 2° ± 10°N	>1400	≈2	1.7–4.8	*g*
1996/10/06	35°±15°W, 75° ±15°N	>1400	4.18	1.7–4.8	*g*
1997/06/28	Pillan, 244°W, 12°S	>1600	≈25	0.4–5.2	*h*
1999/08/02	Gish Bar Patera?, 89.1°W, 15.6°N	1247	24.2	2.38, 3.39, 4.68	*i*
1999/11/13 2000/12/16	Tvashtar Paterae, 120.2°W, 61.5°N	>1300	≤13.7	0.98, clear filter	*j*
2001/02/19	Amirani	990 ± 35	11.5	1.61–2.12	*k*
2001/02/22	Surt	1080–1475	32	1.61–2.12	*k*

1978–1996 data compiled by Spencer and Schneider (1996).

* Longitude range, unless otherwise constrained, is derived from requiring the outburst location to be on the Earth-facing hemisphere of Io throughout the observation. Temperatures in parentheses are assumed.

a U. Fink, L. Lebovsky, H. Larson, unpublished data (Spencer and Schneider, 1996).

b Witteborn *et al.* (1979).

c Sinton (1980b). Consistent with location of Surt.

d Johnson *et al.* (1988) and Blaney *et al.* (1995).

e 1225 K temperature from Blaney *et al.* (1995). Decrease in color temperature implies cooling of lava surfaces. Data are consistent with the location of Loki Patera (long. 310°W, lat. 15°N). 1220–1600 K range from Davies (1996). The changing 4.8- and 8.7-μm fluxes were modeled as a waning fire-fountain event that fed spreading lava flows.

f Spencer *et al.* (1995b). Longitude is constrained by disk-resolved images, which also give a latitude of 30°–60°S. Possibly Arusha Patera, 100.7°W, 39°S.

g Stansberry *et al.* (1997).

h *Galileo* NIMS data; see Davies *et al.* (2001). Eruption also seen by *Galileo* SSI (Keszthelyi *et al.*, 2001a; Williams *et al.*, 2001a).

i Howell *et al.* (2001).

j Milazzo *et al.* (2005).

k Marchis *et al.* (2002). Surt peak power output $= 7.8 \times 10^{13}$ W, or 78% of Io's total thermal emission.

was thought that the cause was possibly a massive vapor explosion, which would account for the short duration of the event. Other outbursts prior to *Galileo* were subsequently reported (Sinton *et al.*, 1983; Johnson *et al.*, 1988; Howell and Sinton, 1989; McEwen *et al.*, 1989; Veeder *et al.*, 1994). Observations of outburst-class events are detailed in Table 2.1.

2.5.2 *Lava fountains and outburst eruptions*

Outburst eruptions proved to be of particular importance to the understanding of Io's volcanism because the scale of these events provided the best opportunities to constrain lower-limit temperatures of erupting material. For example, using the NASA Infrared Telescope Facility (IRTF) on Mauna Kea, Hawai'i, two outburst events were observed on August 7, 1986 and January 9, 1990 (Johnson *et al.*, 1988; Veeder *et al.*, 1994). For the August, 1986 event, data were collected at 8.7, 10, and 20 μm during a 5-hour period, with additional data at 4.8 μm collected during the last 2.5 hours of observing. The initial analysis of those data suggested the addition of a new volcanic thermal anomaly on Io's leading hemisphere with an area of 15 km^2 and a temperature of 900 K, well in excess of sulphur's boiling point (717 K at 10^5 Pa) – strong evidence of silicate volcanism (Johnson *et al.*, 1988). Subsequent analysis with a more refined model suggested temperatures as high as 1500 K, with a source radius of 8 km (Veeder *et al.*, 1994).

The January, 1990 event was observed for 2.6 hours at 4.8, 8.7, 10, and 20 μm, allowing determination of temperature and area change over this period. The location of the outburst was close to the longitude of Loki Patera. The data were initially modeled as a large silicate lava flow that increased its area at a rate of $\approx 1.5 \times 10^5$ m^2 s^{-1} and cooled from 1225 K to 555 K over the 2.6-hour observation period (Blaney *et al.*, 1995). Assuming a flow thickness of ≈ 2 m, the volumetric eruption rate was estimated at 3×10^5 m^3 s^{-1}, an enormous rate by terrestrial standards, perhaps approached only by eruptions of terrestrial continental flood basalts.

The January, 1990 outburst data (4.8 μm and 8.7 μm) (see Figure 2.5) were re-analyzed using a more sophisticated model of volcanic thermal emission, based on the cooling of lava in an ionian environment. The model calculated the distribution of temperatures and areas as a function of time and rate of areal coverage (Davies, 1996). The January, 1990 data were interpreted as having two time-variable components: a large hot area that was reducing in area with time and a cooler component that was conversely increasing in area with time. These trends were interpreted as either silicate lava fountains emanating from a fissure, with the lava fountains feeding clastogenic flows that spread across the surface (Plate 15a), or a lava fountain event in conjunction with the resurfacing or overflowing of a silicate lava lake. In the former case, a mass eruption rate of $\approx 10^5$ m^3 s^{-1} was again derived. The model includes thermal emission from areas at the magma eruption temperature, and magma eruption temperature was constrained to between 1200 K and 1600 K. It was also estimated that ≈ 30 of these events would be sufficient to resurface Io at a depth of ≈ 1 cm year^{-1} (Davies, 1996) if the material erupted were evenly distributed globally.

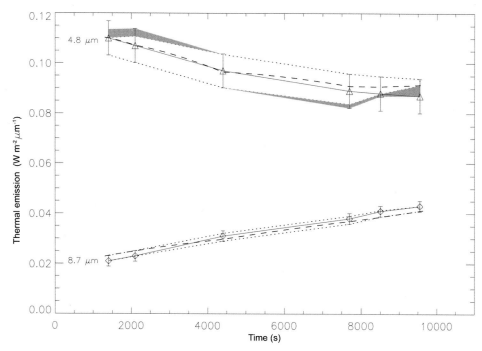

Figure 2.5 During the January, 1990 Io thermal outburst, data were obtained over a 2-hour period at 4.8 μm and 8.7 μm by Veeder *et al.* (1994). The data sequence allowed detailed modeling of eruption dynamics, with modeled magma eruption temperatures between 1200 K and 1600 K, indicative of silicate volcanism, lying within the error bars (gray areas). From Davies (1996). Reprinted with permission of Elsevier.

Whatever the exact eruption mode, these studies concluded that outbursts were infrequent, but cataclysmic, silicate eruptions of short duration. The derived temperatures were conclusive evidence of volcanism at silicate temperatures. The frequency of these events – and the implied volumes of material erupted – showed that this class of eruption represented an important mechanism of heat transfer from the interior of Io.

The implied outburst temperatures were far in excess of anything possible with sulphur volcanism, thus eliminating sulphur as the universal magma on Io. With the confirmation of high-temperature volcanism, a new synthesis of models began to emerge, with both silicate and sulphur volcanism present.

2.6 Stealth plumes

Just before the insertion of *Galileo* into Jupiter orbit, a new class of volcanic plume was proposed for Io: a "stealth" plume (Johnson *et al.*, 1995). These are diffuse plumes caused by the mobilization of sub-surface sulphur dioxide through direct contact with silicate magma. This study was particularly important because it showed that layers of sulphur dioxide frost deposited on the surface could be buried under silicate lavas with relatively little loss of solid sulphur dioxide. Later intrusion of silicates into this buried volatile deposit would initially lead to large, diffuse, high-entropy plumes and, on smaller scales, to geyser and fumarole activity. This could explain the origin of a "patchy" sulphur dioxide atmosphere (e.g., Sartoretti *et al.*, 1994). The model also showed how deep reservoirs of sulphur dioxide could form in the upper crust, ripe for mobilization by contact with hotter fluids.

2.7 Io on the eve of *Galileo*

The scene was thus set for *Galileo*. Io was perceived as a body where heat loss was dominated by active volcanism. Ground-based instruments had constrained Io's heat flow but, except in the cases of the largest eruptions, were not sensitive enough to detect relatively small areas at high temperatures that would constrain composition. A synthesis of proposed composition models had emerged, with volcanism consisting of a mixture of rare, large silicate eruptions, and common, smaller sulphur eruptions.

On December 7, 1995, *Galileo* entered orbit around Jupiter, beginning a new era of exploration and discovery at Io.

3

Galileo at Io

The *Galileo* mission was designed to be a detailed, in-depth investigation of Jupiter's atmosphere, the nature and evolution of its satellites and rings, and its magneto-spheric environment. The *Galileo* spacecraft (Figure 3.1) and instrumentation were already under development by the end of the 1970s. *Galileo* consisted of an orbiter and an atmospheric probe that was destined to plunge into Jupiter's atmosphere. *Galileo* was a large spacecraft, weighing 2717 kg at launch. Of this mass, 925 kg was usable propellant, 339 kg was the atmospheric probe, and 118 kg was dedicated to orbiter science instruments. To maintain stability, the spacecraft rotated about its central axis at rates of up to 10 rpm. A "de-spun" section rotated at the same rate in the opposite direction in order to provide a stable platform for the imaging instruments. Four instruments (Table 3.1) – three imagers and an ultraviolet sensor – covering the electromagnetic spectrum from the extreme ultraviolet to the far infrared were mounted on a movable scan platform that allowed pointing of the instruments and the slews necessary to compensate for blur during fast satellite fly-bys. These instruments were bore-sighted to allow complementary imaging of a target by all of the imagers.

Galileo arrived at Jupiter on December 7, 1995. That it arrived at all was a triumph of human ingenuity over adversity, a decade of stubborn tenacity, and brilliant engineering solutions to some of the most intractable problems ever encountered with a spacecraft. The details of *Galileo*'s many travails, both political and technical, are told elsewhere (O'Neil *et al.*, 1997; Harland, 2000; Fischer, 2001). The biggest problem that faced engineers and scientists was the failure of the High Gain Antenna (HGA) to open after spacecraft deployment. The 4.8-m-diameter HGA was capable of sending back data at a rate of 134 kbps (kilobits per second); however, every attempt to open it failed. Without the HGA, the rate of data return was dramatically reduced. As a workaround, during each satellite encounter, data from the science instruments were written to a digital tape recorder with a capacity of 900 MB. This recorder was originally placed onboard *Galileo* primarily to record spacecraft

Table 3.1 Galileo *scan-platform imaging instruments*

Experiment	Mass (kg)	Range	Objectives	References
Solid State Imaging experiment (SSI)	28	Effective wavelengths from 404–986 nm	Image Jupiter and satellites	Belton *et al.* (1992)
Near-Infrared Mapping Spectrometer (NIMS)	20	0.7–5.2 μm, up to 408 bands (0.01 μm resolution)	Determine surface and atmosphere composition; measure thermal radiation	Carlson *et al.* (1992)
Photo-Polarimeter Radiometer (PPR)	5	410–945 nm and 5 bands from 15–100 μm	Determine physical properties of clouds; measure thermal radiation	Russell *et al.* (1992)
Ultra-Violet Spectrometer/ Extreme Ultra-Violet Spectrometer (UVS-EUVS)	17	113–432 nm (UVS) 0.7–1.3 nm resolution 54–128 nm (EUVS) 1.5–3.5 nm resolution	Study emission and absorption features, airglow and auroras, and plasma torus	Hord *et al.* (1992)

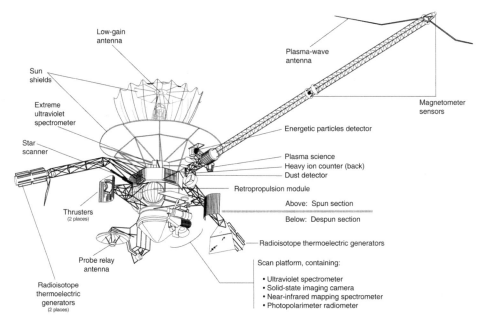

Figure 3.1 The *Galileo* spacecraft showing the likely final configuration of the High Gain Antenna (HGA), which failed to fully deploy. Courtesy of NASA.

engineering telemetry. After each encounter, data were played back and transmitted to Earth using the *Galileo* Low Gain Antenna (LGA) at a rate of 40 bps to 160 bps. Data return continued until the next satellite encounter, when the data remaining on the tape recorder were mostly overwritten – and therefore lost. The use of both tape recorder and LGA were brilliant improvisations that enabled *Galileo* to meet all of its mission science requirements. Three key resources were nevertheless always in short supply: time during a satellite fly-by, space on the tape recorder, and downlink time through the LGA before the next encounter.

3.1 Instrumentation

3.1.1 Solid State Imaging experiment

Visible-wavelength imagery is the mainstay of remote sensing and spacecraft exploration. *Galileo*'s Solid State Imaging experiment (SSI) produced the highest spatial resolution images of Io. At best, SSI images showed Io's volcanic features in unprecedented detail (see Plates 1, 6, and 12). SSI is described by Belton *et al.* (1992) and Klassen *et al.* (1997, 2003). The SSI camera used an 800-pixel × 800-pixel charged-couple device as its detector. Exposure time, gain state, filter, summation mode, and compression mode were all adjustable instrument parameters. SSI had eight filters. At Io, the primary filters used were violet (0.413 μm), green (559 μm), red (665 μm), near-infrared (756 μm), 1 μm (>0.968 μm), and the clear filter, a broad wavelength filter with an effective wavelength of 0.652 μm. Most of the highest resolution images of Io were obtained using the clear filter (Keszthelyi *et al.*, 2001a; Turtle *et al.*, 2004). The range of filters yielded images of Io with accurate visible color balance, an improvement on *Voyager* ISS, which lacked a red filter.

SSI observations of Io were designed to meet a range of scientific objectives, including (a) searching for high-temperature sources and auroral glows in long-exposure eclipse observations, (b) compiling global inventories of active plumes and surface changes (from the *Voyager* epoch, and also from *Galileo* orbit-to-orbit), (c) obtaining images at different resolutions of Io's geomorphology (especially to resolve the question of the dominant crustal component), and (d) by obtaining observations at different phase angles, examining the structure and properties of Io's regolith.

SSI returned 718 images of Io during the *Galileo* mission, with resolutions ranging from 64 km per pixel to 5.5 m per pixel. Primarily a visible-wavelength instrument, SSI was especially sensitive to thermal emission from very-high-temperature sources at wavelengths where *Galileo* NIMS had low signal-to-noise sensitivity.

Apart from returning high-spatial-resolution, visible-wavelength images of Io, SSI was also capable of detecting auroral glows caused by charged particles interacting with gases in Io's tenuous atmosphere and volcanic plumes. In eclipse observations (Plate 6f), SSI was capable of seeing the hottest areas (in excess of 700 K) of active lava flows, lava lakes, and lava fountains glowing in the dark (McEwen *et al.*, 1997). Later in the mission, SSI would on occasion detect volcanic thermal emission in dayside observations, for example, at the volcanoes Pele (Davies *et al.*, 2001) and Tvashtar Paterae (Keszthelyi *et al.*, 2001a).

3.1.2 Near-Infrared Mapping Spectrometer

The Near-Infrared Mapping Spectrometer (NIMS) was the first of a new class of spacecraft instruments, a hyperspectral imager capable of obtaining images of a target at more than 100 discrete wavelengths, therefore producing not just a spectrum of the field of view but a spectrum for each pixel in the image. Data are stacked into a "cube," with each layer being an image at a different wavelength. NIMS is described in detail by Carlson *et al.* (1992). NIMS was designed to measure the spectra of reflected sunlight at high spectral resolution. Absorption bands in the spectra reveal the composition of materials making up the surface of planets, satellites, and planetary atmospheres.

NIMS images, obtained during day and night (or eclipse), fell into one of three categories: global images obtained at low spatial resolution, often >200 km pixel^{-1}; medium-resolution images at 50 km to 100 km pixel^{-1}; and high-resolution images at resolutions down to 0.5 km pixel^{-1}. Observations in the latter category are small in number but make up for this scarcity with unprecedented spatial detail of select targets, such as Loki Patera, Prometheus, and Pele.

Additionally, NIMS was capable of detecting and measuring thermal emission from active volcanism on the surface of Io, even in daylight (Plate 7). NIMS had a spectral range from 0.7 μm to 5.2 μm and was therefore capable of detecting thermal emission from sources at temperatures from ≈180 K to more than 1200 K (Smythe *et al.*, 1995; Davies *et al.*, 2001). This range made the instrument ideal for observing thermal emission from active or recently active volcanoes (see Chapter 6). NIMS operated in different modes, which determined the number of wavelengths at which data were obtained. At best, NIMS obtained data at 408 wavelengths for each pixel. Free from the effects of a dense, infrared-absorbing atmosphere like Earth's, the NIMS Io data are the least adulterated spectral data of volcanic thermal emission ever obtained. Because the NIMS wavelength range covered shorter wavelengths than *Voyager* IRIS, NIMS was much more sensitive to higher temperatures than IRIS.

As with most datasets from spacecraft instruments, the processing of NIMS data to the point at which quantitative analysis could be performed was complex. The steps involved are described in numerous publications (see Davies *et al.*, 2000a, 2001; Lopes *et al.*, 2004).

NIMS contained a scanning grating spectrometer that formed spectra as the mirror swept across 17 detectors. The 17 wavelengths thus obtained for each of 24 grating steps were acquired simultaneously. At the end of the mirror sweep, the grating position changed, and the next sweep obtained data at a different set of 17 wavelengths. Each sweep was about 0.33 second apart. With 24 grating positions, the most complete spectrum consisted of the addition of 24 mirror sweeps of up to 17 wavelengths per grating position, yielding a spectrum of 408 wavelengths. An image of the target was constructed by adding together adjacent strips across a target, with a swath overlap of about 50%, which were obtained by slowly slewing the spacecraft scan platform.

NIMS data were processed into two formats: "tubes" and "cubes." Tube format products contained spectra before spatial averaging took place to make a cube product. A cube had post-observation navigation applied, which was more accurate than the tube navigation. Cubes are therefore more useful for mapping and geo-location purposes. Cubes also had the 50% swath overlap present in the tube removed through an averaging process. Importantly, the averaging used to remove the mirror swath overlap in cube formation led to a smearing of isolated hot pixels into adjacent pixels, altering the shape of the spectra in all pixels adjacent to the hot pixel as well as the hot pixel spectrum itself. The tube, where each spectrum corresponds to a *specific* location on Io's surface, thus contains the most robust NIMS spectral data for the purposes of modeling volcanic thermal emission.

Small motions of the scan platform during the 24 steps of the observation sometimes led to movement of the instrument field of view. This jitter led to patterning in the data (see Figure 3.2 and Soderblom *et al.*, 1999) and was exacerbated by the unique Io case, where small, intense hot spots resided within some pixels. Analysis showed that the best way to treat NIMS hot-spot data was to separate the full spectrum back into the individual grating position spectra and treat these individually (Davies *et al.*, 2000a).

Because the energy from a point source was spread between adjacent NIMS pixels (a characteristic resulting from the way in which data were sampled), another necessary step in data processing was the addition of intensities from adjacent pixels in the mirror-sweep direction, to obtain the full hot-spot intensity.

For low-spatial-resolution dayside data, the intensity measured by NIMS was a mixture of thermal emission from a hot sub-pixel area, often of low albedo and high emissivity, and reflected sunlight from the often higher-albedo, non-thermal remainder of the NIMS pixel. At NIMS wavelengths, reflected solar flux decreased with increasing wavelength, often as the thermal emission from the hot area increased. In daylight observations, thermal emission was almost always only detectable at NIMS wavelengths longer than ≈3 μm (see Figure 6.3). The loss of wavelength range limits temperature fits to only long-wavelength daytime data, but nevertheless

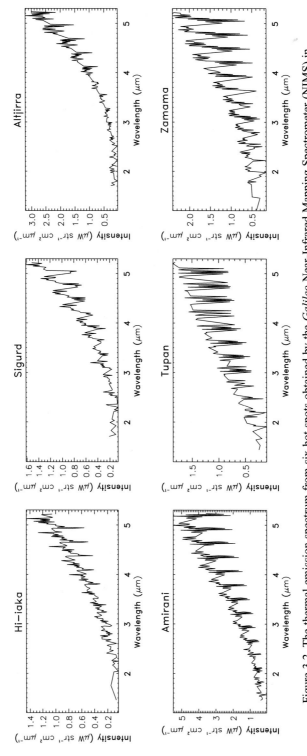

Figure 3.2 The thermal emission spectrum from six hot spots obtained by the *Galileo* Near-Infrared Mapping Spectrometer (NIMS) in June, 1996 (Orbit G1). The patterning (or "jitter") in the data is a result of the movement of the instrument field of view as the hyperspectral cube is constructed. From Davies *et al.* (2000a). Reprinted by permission of Elsevier.

two or more wavelengths could be used to derive a single-temperature fit to the data (e.g., Lopes-Gautier *et al.*, 2000). Such fits set a minimum temperature for the hot spot. Subtraction of the spectrum from a nearby pixel deemed to be non-thermal also isolated the thermal component (e.g., Lopes *et al.*, 2004). In some high-spatial-resolution data, the instrument "jitter" was used to separate the thermal component from the reflected sunlight component, yielding a full spectrum suitable for further analysis (Soderblom *et al.*, 1999). An example of such a "deconvolved" spectrum for Pele is shown in Figure 10.1.

These steps produced spectra ready for analysis using models of thermal emission. The analytical methods and results are described in later chapters.

NIMS obtained 190 observations of Io at resolutions ranging from hundreds of kilometers per pixel to 90 m pixel^{-1}, the latter obtained during one of the close Io fly-bys (Lopes *et al.*, 2004). Many of these observations consisted of hundreds of images, each obtained at a different wavelength; however, it was found that in nighttime data at short wavelengths, often, a very low signal-to-noise ratio was obtained. Therefore, data from detectors 1 and 2 (covering wavelengths from 0.7 μm to 1.24 μm) were not generally used in quantitative analyses.

Spectral resolutions ranged from 10 nm (at best), obtained with 408-wavelength observations, to observations with only 8 wavelengths spread across the NIMS wavelength range. The latter, theoretically, yielded spectra suitable for modeling thermal emission. In practice, however, modeling these particular products proved to be difficult because radiation spikes were hard to identify without comparison with adjacent, non-noisy data.

3.1.3 Photo-Polarimeter Radiometer

The *Galileo* Photo-Polarimeter Radiometer (PPR) is described by Russell *et al.* (1992). PPR was the only instrument on *Galileo* capable of detecting thermal emission from surfaces colder than ≈200 K, such as low-temperature hot spots (including cooled flow surfaces) and the non-volcanic, passive (sun-heated) surface. PPR was a single-aperture photometer and had the lowest spatial resolution of the imaging instrument suite, a function of the relatively large instrument field of view (2.5 mrads). PPR was sensitive from visible wavelengths to 100 μm. Difficulties with the filter wheel for much of the *Galileo* mission greatly reduced instrument utility, but, between 1999 and 2000, PPR obtained observations during six close fly-bys of Io (Spencer *et al.*, 2000b; Rathbun *et al.*, 2004), that provided thermal infrared fluxes for many volcanoes and mapped much of Io's passive surface from pole to pole. At Io, PPR was used most often in radiometry mode, measuring infrared thermal emission (see Chapter 6). Many PPR observations were "ride-along," taken in conjunction with NIMS or SSI, yielding single-aperture scans across a target. Three filters were used for Io radiometry: a "17-μm" filter, which covered the range 14.7 μm to 18.9 μm and was used during the ride-along observations; a "27-μm" filter, covering the range 24.1 μm to 32.3 μm, usually used for dayside observations; and the open filter, covering the range from visible to 100 μm (Rathbun *et al.*, 2004). PPR obtained more than 100 observations of Io from 1999 through 2002.

3.1.4 *Ultra-Violet Spectrometer*

The fourth instrument mounted on the *Galileo* scan platform was the Ultra-Violet/Extreme Ultra-Violet Spectrometer (UVS-EUVS), which was designed to study the upper atmosphere of Jupiter, the Io plasma torus, and gases escaping from satellites (Hord *et al.*, 1992). As such, the instrument was not suited to study volcanic activity on the surface of Io.

3.2 *Galileo* observations of Io

Galileo was placed in a highly elliptical orbit around Jupiter, with data being collected at perigee, when the spacecraft swept through the satellite system. Observations were naturally grouped by orbit number. Each orbit was further designated after the satellite of particular interest during the orbit, identified by the satellite initial in front of the orbit number. Hence, the first orbit, which included a fly-by of Ganymede, became Orbit G1. Only on the twenty-fourth orbit would Io's initial be found in an orbital encounter identifier.

The *Galileo* mission at Jupiter was divided into three phases: the Prime (or nominal) Mission, the *Galileo* Europa Mission (the first mission extension), and the *Galileo* Millennium Mission (the second and final extension).

3.2.1 *Orbit insertion*

Only one close encounter with Io by *Galileo* was planned in the Prime Mission. This encounter was to occur during the insertion of *Galileo* into orbit around Jupiter in December, 1995. Orbit insertion was designated J0. During J0, *Galileo*'s point of closest approach to Io was only 180 km above the surface, promising spatial resolutions with SSI of as high as a few meters per pixel. Unfortunately, because of concerns about the health of the data recorder, no imaging data of Io could be obtained. Recorder use was minimized to safeguard, as much as possible, the data from the atmospheric probe that plunged into Jupiter's atmosphere. Delivery of the atmospheric probe was the primary *Galileo* science objective. For Io scientists, this was a crushing blow. Io lies deep in the jovian radiation belts – a very unhealthy environment for a spacecraft and its instrumentation. During this single Io fly-by,

Galileo received more than a third of the expected total mission radiation dose. J0 was the only planned close fly-by of Io because the spacecraft was not expected to survive too many such encounters. Scientists primarily interested in Io had to wait until after other mission science objectives were fulfilled before *Galileo* could again be exposed to the hostile radiation environment around Io.

Even so, subjected to the degenerative effects of high radiation, *Galileo* slowly deteriorated over the years. Not only did the main spacecraft bus and instruments suffer physical damage, affecting overall performance (there was, for example, a progressive deterioration of the spacecraft gyroscopes that occasionally caused severe wobble), but observations were also lost as a result of computer resets and spacecraft electrical failures (e.g., see Keszthelyi *et al.*, 2001a). Nevertheless, *Galileo* would eventually exceed its planned lifetime radiation dose of 150 kRad by a factor of 5, which is a testament to the ruggedness of the spacecraft, the skill of its builders, and the innovation and tenacity of its controllers on Earth.

3.2.2 Prime Mission

The Prime Mission covered the first 11 orbits of Jupiter, with close fly-bys of Ganymede, Callisto, and Europa (see Table 3.2) from June, 1996 to November, 1997. Observations of Io during the Prime Mission were limited to global observations at relatively low spatial resolutions (mostly from 5 to 20 km pixel^{-1} with SSI, and 122 to 725 km pixel^{-1} with NIMS), but that nevertheless allowed (a) monitoring of Io's surface volcanic activity and plume activity; (b) comparing the new imaging data with data obtained by *Voyager* and the Hubble Space Telescope (Sartoretti *et al.*, 1995; McEwen *et al.*, 1997; Spencer *et al.*, 1997a; McEwen *et al.*, 1998a; Geissler *et al.*, 1999); (c) mapping sulphur dioxide concentrations on Io's surface at regional scales with NIMS (Carlson *et al.*, 1997; Douté *et al.*, 2001); and (d) modeling the integrated thermal emission spectra from Io's volcanoes (Davies *et al.*, 1997; Lopes-Gautier *et al.*, 1997, 1999; Davies *et al.*, 2000a, 2001; Davies, 2001).

3.2.3 Galileo *Europa Mission*

The *Galileo* Europa Mission (known as GEM) covered December, 1997 (E12) through November, 1999 (I25) and included an additional mission extension (unofficially called GEM-squared) to May, 2000 (G28). Monitoring of Io was continued to provide context for closer approaches as well as allowing monitoring of changes since the last reasonably close approach on March 29, 1998 (E14). During GEM, the anti-jovian hemisphere of Io was observed, at best, at a resolution of 1.4 km pixel^{-1}. Equatorial and northern passes of the anti-jovian hemisphere yielded resolutions

Table 3.2 Galileo *encounters and Io observations*

Orbit designation	Focus	Satellite fly-by date	Typical distance to Io when data obtained (× 10³ km)	SSI resolution (km pixel⁻¹)	NIMS resolution (km pixel⁻¹)	Io data obtained
J0	Io	Dec. 7, 1995	–	–	–	No observations
G1	Ganymede	June 27, 1996	696	9–23	348	SSI, NIMS
G2	Ganymede	Sept. 6, 1996	273	5–10	137	SSI, NIMS
C3	Callisto	Nov. 4, 1996	243	2.5–23	122	SSI, NIMS
E4	Europa	Dec. 19, 1996	312	6–18	156	SSI, NIMS
J5	–	–	–	–	–	No observations
E6	Europa	Feb. 20, 1997	402	4.1–21	201	SSI, NIMS
G7	Ganymede	Apr. 5, 1997	530	5.7–12	265	SSI, NIMS
G8	Ganymede	May 7, 1997	955	10–13	478	SSI, NIMS
C9	Callisto	June 25, 1997	1449	6–12	725	SSI, NIMS
C10	Callisto	Sept. 17, 1997	1330	3.8–64	665	SSI, NIMS
E11	Europa	Nov. 6, 1997	828	7.9–19.2	414	SSI, NIMS
E12	Europa	Dec. 16, 1997	514	–	257	NIMS
E13	Europa	Feb. 10, 1998	–	–	–	No observations
E14	Europa	Mar. 29, 1998	250	2.6–11.3	125	SSI, NIMS
E15	Europa	May 31, 1998	336	10.8–14	168	SSI, NIMS
E16	Europa	July 21, 1998	704	–	352	NIMS
E17	Europa	Sept. 26, 1998	–	–	–	No observations
E18	Europa	Nov. 22, 1998	–	–	–	Io data lost
E19	Europa	Feb. 1, 1999	–	–	–	Io data lost
C20	Callisto	May 5, 1999	780	–	390	NIMS
C21	Callisto	June 30, 1999	129	1.3–30	65	SSI, NIMS
C22	Callisto	Aug. 14, 1999	786	10–16	393	SSI, NIMS
C23	Callisto	Sept. 16, 1999	–	–	–	No observations
I24	Io	Oct. 11, 1999	0.611	0.009–6.5	0.31	SSI, NIMS
I25	Io	Nov. 26, 1999	0.301	0.147–0.265	0.15	SSI, NIMS
E26	Europa	Jan. 4, 2000	331	3.4	166	SSI, NIMS
I27	Io	Feb. 22, 2000	2.013	0.0055–17	1.01	SSI, NIMS, PPR
G28	Ganymede	May 20, 2000	–	–	–	No observations
G29	Ganymede	Dec. 28, 2000	968	10–17	484	SSI, NIMS, PPR
C30	Callisto	May 25, 2001	382	–	191	NIMS, PPR
I31	Io	Aug. 6, 2001	0.197	18–19.6	0.10	SSI, NIMS, PPR
I32	Io	Oct. 16, 2001	0.182	0.009–9.8	0.09	SSI, NIMS, PPR
I33	Io	Jan. 17, 2002	0.193	–	–	PPR
A34	Amalthea	Nov. 5, 2002	–	–	–	No observations
J35	Jupiter impact	Sept. 21, 2003	–	–	–	No observations

from 15 km to hundreds of meters per pixel with SSI. On June 30, 1999 (C21), *Galileo* passed Io at a range of 129 000 km, the closest encounter since orbit insertion. Data were obtained at spatial resolutions as high as 1.3 km pixel^{-1} by SSI, and 65 km pixel^{-1} by NIMS. The highlights of GEM were close Io fly-bys in October, 1999 (I24, closest approach 611 km), 1 month later (I25, closest approach 301 km), and in February, 2000 (I27) – all with data obtained at high resolutions (Table 3.2).

SSI, NIMS, and PPR were often impacted by radiation. After a nerve-wracking recovery from the spacecraft going into a radiation-induced safe (non-operating) mode just before the I24 closest approach to Io, the NIMS grating stuck. It would never move again. At a stroke, the ability to obtain high-spectral-resolution data was lost, although the data that were collected could still be used for determining temperatures and mapping the distribution of sulphur dioxide. This was, again, highly frustrating for scientists who were waiting for detailed spectra of the surface at spatial resolutions high enough to distinguish individual flow units and eruption centers too hot (>450 K) to have coatings of condensed volatiles.

Many of the I24 SSI observations were scrambled, the result of radiation damage to the camera, although some of these were successfully unscrambled (Keszthelyi *et al.*, 2001a). NIMS had better luck with I24 data collection, albeit with the loss of spectral resolution. During I25, another spacecraft "safeing" led to the loss of many planned observations, although an observation of a series of nested calderas in the northern hemisphere of Io at Tvashtar Paterae caught, to general astonishment, kilometer-high lava fountains erupting from a 25-km-long fissure (Keszthelyi *et al.*, 2001a; Wilson and Head, 2001).

3.2.4 Galileo *Millennium Mission*

The final phase of Io observations was during the *Galileo* Millennium Mission, the last extension of the *Galileo* mission. At the end of December, 2000 (G29), *Galileo* observed Io at the same time as the *Cassini* spacecraft, which obtained a gravity-assist from Jupiter on its way to Saturn. Three close fly-bys of Io took place in 2001 and 2002, allowing high-resolution observations of selected targets. Orbit I31 in August, 2001 included a low-altitude pass ($\approx$200 km) over the northern hemisphere, but radiation halted SSI and high-spatial-resolution observations were lost. Prudently, contingency observations planned for such a possibility were obtained at lower spatial resolutions. Orbit I32 in October, 2001 was a triumph. Almost all planned observations were obtained at resolutions down to 9 m pixel^{-1} for SSI and 90 m pixel^{-1} for NIMS. Additionally, PPR obtained dozens of high-resolution profiles across several volcanoes, supplying thermal infrared (long-wavelength) constraints on power emission (Rathbun *et al.*, 2004).

During the final Io encounter (I33) in January, 2002, spacecraft anomalies meant that only PPR obtained data. Unfortunately, the lost observations included high-resolution SSI dayside observations of volcanoes on the jovian-facing hemisphere of Io, including Loki Patera and the source of the Pele plume. These lost observations are described by Turtle *et al.* (2004). Io would not be imaged again by *Galileo*. On September 21, 2003, to avoid any possibility of contaminating Europa, *Galileo*, its propellant almost depleted, was deliberately steered into the atmosphere of Jupiter and destroyed.

Section 2

Planetary volcanism: evolution and composition

4

Io and Earth: formation, evolution, and interior structure

Volcanism has shaped the surface of the terrestrial planets. Earth, the Moon, Venus, and Mars have all been heavily modified by volcanism during at least some part of their history. *Active* high-temperature volcanism, with magma at temperatures in excess of 1000 K issuing onto the surface, however, has been observed on only two planetary bodies – Earth and Io[1] (compared in Table 4.1). This high-temperature volcanism is clear evidence of the triumph of interior heating processes over planetary heat loss mechanisms. Interior heating has melted at least part of the planetary mantle to form silicate magmas. The forces primarily responsible for heating within these two bodies, however, are very different.

Why, then, are Earth and Io currently volcanically active? What are the origins of the heat that is generating and being lost through volcanism, and why is Io unique in the jovian system in having high-temperature volcanism? How have these bodies evolved over time?

4.1 Global heat flow

First and foremost, it is necessary to quantify surface heat flow in order to model heating and heat transport mechanisms. Determination of heat flow from remote-sensing data is complex, being dependent on the thermal properties of surface and subsurface materials. In Io's case, these properties and materials are not known with any certainty. In comparison, measurement of heat flow on Earth is easily accomplished with *in situ* measurements.

---

[1] Observations in 2005 and 2006 of the saturnian moon Enceladus by the *Cassini* spacecraft made the exciting discovery of active water geysers (Porco *et al.*, 2006, *Science*, **311**, 1393–401) and surface thermal anomalies at ≈180 K (Spencer *et al.*, 2006, *Science*, **311**, 1401–5). The possibility exists that these anomalies might be driven by sub-surface volcanic activity (Matson *et al.*, 2006a). The current uncertainty as to the cause of the geysers and hot spots, however, means that Enceladus is beyond the scope of this book, which concentrates on Earth and Io.

Table 4.1 *Earth and Io constants*

	Units	Earth	Io
Equatorial radius	km	6378.3	1826.5 ± 0.12^b
Polar radius	km	6356.9	1812.2 ± 0.5^b
Mean radius	km	6371.0	1821.6 ± 0.5^c
Area	10^6 km^2		
Total		510	41.5
Land		149	–
Oceans		361	–
Mass	kg	5.976×10^{24}	8.935×10^{22}
Volume	m^3	1.08×10^{21}	2.52×10^{19}
Mean density	kg m^{-3}	5517	3527.5 ± 2.9^c
Acceleration due to gravity	m s^{-2}	9.812	1.79
Moment of inertia (C/MR^2)^a		0.334	0.37685 ± 0.00035^c
Heat flow (mean)	W m^{-2}	0.07^d	$>2^e$
Heat flow (mean, continents)	W m^{-2}	0.0565	–
Heat flow (mean, oceans)	W m^{-2}	0.0782	–
Sidereal rotation period	hrs	23.9345	42.456
Obliquity		23.45°	0.04°
Relief			
Land: greatest height	m	8848 (Mt. Everest)	18 200 (Boösaule Montes)^f
Land: mean height	m	840	–
Ocean: greatest depth	m	11 035 (Mariana Trench)	–
Ocean: mean depth	m	3808	–
Patera: greatest depth	m	–	>2500 (e.g., Bochica Patera)^g

[a] This is the normalized moment of inertia. C = axial moment of inertia, M = mass, R = radius.
[b] Davies *et al.* (1998).
[c] Anderson *et al.* (2002).
[d] Turcotte and Schubert (1986).
[e] Veeder *et al.* (1994). See Table 4.2.
[f] Schenk *et al.* (2001).
[g] Davies and Wilson (1987). Few patera depths have been calculated and published.

Earth's average surface heat flow is ≈0.07 W m^{-2} (Turcotte and Schubert, 1986). Significant differences on regional and local scales (Table 4.1) exist. Continents have a mean heat flow of 0.0565 W m^{-2} (a total of 1.13×10^{13} W). Ocean floors and margins have a mean heat flow of 0.0782 W m^{-2} (a total of 3.55×10^{13} W). Africa's average heat loss is 0.0498 W m^{-2}, expected for old, thick, relatively cold archean crust. At the other end of the heat flow scale is the North Pacific: young, hot oceanic crust with a heat flow of 0.0954 W m^{-2}. Global heat loss is 4.68×10^{13} W (or 1.48×10^{21} J yr^{-1}). Most of this heat is brought to the surface by conduction through continental and oceanic crust and through recycling of oceanic plates. The quantity of heat brought to the surface through effusive volcanism is extremely small in comparison (discussed in detail in Chapter 17).

Several estimates of Io's heat flow have been made over the past 25 years (Table 4.2). The first estimate of Io's heat flow was by Matson *et al.* (1981), who produced a number of 2.1 ± 0.7 W m^{-2}. The analysis of 12 years of ground-based photometric observations of Io at wavelengths from 5 to 20 μm by Veeder *et al.* (1994) estimated average heat flow at ≈ 2.5 W m^{-2}, with a global thermal output of 10^{14} W. Veeder *et al.* proposed that much of this heat came from large areas at temperatures lower than 200 K, with the remainder from smaller, hotter areas (i.e., active volcanoes). The heat flow estimate was considered to be a lower limit, as it did not include (a) the conduction of heat through Io's lithosphere (probably insignificant: see O'Reilly and Davies, 1981) or (b) a heat contribution from theoretical endogenically warmed polar regions (see Veeder *et al.*, 2004). *Voyager* IRIS determined the location of several hot spots and this helped constrain background thermal emission, yielding a minimum value of 1.85 W m^{-2} (McEwen *et al.*, 1996). *Galileo* PPR data yielded a heat flow of 2–2.6 W m^{-2} (Rathbun *et al.*, 2004), confirming the original estimates of Matson *et al.* (1981).

Nothing demonstrates the magnitude of the processes heating Io more than the average global heat flow, which is 25 times greater than Earth's. In contrast to Earth, most heat on Io is transported to the surface by active volcanism through a cold lithosphere (O'Reilly and Davies, 1981).

The internal heat lost from the surfaces of Io and Earth has origins that stretch back to the dawn of the Solar System and the formation of the planets and their satellite systems. Formation and heating mechanisms are discussed in the next sections of this chapter.

4.2 Planetary formation

4.2.1 Formation and differentiation of Earth

The generally accepted model for Earth formation is that the Earth formed from part of a primordial solar nebula of dust and gas that underwent gravitational collapse. The Sun formed at the center of the system and the planets and their satellites formed from the material left over from solar formation. Quite rapidly, matter coalesced into dust, into larger aggregations, and into bodies with diameters of 1 km or more – the latter called planetesimals. The proto-Earth formed over 4.5 billion years (Gyr) ago through the accretion of planetesimals, a process that became self-perpetuating and accelerated as the gravitational pull of the growing Earth increased.

Recent computer simulations and analyses of extinct radioisotope lines show that the larger planetary bodies would have experienced runaway growth at a rate

Table 4.2 *Theoretical and observational estimates of Io's internal heating*

Source of estimate	Notes	Equivalent surface heat flow (W m^{-2})	Total power (W)	Reference
Estimates of radioisotope decay	Chondritic (present day)	0.011	4.5×10^{11}	Cassen et al. (1982)
	Lunar (present day)	0.015	6.1×10^{11}	Cassen et al. (1982)
Tidal dissipation model	Theoretical upper limit	0.46–1.6	1.9–6.6×10^{13}	Cassen et al. (1982)
		0.79	$<3.3 \times 10^{13}$	
Tidal dissipation model	Steady-state over 4.5 Gyr	0.8	3.3×10^{13}	Peale (1986)
Infrared observations	Ground-based photometry	≈2	$\approx 9 \times 10^{13}$	Pearl and Sinton (1982)
Infrared observations	Ground-based photometry	1.8 ± 0.6	$7.5 \pm 2.5 \times 10^{13}$	Sinton (1981)
Infrared observations	Ground-based photometry	2.1 ± 0.7	$8.3 \pm 4.2 \times 10^{13}$	Matson et al. (1981)
Infrared observations	Ground-based photometry	1.5–2.0	8×10^{13}	Johnson et al. (1984)
Infrared observations	Ground-based photometry; large areas at $T < 200$ K, and hotter volcanic sources	2.5	10^{14}	Veeder et al. (1994)
Spacecraft observations	*Voyager* IRIS data, extrapolated globally, excluding Loki Patera	>1.85	$>7.7 \times 10^{13}$	McEwen et al. (1996)
Spacecraft observations	Analysis of *Galileo* PPR nighttime data	1.7	7×10^{13}	Spencer et al. (2000b)
Theoretical modeling	Establishing theoretical upper limit	<13.5	n/a	Matson et al. (2001)
Spacecraft observations	*Galileo* PPR data	2–2.6	0.8–1.1×10^{14}	Rathbun et al. (2004)
Theoretical modeling	Includes warm polar units	3	1.25×10^{14}	Veeder et al. (2004)

sufficient to create bodies approaching the size of the terrestrial planets in about 10 million years, with the Earth attaining about 63% of its mass in this time (Jacobsen, 2005).

Subsequently, these proto-planets swept up most large planetesimals over another 20 million to 40 million years (see reviews by Jacobsen and Yin, 2003; Jacobsen, 2005). The planetesimals appear to have been partly molten as a result of the decay of short-lived radioisotopes (Urey, 1955), if the bodies were large enough to insulate the interior against heat loss (Hevey and Sanders, 2006). Subsequent to the rapid planetary formation process (which included a major impact between Earth and a Mars-sized object that created the Moon; see Jacobsen [2005] for dating of this event), bombardment continued at a lower, but still intense, level for some hundreds of millions of years.

The formation of the Earth, as noted by Hamilton (2003), occurred "hot, fast and violently." Collisional heating (the transfer of kinetic energy), heating from the decay of short-lived radioisotopes, and heat liberated through gravitational potential energy release as the interior of the Earth self-compressed would have rapidly raised internal temperatures to form – and sustain – a global magma ocean.

The interior of Earth underwent core separation when sufficient interior temperatures ($\approx$2000 K) had been reached to allow melting and for dense iron and siderophilic materials to rapidly sink, and lighter silicates to rise. The onset of differentiation into core and mantle appears to have taken place very early in Earth's history, perhaps within a few tens of millions of years after initial formation (Jacobsen, 2005). A primitive crust, rich in lithophilic radioisotopes, would also have formed. Differentiation continued even as the Earth grew through further accretion (Jacobsen, 2005).

Additional heating would have taken place from the exothermic formation of serpentine in the mantle (Cohen and Coker, 2000). Unlike the heat of accretion, all of this energy was initially retained inside the planet. The downward mass movement of material would have raised the core temperature by over 2000 K to more than 4000 K (Solomon, 1981).

4.2.2 Formation and differentiation of Io

Little is known of the accretion of Io and the other Galilean satellites and the role that accretion played in the heating of Io. There are many uncertainties with regard to the timing of formation and therefore the role that may have been played by short-lived radioisotopes in heating the early Galilean satellites (e.g., Consolmagno, 1981).

Again, the generally accepted model for satellite formation has the jovian satellites forming from gas and planetesimals in a sub-nebula around a forming Jupiter (see reviews by Stevenson *et al.*, 1986; Schubert *et al.*, 2004). As the satellites formed, internal heating took place as each body compressed under its own weight, raising the interior temperature of Io to 1000 K or more (Schubert *et al.*, 1986). Impacts from other debris caught in the gravitational field of the forming satellite, often at accelerated velocities caused by Io's proximity to the mass of Jupiter, would have added to heating by the transfer of kinetic energy. Nevertheless, as with Earth, much of this heat would have been lost during the accretion process (Consolmagno, 1981) and would not have persisted over aeons of time. Enough uncertainty of the duration of satellite formation also exists (see Schubert *et al.*, 2004) to make quantification of internal heat generation difficult. What is known is that there was sufficient heating to melt and drive off any water remaining after accretion, and that the average density of Io (3527 kg m^{-3}) allows for an iron or iron sulphide core (Cassen *et al.*, 1982) – meaning there was sufficient heating from all sources at some point for melting, leading to core–mantle separation. Because the release of heat of differentiation by the formation of the ionian core was insufficient to melt silicate rock or an iron–sulphur alloy (Schubert *et al.*, 1986), another source of energy is needed to generate the active volcanism occurring on Io.

4.3 Post-formation heating

4.3.1 Radiogenic heating

The bulk of Earth's internal heat comes from the decay of radioisotopes. Earth is volcanically active today mostly as a result of past heating by radioisotope decay. Because unstable radioactive material was originally distributed homogeneously within the forming planetary bodies, early radiogenic heating was also homogeneous, unlike accretionary heating, which was initially concentrated in near-surface layers.

Radiogenic heating within a planet depends on the amount of radioisotopes present, which amount is proportional to the volume of the planet. Assuming homogeneity after planetary formation, a small planet has fewer radioisotopes than a larger one. Heat generated is opposed by heat loss, which is primarily a function of the surface area of the planet. First, the larger the body, the greater the distance heat has to be transferred to reach the surface. Second, the ratio of surface area to volume increases as planetary radius decreases (surface area is a function of radius^2 and volume is a function of radius^3); as a result, small bodies cool faster than large bodies. Io is the same size as Earth's Moon, with a volume slightly more than 2% that of Earth; Io's surface area is 8% that of Earth. The Moon and Mars have lost most of the heat that could drive active volcanism as a result of their relatively small size. Currently, surface heat loss is sufficient to prevent internal heat buildup to the point where large-scale melting takes place. The Earth and Venus have sufficient heat content and internal heating to drive active volcanism, although volcanism on

Venus may be dormant at the present time. Io would be a volcanically dead world if the heat of accretion, differentiation, and radioisotope decay were the only sources of heat. Other processes, however, are contributing to Io's thermal budget.

Short-lived radioisotopes (SLRIs) played a major role in early heating of planetesimals and planets, leaving daughter products from radioactive decay that can be used to date early Solar System events with great precision. The most important SLRI was Aluminum-26 (^{26}Al) because of its relative abundance and short half-life (0.78 Myr). ^{26}Al is, to all intents and purposes, extinct in the Solar System, but its daughter product, Magnesium-26 (^{26}Mg), remains. SLRI daughter products have been found in meteorite calcium-aluminum-rich inclusions (CAIs), the oldest objects in the Solar System. CAI genesis, condensing out of the solar nebula, is taken to be the chronological origin event of the Solar System. CAIs have estimated ages exceeding 4.567 Gyr (Amelin *et al.*, 2002; Zinner and Gopel, 2002).

Heat release from ^{26}Al decay would have decreased markedly after 3–4 Myr (Cohen and Coker, 2000), although the heat already produced would take much longer to escape from large planetesimals and from the proto-Earth these bodies were joining. The young Earth was also heated by other SLRIs (see list in Cohen and Coker, 2000), especially Iron-60 (^{60}Fe, half-life 1.5 Myr), and including Calcium-36 and -41 (^{36}Ca and ^{41}Ca, half-lives of 0.3 Myr and 0.1 Myr, respectively). These radioisotopes were so short-lived that it is postulated that the proto-Solar System was either seeded with them as the result of a nearby supernova or highly active star, which would mean that there was a uniform distribution throughout the Solar System, or the SLRIs were formed by irradiation close to the young Sun and transported by the solar wind out to the asteroid belt, yielding a heterogeneous distribution (see review in Russell *et al.*, 2001; see also Gounelle and Russell, 2005; Gounelle *et al.*, 2006).

The products of CAI SLRI decay have been used as an early Solar System chronometer. For example, the extinct Hafnium–Tungsten (^{182}Hf–^{182}W) radionuclide system is a favorable chronometer of the core formation process because hafnium is strongly lithophilic, remaining in the silicate mantle, and tungsten is moderately siderophilic, partitioning into the metal core. A study of the decay of ^{182}Hf (half-life 9 Myr) by Jacobsen (2005) yielded an early date of the Earth's core formation. The relative abundances of ^{182}W in terrestrial and lunar rocks showed that the Moon formed shortly afterward, at about 30 Myr after Solar System origin.

SLRIs were not the only contributors to heating of the Earth. The effects of SLRIs were impressive but short-lived. The heating and melting processes they began, however, were continued by the decay of long-lived radioisotopes (LLRIs). Four of these LLRIs are still prime contributors to heating of the Earth, having played a major role in heating the Earth over geologic time. These are Potassium-40 (^{40}K), Uranium-235 (^{235}U), Uranium-238 (^{238}U), and Thorium-232 (^{232}Th). Four and a half Gyr ago, the contribution of radiogenic heat flow was close to 400 mW m^{-2} at the Earth's surface, with the greatest contributors being ^{40}K (150 mW m^{-2}) and ^{235}U (110 mW m^{-2}) (Turcotte and Schubert, 1986). Currently, average global heat flow from all sources is $\approx$70 mW m^{-2} (3.6×10^{13} W globally), somewhat more than the heat currently being generated in the mantle by LLRI decay (currently $\approx$2.5–3.0 $\times$ 10^{13} W, or 1.3–1.6 $\times$ 10^{20} J yr^{-1}). ^{232}Th and ^{238}U are the largest contributors (Turcotte and Schubert, 1986). Over the age of the Earth, 5×10^{32} W of heat resulting from LLRI decay has been lost at the surface.

Io today

For Io, not enough is known about the time of formation of the jovian satellite system, the speed of accretion, and the local abundances of radioisotopes to confidently determine any effects SLRIs may have had on early satellite evolution. LLRIs certainly played a role in heating Io's interior. By assuming that radiogenic heating on Io would be comparable with that on the Moon (a body of similar size and density), radioisotopes within Io would liberate about 600 GW of radiogenic heat at present (Cassen and Reynolds, 1974). This is an upper limit. Current radioactive heating of Io is estimated to contribute only 0.011 to 0.015 W m^{-2} of the surface heat flow (Table 4.2), depending on whether Io has a chondritic (more likely) or lunar composition (Cassen *et al.*, 1982). Both numbers are far below the estimate of 2.5 W m^{-2} for Io's heat flow derived by Veeder *et al.* (1994).

　　Further research on (a) the effects of radiogenic heating elsewhere in the Solar System, (b) terrestrial and lunar radioisotope abundances and analysis of CAIs, and (c) the modeling of other Solar System bodies, such as Callisto (McKinnon, 2006) and the saturnian moon Enceladus (Castillo *et al.*, 2005), may well yield constraints that can be applied to Io.

4.3.2 Tidal heating

The importance of tidal heating of Io was first recognized by Peale *et al.* (1979). Io is caught in a gravitational tug-of-war with Jupiter, Europa, and Ganymede. Acting to speed Io in its orbit is Jupiter's gravitational pull. Io's gravity raises a bulge on Jupiter. Because Jupiter rotates faster than Io orbits Jupiter, the bulge on Jupiter always leads Io and acts to accelerate Io and expand its orbit. This process is countered by means of the forced orbital resonance lock, discovered by Laplace in 1805, among Ganymede, Europa, and Io. The resonance is such that, for every revolution Ganymede makes around Jupiter, Europa makes two revolutions and Io four, such that

$$n_{Io} - 3n_{Europa} + 2n_{Ganymede} = 0, \tag{4.1}$$

where n_i is the satellite mean angular velocity, also known as the "mean anomaly." Whereas Io orbits Jupiter every 42 hours and 27 minutes, Europa takes 85 hours, almost exactly double Io's period. Each time Io catches up to and speeds past Europa, the mutual gravitational attraction pulls both satellites away from perfectly circular orbits, increasing Io's orbital eccentricity. Without the resonance, tidal dissipation within the satellites would rapidly damp down orbital eccentricity to zero. With the resonance, the effect on Io is profound. The small eccentricity has a massive effect gravitationally, resulting in the flexing of Io on a high-frequency cycle through the raising of tides. A mostly silicate body, Io resists this deformation and energy is dissipated internally as heat. Dissipation within Io is at a maximum closest to Jupiter (periapsis) and at a minimum when farthest away (apoapsis). Here, therefore, is the source of the energy that currently drives Io's volcanism – the tapping of the mechanical energy of the orbital system (Peale *et al.*, 1979; Yoder, 1979).

Tidal heating also plays a role on Earth, which by Solar System standards is orbited by a very large moon (compared with parent body size). Gravitational forces result in tides in Earth's oceans on the scale of meters, but the lunar tidal effect on the rigid lithosphere and mantle does not greatly exceed some tens of centimeters (Bartha, 1981; Melchior, 1983). Tidal heating of the Earth has been estimated at approximately 3×10^{19} J year^{-1} (Francis and Oppenheimer, 2004), only about 2% of current global heat loss. Over geologic time, tidal forces exerted on the Moon by Earth have had the effect of increasing the Moon's orbital velocity, leading to a corresponding increase in the Earth–Moon distance, a process first investigated by Darwin (1880). At the same time, the transfer of angular momentum to the Moon slowed the Earth's rotation rate. Orbital evolution also led to phase-locking so that the same side of the Moon always faces Earth. Io also is similarly phase-locked with respect to Jupiter.

4.3.3 Orbital evolution and the dissipation of heat

The dissipation of tidal energy within Io, and the consequences of tidal heating on interior structure and composition, is a topic that has been closely studied and widely discussed. Few concrete conclusions have been reached, however. Reviews have been published post-*Voyager* and prior to *Galileo* (e.g., Nash *et al.*, 1986; McEwen *et al.*, 1989; Spencer and Schneider, 1996), before high-resolution passes of Io by *Galileo* (McEwen *et al.*, 2000a; Schubert *et al.*, 2001), and after the *Galileo* mission (Keszthelyi *et al.*, 2004a; McEwen *et al.*, 2004; Schubert *et al.*, 2004). Further analysis of *Galileo* data continues to reshape ideas of Io's interior and possible evolution.

How long Io has been volcanically active depends on several mutually dependent factors, none of which are completely understood. The magnitude and location within Io of tidal dissipation depends on orbital parameters (in particular, orbital eccentricity and the mean anomaly, and the rates at which these factors may be changing) and on the rheology of Io's interior, which determines how much – and where – energy is dissipated within Io.

The degree of dissipation is designated Q, the "quality factor," a quantity that is inversely proportional to the rate at which energy is dissipated in a body per cycle. The more rigid (and viscous) the body, the higher the value of Q. For example, a bell has a high Q, a lump of putty a low Q.

Io's heat flow of 2.5 W m^{-2} (Veeder *et al.*, 1994) is not matched by what is expected from dissipation for steady-state heating of Io, given what is known of Io's and Jupiter's interiors. Io's observed heat flow is three times as large as the upper limit of ≈ 0.8 W m^{-2} (Peale, 1986, 1989) thought possible through steady-state tidal heating over the lifetime of the Solar System (4.56×10^9 Gyr).

Coupled thermal and orbital evolution models (Ojakangas and Stevenson, 1986; Greenberg, 1989; Fischer and Spohn, 1990; Hussmann and Spohn, 2004) predict periods of intense heating that result in widespread volcanic activity, followed by longer periods of relatively little or no activity. Such models suggest that Io may currently be at or close to the peak of a heating cycle, in a hot and highly dissipative state. Hussmann and Spohn (2004) considered Europa and Io as a single system and modeled their mutual thermal and dynamic evolution. Several dissipative models, dependent on heat transport mechanism, were studied. For example, in the model where Io has convection in the entire mantle, high heat flow spikes are produced about every 140 Myr (Hussmann and Spohn, 2004). Confirmation of this and other theories depends on a determination of how Io's orbit may be changing. The key measurement to make is determination of the rate of change of Io's mean anomaly.

The acceleration of a satellite's mean anomaly n is designated $\dot{n}$. If Io's mean anomaly is increasing with time (positive $\dot{n}$), Io is slowly spiraling toward Jupiter. If this is the case, then Io is losing more tidal energy from tidal dissipation within the satellite than it is gaining from tidal torque from Jupiter (Aksnes and Franklin, 2001). If Io is migrating inward, then the amount of heating caused by the orbital resonance with Europa is decreasing, and the amount of heating within Io by tidal dissipation is also decreasing (Greenberg, 1987). If at or just past a peak in a heating/cooling cycle, Io is therefore cooling down, consistent with models of periodic heating (Ojakangas and Stevenson, 1986; Fischer and Spohn, 1990). Because heat loss lags behind the heating process, Io's heat flow remains above the heating rate averaged over the full cycle, which is consistent with observations. In this scenario, the tidal torque from Jupiter would most likely eventually arrest the inward movement, and Io would migrate outward again, heating up as it did so and beginning another cycle. A negative value of $\dot{n}$ implies that Io is moving outward, driven by Jupiter's tidal torque. Tidal heating would therefore be on the rise.

A determination of $\dot{n}$ is essential, therefore, to correctly extrapolate the past evolution of the Galilean satellites and to model the system cyclicity induced by orbital resonances (Lainey and Tobie, 2005).

Studies of occultation observations dating back to the fifteenth century have attempted to quantify $\dot{n}$ (e.g., de Sitter, 1931; Goldstein and Jacobs, 1986; Greenberg, 1987; Goldstein and Jacobs, 1995; Aksnes and Franklin, 2001; Lainey and Tobie, 2005). An appraisal of these results, determining the errors in observation timings, concluded that it was not possible to determine from existing data how $\dot{n}$ was changing (Lainey and Tobie, 2005).

An understanding of Io's tidal heating and evolution will probably be reached through sophisticated modeling, combining the thermal and dynamic evolution of Io using improved post-*Galileo* interior constraints. Additionally, more appropriate values of Q for Io (current values typically assume a homogenous body, which Io is not) and for Jupiter will be needed to constrain models. Q for Jupiter is poorly constrained between 10^5 and 10^6 (Goldreich and Soter, 1966). Accurate measurements of satellite positions over a suitable time period should establish the value of $\dot{n}$ (Lainey and Tobie, 2005).

4.4 Interior structure

4.4.1 Interior structure of the Earth

Recent summaries of the current state of knowledge of Earth's interior structure can be found in Schminke (2004) and Anderson (1999). The structure of Earth's interior (Figure 4.1 and Table 4.3) has been revealed by the behavior of seismic waves through the planet (seismic tomography) and meets the requirements of Earth's moment of inertia, which constrains interior mass distribution. A homogeneous sphere has a moment of inertia of 0.4. Earth's moment of inertia is 0.344, indicative of a dense core.

At Earth's center, at temperatures exceeding 6300 K, is a solid inner core of iron–nickel (density $\approx 14\,000$ kg m^{-3}). At a depth of 5150 km, the reduction in pressure results in a phase change to a liquid outer core (density $\approx 11\,000$ kg m^{-3}). Convection in this molten layer is the likely source of the dynamo effect that generates Earth's magnetic field. At a depth of 2890 km, there is an abrupt transition and steep temperature drop between the outer core and the solid but plastic mantle. This turbulent region, known as the D″ layer, is thought to be the source of some of the convection patterns and a postulated source of mantle plumes that drive both plate tectonics and hot-spot volcanism. The mantle consists primarily of iron and magnesium-rich silicates and is divided into sub-zones determined by mineral assemblages that form as a consequence of changing temperature and pressure. The lower part of the lower mantle is made up of high-temperature-phase minerals

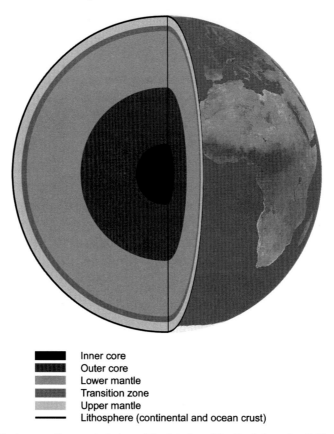

Inner core
Outer core
Lower mantle
Transition zone
Upper mantle
Lithosphere (continental and ocean crust)

Figure 4.1 A generally accepted view of the interior of the Earth (see also Table 4.3). The separation of core and mantle appears to have taken place very rapidly (≤ 10 Myr) after initial formation and continued during additional accretion. Melting of the upper mantle produces basalt magma that mostly erupts at ocean spreading centers.

such as perovskite and stishovite, and the upper 650 km is primarily in the form of olivine and pyroxene. Toward the top of the lower mantle, spinel and garnet are found. Between depths of 650 km and 400 km, there is a transition zone, possibly because of abrupt changes in mineral structure as a result of changes in temperature and pressure, between what can be regarded as the lower and upper mantle. Above this transition zone is the upper mantle, also known as the aesthenosphere. Decreasing pressure in this viscous layer, from depths of 200 km to 100 km, causes some mineral assemblages to melt, forming a partially molten layer known as the *Low Velocity Zone*, so called because seismic waves cross this "mushy" zone at a slightly reduced velocity ($\approx 2\%$) compared with transmission velocities through the more rigid material on either side. An increase in temperature in this zone approaches the melting temperature of mantle peridotite.

Table 4.3 *Interior structure of the Earth*

	Radius (km)	% of Earth mass	Mass (kg)	% of Earth volume	Volume (m³)	Mean density (kg m⁻³)	Temperature (K)	Dominant composition
Inner core	1220	1.7	1.016×10^{23}	0.66	7.239×10^{18}	14 034	>6300–6800	Fe-Ni
Outer core	3480	30.8	1.841×10^{24}	15.53	1.693×10^{20}	10 871	5300–6300	Fe-Ni
Lower mantle	5720	49.2	2.940×10^{24}	55.73	6.075×10^{20}	4 840	3000–3800	Peridotite
Transition zone	5970	7.5	4.482×10^{23}	9.85	1.074×10^{20}	4 175	2800	Peridotite
Upper mantle	≈6350	10.3	6.155×10^{23}	16.63	1.813×10^{20}	3 395	1570–1770	Peridotite
Continental crust[a]	6371	0.374	2.235×10^{22}	0.76	8.278×10^{18}	2 700	<1300	Granite/basalt
Oceanic crust[b]	6371	0.099	5.916×10^{21}	0.19	2.040×10^{18}	2 900	<1300	Basalt

Note: Temperature and radius data are after Anderson (1999).
[a] ≈30 km thick over 30% of the Earth's surface.
[b] 10 km thick over 70% of the Earth's surface.

This region of partial melting is generally regarded as the source of the basaltic magma that erupts at mid-ocean ridges, the most abundant material erupted from the mantle. Above this region, the mantle recrystallizes to a solid, above which is found a crust of sedimentary, metamorphic, and igneous rock. This uppermost solid region is the lithosphere, which is broken up into a number of tectonic plates.

This description of the Earth's interior is somewhat simplified. Continental and oceanic crustal thicknesses vary greatly around the world. The mantle zones exhibit great variability in depth and location, and even the inner core exhibits inhomogeneities. The mantle composition is varied enough that rocks from around the world exhibit unique radioisotope contents and compositions (Anderson, 1999; Head, 1999). Basalt makes up all of the oceanic crust, covering 70% of Earth's surface. Islands such as Hawai'i and Iceland are formed of great basaltic piles that have accumulated over time. Hidden under the oceans is a 40 000-km-long network of ridges where new basalt is erupted, forming new ocean floor. Ocean floor is recycled through subduction zones, where cold, dense plates sink back into the mantle.

At depths greater than 60 km in the mantle, cold basalt converts to the mineral eclogite, which is denser than the shallow mantle. Large bodies of eclogite can sink through the mantle, a possible explanation for why the crust of the Earth is no greater than $\approx$60 km thick.

4.4.2 Interior structure of Io

Io's inferred global structure (Figures 4.2 and 4.3) at the end of the *Galileo* mission was summarized by Schubert *et al.* (2004) and Keszthelyi *et al.* (2004a). What is known about Io's interior structure is derived from estimates of the normalized moment of inertia (0.37685 ± 0.00035) derived from four *Galileo* fly-bys (Anderson *et al.*, 2001) and mean density. Io's mean density, 3527.5 kg m^{-3}, is consistent with an iron and silicate composition. For a three-layer planet, mean density is related to internal structure using the relationship

$$\bar{\rho} = \rho_s + (\rho_c - \rho_m)\left(\frac{r_c}{R}\right)^3 + (\rho_m - \rho_s)\left(\frac{r_m}{R}\right)^3, \tag{4.2}$$

where $\bar{\rho}$ is mean density; ρ_s, ρ_m, and ρ_c are densities of the outer shell (lithosphere), mantle, and core; and r_m and r_c are the radii of mantle and core. R is Io's radius. The relationship between the mean density and moment of inertia is given by

$$\bar{\rho}\left(\frac{C}{MR^2}\right) = \frac{2}{5}\left[\rho_s + (\rho_c + \rho_m)\left(\frac{r_c}{R}\right)^5 + (\rho_m + \rho_s)\left(\frac{r_m}{R}\right)^5\right], \tag{4.3}$$

where C is the satellite axial moment of inertia and M is the mass of the body. It should be stressed that Equations 4.2 and 4.3 do not take into account the effects

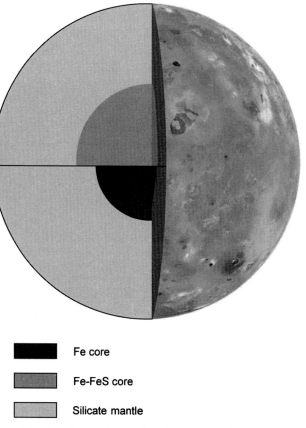

Fe core

Fe-FeS core

Silicate mantle

Figure 4.2 A post-*Galileo* view of the interior of Io. The distribution of mass inside Io is constrained by Io's moment of inertia, which points to an Fe-rich core. If pure Fe, the core has a radius of ≈650 km, about 10% of Io's mass. If the core is an Fe–FeS alloy, it has a radius of about 950 km (see Anderson *et al.*, 1996, 2001; Kuskov and Kronrod, 2001a, 2001b; Sohl *et al.*, 2002; Schubert *et al.*, 2004). Because these estimates are dependent on Io's interior temperature, which controls density, uncertainties are ≈10%. Surrounding the core is a silicate mantle, possibly an enstatite-rich peridotite (Keszthelyi *et al.*, 2004a).

of pressure on density, so these are approximations. Equations 4.2 and 4.3 can be solved for the case where $\rho_m = \rho_s$ (e.g., Schubert *et al.*, 2004). Results are shown in Table 4.4. Io's moment of inertia indicates the presence of an iron-rich core, although uncertainties remain about the exact composition (possibly pure iron or an iron–iron sulphide alloy) and radius. Estimates for the radius range from ≈650 km if the core is pure Fe ($\rho_c = 8090$ kg m^{-3}), making up ≈10% of Io's mass and ≈5% of the volume, to ≈950 km for Fe–FeS ($\rho_c = 5150$ kg m^{-3}), taking up ≈21% of Io's mass and ≈14% of its volume. These compositions and mass distributions also satisfy cosmo-chemical considerations (e.g., Consolmagno, 1981) that point to Io's having an iron-rich core that is most likely sulphur-rich. In studies of the Fe–S system under high pressure, Sanloup *et al.* (2000, 2001) found that, as sulphur

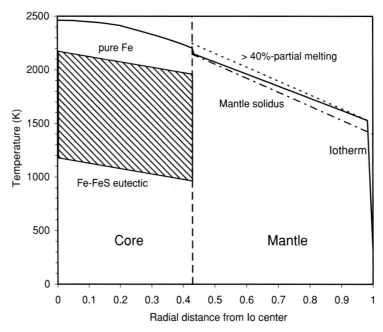

Figure 4.3 The thermal structure within Io, proposed by Schubert *et al.* (2004). Plotted are a simplified mantle solidus based on the solidus of dry peridotite (Takahashi, 1986, 1990); a line of constant 40% melt fraction after Wylie (1988); a range of liquid core compositions (shaded area) from Fe to a eutectic Fe–FeS mixture (Usselmann, 1975a, 1975b; Boehler, 1986, 1992); and an Io thermal profile (Iotherm), which follows the wet adiabat in the lower mantle, a constant degree of partial melting in the aesthenosphere, and the adiabat in the core. © Cambridge University Press.

content increased, incompressibility decreased, leading to a denser core. This density increase with increasing sulphur content is another factor that has yet to be included in core density and size estimates.

Kuskov and Kronrod (2001a) found that the correspondence between the density and moment of inertia values for bulk composition of Io may be described as close to those of ordinary L and LL chondrites. If the core composition is at the Fe–FeS eutectic, the core density ρ_c is ≈ 5150 kg m^{-3}, with a melting temperature of ≈ 1200 K to 1300 K (Brett, 1973; Anderson *et al.*, 2001). The base of the mantle is expected to be considerably hotter than this, so Io's core is partially or wholly molten. A wholly or mostly molten core would explain the absence of a detectable intrinsic magnetic field (see Chapter 9).

Assuming an original chondritic composition and an iron-rich core, the composition of the silicate remainder in the mantle can be calculated. Keszthelyi *et al.* (2004a) determined that the most common likely mineral in the Io mantle is enstatite, although geochemical modeling does not reveal whether the mantle is solid, partially molten, or liquid.

Table 4.4 *Interior structure of Io*

	Radius (km)	% of Io mass	Mass (kg)	% of Io volume	Volume (m^3)	Mean density (kg m^{-3})	Temperature (K)	Dominant composition
Fe core								
Core	650	10.42	9.308×10^{21}	4.54	1.150×10^{18}	8090	2100–2400	Fe
Mantle	1791	89.58	8.003×10^{22}	95.46	2.417×10^{19}	3311	1600–2100	Peridotite?
Fe–FeS core								
Core	950	20.71	1.850×10^{22}	14.18	3.592×10^{18}	5150	2100–2400	Fe–FeS
Mantle	1791	79.28	7.082×10^{22}	85.82	2.173×10^{19}	3259	1600–2100	Peridotite?
Lithosphere	1821	4.06	3.630×10^{21}	4.86	1.231×10^{18}	≈3000	130–1500	Silicate

Note: Fe and Fe–FeS cores are end-member models.

4.5 Volcanism over time

4.5.1 Earth

It is generally accepted that Earth has been volcanically active for most of its history (see summary in Head, 1999). The large size of the planet means that internal heating sources are greater than those of smaller terrestrial planets and moons. As a result of extreme recycling of the crust, most of Earth's surface is less than 200 Myr old. Over 70% of the surface of the Earth has formed in the last 5% of the history of the planet. The oldest continental rocks date back 3.8 Gyr, a large proportion of the age of the planet. Earth's volcanic history appears to have been of a steady-state nature (e.g., Head, 1999), punctuated by events that may each have been unique. An example is the formation of the continents, which appears to have taken place over a relatively short period of time. Another period of activity, unlikely to be repeated, involved extensive eruptions of ultramafic magmas (komatiites), mostly 3.5 Gyr to 2.5 Gyr ago. Komatiites, evidence of a high degree (>40%) of mantle melting (see Chapter 5), may have been produced in deep-origin mantle plumes. They have been used to illustrate the "cooling Earth" model, where a sequence of eruptions over 1 Gyr shows a decreasing magnesium content, possibly due to mantle cooling (Arndt and Nisbet, 1982).

What caused the onset of plate tectonics as we see it today is open to debate. A possibility is that internal heating led to convection in the mantle, which fractured the lithosphere. The surface of the Earth is now divided into 12 rigid lithospheric plates. The movement of these plates; their creation at divergent centers (such as mid-ocean ridges); the effects of their destruction in subduction zones where cold, sediment-laden, and water-rich plates plunge into the mantle; and the results of plate collisions have created the topography of the Earth as seen today.

The recycling of oceanic plates dominates heat loss from the Earth; the conveyor belt of plate formation, movement, and destruction is an efficient way of removing heat from the Earth's interior while maintaining a thin lithosphere, which over most of the planet has never had a chance to thicken and impede heat loss by conduction. Plate tectonic boundary zones are where the majority of volcanoes are found. Most magma reaches the top of the lithosphere deep under oceans at mid-ocean ridges. Deeper-seated manifestations of heat transport also result in volcanic activity. Thermal anomalies in the lithosphere, possibly resulting from mantle plumes, form intra-plate volcanoes over hot spots far away from active margins and rifts. Hawai'i is a mid-ocean-plate hot spot; Yellowstone, in Wyoming, is a hot spot in a continental plate.

The last period of global volcanic activity was relatively recent: the emplacement of continental flood basalts (described in more detail in Chapter 11). These are the largest eruptions of lava that survive, with known volumes of individual flows exceeding 2000 km^3 (Rampino and Stothers, 1988; Mahoney and Coffin, 1997). The ages of these lavas date from the Siberian flood basalts (250 ± 1 Myr ago) to the relatively recent Columbia River Flood Basalts (CRBs) (16 ± 1 Myr ago). The scale of the eruptions staggers the imagination. The total volume of the Siberian lavas is over 2×10^6 km^3, erupted at an average rate of about 2 km^3 yr^{-1} (Mahoney and Coffin, 1997).

To summarize, heat loss from the Earth is dominated by conduction through the lithosphere and through plate recycling. Volcanism currently plays little part in the heat loss process (discussed in Chapter 17). Over time, the dynamic level of plate tectonics will wane as the Earth continues to lose heat from the mantle and as internal heating mechanisms decrease. Eventually, mantle convective forces will decline to the point where plate tectonics will cease, and the lithosphere will thicken and become more rigid. Earth, like the Moon, Mars, and Venus, will become a one-plate planet.

4.5.2 Io

How long has Io been volcanically active? If satellite resonances formed recently, then the onset of volcanism on Io may have been relatively recent (Peale, 2003), and we may therefore be witnessing Io at a peak in volcanic activity. If this is the case, then the mantle of Io may be chemically undifferentiated (not be confused with the differentiation of a planet into core and mantle) and the composition of the magmas produced through partial melting of the mantle would be primitive, ranging from mafic to ultramafic compositions, depending on the degree of mantle melting (discussed in Chapter 9).

If the satellite resonances formed during the early stages of Solar System formation, then Io's volcanism may have been constant or, more likely, intermittent for billions of years, and Io's interior would have been reworked many hundreds of times (Carr, 1986; Keszthelyi and McEwen, 1997a). Magmas might therefore be expected to be highly fractionated, the leavings of many episodes of melting. Melts of ultramafic composition would be expected to sink, leaving more evolved silicates that have lower melt temperatures (Keszthelyi and McEwen, 1997a). It may be that, after the quiescent period of cooling during an orbital cycle, the onset of heating and subsequent heat buildup within Io is so rapid that mantle melting creates a magma ocean of primitive composition into which the crust is recycled. Alternatively, solid-state convection within the mantle might be vigorous enough to effectively recycle material that delaminates from the base of the crust (see Hussmann and Spohn, 2004).

4.6 Implications

After *Galileo*, the model of Io's interior bears some resemblance to the "liquid aesthenosphere" model proposed in the aftermath of *Voyager* (see Figure 1.9 and Spencer and Schneider, 1996). Some degree of partial melting at the top of the mantle produces magmas that are erupted at the surface. The derivation of magma temperature therefore becomes a crucial measurement, at one end of a chain of causes and effects. Magma temperature is a function of the degree of partial melting in the mantle. The degree of melting in turn constrains where most tidal heating is dissipated. Dissipation location constrains the mechanisms driving heat transport within Io. The sum of these processes controls how the interior of Io continues to evolve.

It is also important to quantify other processes that might superheat magma during ascent to the surface. Unlike *Voyager*, the *Galileo* spacecraft had instruments (NIMS and SSI) that were capable of measuring thermal emission and constraining the temperature from newly erupted lava. Methods for determining minimum lava eruption temperatures from remote-sensing data are described in later chapters.

5

Magmas and volatiles

The surface of Io is covered in sulphur dioxide (SO_2), sulphur, and silicates. Based on images of lava flows, estimates of lava eruption temperatures, and observed topography, silicate flows are common. The absence of steep-sloped ($>10°$) volcanic edifices (e.g., Clow and Carr, 1980; Schenk *et al.*, 2004) suggests that lavas are fluid and of low viscosity, which indicates they are low in silica and therefore more akin to basalt or ultramafic composition lavas than silica-rich, high-viscosity lavas such as terrestrial rhyolite and dacite.

Volcanogenic elemental sulphur forms deposits on the surface of Earth and Io. Secondary sulphur flows, comprised of re-mobilized deposits of volcanogenic sulphur, are found on Earth and are probably quite common on Io as well.

SO_2 dissolves in silicate and sulphur magma under sufficient pressure. It is a common volatile in terrestrial magma and is ubiquitous on the surface of Io (Carlson *et al.*, 1997). SO_2 plays an important part in Io's volcanism, driving most plume activity. SO_2 in the lithosphere and on the surface of Io is readily mobilized by contact with hot silicates and sulphur.

This chapter reviews the physical and thermodynamic properties of these materials and their role in Io's volcanism. The first part of the chapter considers silicate magmas. The second considers sulphur and sulphur dioxide.

5.1 Basalt

Basalt is a general term for fine-grained, dark volcanic rocks comprised of primarily plagioclase feldspar and pyroxene in roughly equal abundances with less than 20% of other minerals (olivine, calcium-poor pyroxene, and iron–titanium oxides) (BVSP, 1981). The canonical silica content range for basalt is 45% to 52%, although more recent definitions include other varieties of basalt: low-titanium lunar mare basalts and eucrites are examples in which a low-calcium pyroxene is the dominant mafic mineral. Basalts are erupted partial melts of planetary mantles and, as such, provide important information about planetary composition and evolution. All terrestrial planets have silicate mantles, with compositions similar to those of chondritic meteorites, the basic building blocks of the Solar System. Partial melting of chrondritic silicates produces basalts (BVSP, 1981). Greater degrees of melting produce more magnesium-rich melts, described later. Basalt is the most common igneous rock on Earth, making up the oceanic crust, and has been found in large quantities on the Moon, Mars, and Venus.

On Io, temperatures derived from ground-based observations before *Galileo* reached Jupiter indicated the presence of liquid silicates at temperatures in the basalt range ($\approx$1300–1500 K) (e.g., Johnson *et al.*, 1988; Veeder *et al.*, 1994; Blaney *et al.*, 1995; Davies, 1996). The low viscosity of basalt means that volatile degassing takes place easily. Basalt erupted onto planetary surfaces forms flows with low aspect ratios (the ratio of thickness to area) that can extend for hundreds of kilometers.

Flow forms are generally either blocky, massive flows called "a'a" or relatively smoother-surfaced flows, called "pahoehoe." In both cases, the interiors of the flows are kept hot by the insulating properties of the crust. Pahoehoe flows have a lower viscosity than a'a flows. High-volume eruptions often begin with lava fountaining (also known as "fire fountaining") as gas-rich lava erupts when elastic stress on the magma supply system is at a maximum (Wadge, 1981). In a vacuum, basalt and other silicates will undergo evaporative loss of volatiles, a process that increases with increasing melt temperature. The loss of alkalis (K, Na) may explain the presence of these elements in Io's atmosphere (see review in Schaefer and Fegley, 2004).

Three modes of generating basalt by partial melting of the Earth's mantle (peri-dotite) are illustrated in Figure 5.1: (a) increasing temperature, (b) decreasing pressure, and (c) lowering the melting curve by addition of volatiles (such as H_2O, CO_2, or SO_2). Increasing the melt fraction generates ultramafic composition fluids.

5.2 Ultramafic magma

Ultramafic magmas are almost entirely comprised of magnesium and iron silicates (Lesher *et al.*, 1984). Olivine is the dominant mineral, but minor amounts of pyroxene and calcium plagioclase may be present. The Earth's mantle is ultramafic, composed of peridotite. Xenoliths of peridotite are often found in basaltic lava. Early in Earth's history, higher interior temperatures allowed greater degrees of partial melting, producing voluminous, high-temperature, low-viscosity ultramafic magmas, such as komatiites. Once common on Earth, massive ultramafic lava flows exist today as altered or metamorphosed greenstones found almost exclusively in Precambrian terrains (Arndt and Nisbet, 1982). As summarized by Williams *et al.* (2001a), there is considerable debate over maximum MgO content, volatile content, eruption temperature, and style of flow emplacement.

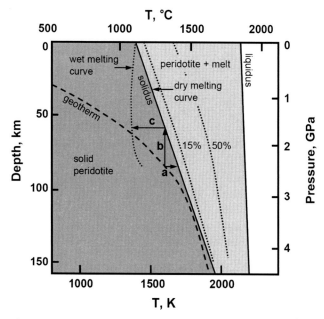

Figure 5.1 Three ways of generating basaltic magma from partial melting of the Earth's peridotite mantle by bringing together the mantle solidus and the geotherm: (a) increasing temperature (e.g., by mechanical friction); (b) decreasing pressure in a dry system, the result of mantle convection, where material rises fast enough that heat cannot be lost through conduction; and (c) the addition of volatiles, which lowers the melting point of rock. Also shown are the 15% and 50% melt curves for mantle peridotite. Large melt fractions produced in a younger, hotter Earth yielded ultramafic magmas, including komatiites.

The komatiites of the Commondale greenstone belt of South Africa are excellent examples of ultramafic lavas and are noted for their large spinifex olivine crystals. Like all ultramafic lavas, komatiites are inferred to have erupted at high temperature (in excess of 1800 K) and low viscosity, possibly resulting in turbulent flows and the production of thermal erosion channels (Williams *et al.*, 2001a). Komatiite flows appear to have been emplaced as fields of multiple flow units, indicative of episodic activity. A wide range of facies is found, including flood or sheet flows; sheet flows with a dominant, centralized pathway; flows confined in channels and lava tubes; compound lobed flows; pillowed and ponded flows; and sub-volcanic intrusives (Arndt *et al.*, 1979; Lesher *et al.*, 1984; Hill *et al.*, 1995; Parman *et al.*, 1997). Flow thicknesses range from tens of centimeters to tens of meters. All surviving terrestrial komatiites appear to have been emplaced in sub-aqueous environments, and most are thought to have been emplaced on the deep ocean floor. Although emplacement as turbulent, channeled flows is the conventional paradigm for komatiites (Huppert *et al.*, 1984; Huppert and Sparks, 1985; Jarvis, 1995), it has also been suggested that emplacement may have been laminar. With lava insulated under crusts, the injection of new lava causes inflation of the flow lobe, which then advances when the lava breaks out (e.g., Hill *et al.*, 2002).

5.3 Lava rheology

The rheological properties of magmas are expressions of the internal structure of molten silicates. Silicate rheology is reviewed by McBirney and Murase (1984). The two most important factors are viscosity and yield strength, which influence flow surface morphology, the size and shape of a flow, flow thickness and width, length of flow, velocity of flow, and effusion rate.

5.3.1 Viscosity

Viscosity is the ratio of shear stress to strain rate, expressed in units of Pascal seconds (Pa s). Magma viscosity is strongly controlled by silica content (the higher the silica content, the higher the viscosity) and crystal content (a function most strongly controlled by temperature) and is also affected by pressure, volatile content, and the thermal and mechanical histories of the melt. Ultramafic magmas, at temperatures in excess of 1800 K, have no silica polymerization and therefore have very low initial viscosity. Viscosity is important because of the role it plays in factors that control the style of volcanic eruption and the physical nature of generated products. The viscosity of an undisturbed magma may increase with time before leveling off, the result of an increasing ordering and polymerization of silica tetrahedra.

Lava viscosity η can be estimated using the Einstein-Roscoe equation

$$\eta = \eta_0 (1 - R\phi)^{-2.5}, \tag{5.1}$$

where η_0 is the viscosity of the liquid alone, ϕ is the crystal content, and R is a constant based on the volumetric ratio of solids at maximum packing, which has been found to be 1.67 (Marsh, 1981).

Viscosity as a function of temperature for basalt and ultramafic lavas is shown in Figure 5.2. Basalts at Pu'u 'O'o, an active vent of Kilauea, Hawai'i, typically erupt at a temperature of $\approx$1420 K with a viscosity of $\approx$300 Pa s, although this viscosity increases rapidly with increasing heat loss (Hon *et al.*, 2003). Because of their high temperatures, komatiitic lavas would have had even lower viscosities, perhaps as low as 0.1 Pa s (Arndt and Nisbet, 1982).

A high-temperature tachylite (vitreous basalt) flow at Mauna Iki, Hawai'i, covered a viscosity range from 7.6 Pa s at 1587 K to 495 Pa s at 1347 K, and an olivine nephelinite (an alkali basalt) in Japan covered a range from 8 Pa s at 1673 K to 19 Pa s at 1473 K (Clark, 1966). More common tholeiitic basalts, such as those from Mauna Loa, span the range 1.4×10^2 to 5.6×10^6 Pa s (Moore, 1987). For

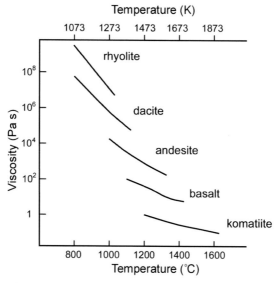

Figure 5.2 Viscosity vs. temperature for silicates ranging from high-silica-content rhyolite to fluid komatiite.

lunar basalts, it is thought that viscosity values are typically an order of magnitude lower than for terrestrial lavas (Murase and McBirney, 1970). If superheating of ascending magma by more than 100 K (discussed in Chapter 9) is taking place on Io (Keszthelyi *et al.*, 2005a), then basalts would be erupted with very low viscosities, perhaps comparable to those inferred for ultramafic lavas. Superheated lavas with no crystals behave as Newtonian fluids.

Viscosities η for moving lava flows have been crudely estimated using the Jefferies equation for laminar flow of an extensive sheet on a uniform slope where

$$\eta = \frac{g\rho(\sin\,\alpha)\,d^2_{(\eta,\rho,g)}}{3\,V_{(\eta,\rho,g)}}, \qquad (5.2)$$

where g is gravity, ρ is density, α is slope angle, $d_{(\eta,\rho,g)}$ is flow thickness, and $V_{(\eta,\rho,g)}$ is mean flow velocity. For channeled flows, the denominator 3 is replaced by 4. Flow velocity therefore increases with increasing slope and gravity and decreasing viscosity.

5.3.2 Yield strength

Cooling and crystallizing magmatic liquids behave as Newtonian fluids until they contain about 20% crystals. At that point, yield strength is the dominant rheological parameter. Yield strength is an important physical boundary for interpreting how lava will behave. A highly polymerized liquid such as lava acts as a Bingham fluid and deforms permanently (flows) only when its yield strength is exceeded. As noted by McBirney and Murase (1984), lava can deform elastically before or after yield strength is exceeded, but such deformation is recovered as the deformational stress is removed.

As temperature falls and crystal content increases, yield strength increases. As a result of cooling at flow margins, yield strength soon greatly exceeds shear strength, and levee formation may begin.

5.3.3 Other properties

Density, specific heat capacity, latent heat capacity, and other physical properties of basalt and other materials are shown in Table 5.1. The density and thermal conductivity of solidified basalt lava depend on initial volatile content, the degree to which the lava loses volatiles, and the speed at which the lava solidifies. Basalt is a relatively low-viscosity lava. Therefore, exsolving gases can escape easily, yielding relatively volatile-free lava and effusive, rather than explosive, volcanic activity. On airless bodies like Io and the Moon, however, even a low residual volatile content (<0.01 wt%) may, under the effects of erupting into a vacuum, result in explosive activity (Wilson and Head, 1981) or the emplacement of lava with a low-density, void-filled (vesicular) crust (Matson *et al.*, 2006b). A porous crust on an airless body has a lower thermal conductivity than solid lava. The lava surface cools rapidly, and heat is transported from the hot interior at a lower rate than through a higher density crust.

The amount of volatiles that may be dissolved in magma depends on several factors, primarily the composition of the melt, temperature, and pressure. The solubility of most volatiles increases with increasing pressure. On Earth, the most common volcanic volatiles are, in decreasing amounts, H_2O, CO_2, SO_2, HCl, F, Cl, and a small amount of other gases. On Io, volcanic volatiles are dominated by SO_2 and sulphur. The total amount of sulphur that can be dissolved in magma depends most strongly on the magma's iron content. The solubility of sulphur and HCl in magma is further complicated by the formation of non-volatile compounds. On Io, it is thought that most sulphur was incorporated into an FeS-rich core during Io's formation. Remaining elemental sulphur is found on the surface and probably throughout much of the lithosphere, where it is constantly recycled.

5.4 Sulphur

Sulphur is a complex element that forms many allotropes exhibiting a variety of physical properties. Many allotropes of sulphur have been identified, but only

Table 5.1 *Physical characteristics of volcanic materials proposed for Io*

Property	Units	Basalt	Ultramafic	Sulphur (S_8)	SO_2
Density solid	$kg\,m^{-3}$	2800	2845	2000	1900
Density liquid	$kg\,m^{-3}$	2600	2680	1800 (392 K) 1610 (717 K)	1500 (263 K)
Liquidus temp	K	1400–1600	1800–1900	400	198
Vapor temp	K	1700+ (SiO_2)[a]	1800–1900 (Al_2O_3)[b] 1900 (MgO)[c]	717 K (10^5 Pa)	200 (10^{-5} Pa)
Specific heat capacity (solid)	$J\,kg^{-1}\,K^{-1}$	1200–1500	1500–1700	606 (200 K) 800 (376 K)	$\approx$1000
Specific heat capacity (liquid)	$J\,kg^{-1}\,K^{-1}$	1200	1750	1012 (400 K) 1861 (431 K) 1185 (700 K)	$\approx$1000
Specific heat capacity (vapor, 10^5 Pa)	$J\,kg^{-1}\,K^{-1}$	–	–	1193 (717 K) 15 115 (922 K) 1047 (1422 K)	592 (250 K)[d] 816 (800 K)[d] 884 (1500 K)[d]
Latent heat of fusion	$J\,kg^{-1}$	3–4×10^5	8×10^5–10^6	3.81×10^4	10^5
Latent heat of vaporization	$J\,kg^{-1}$	–	–	3.26×10^5	3.52×10^5 [e]
Thermal conductivity	$W\,m^{-1}\,K^{-1}$	0.9–4.0	0.5–1.0	0.2 (300 K) 0.16 (500 K)	3.5 (solid) 0.20 (liquid, 290 K) 0.13 (liquid, 400 K)

(continued)

Table 5.1 *(continued)*

Property	Units	Basalt	Ultramafic	Sulphur (S_8)	SO_2
Yield strength	Pa	Io ≈ 10–10^3 [f] Earth 10^2–10^3 Moon/Mars $\approx 10^2$–10^4 [g]	10–100?	≈ 1	≈ 0.1?
Coefficient of thermal expansion	$K^{-1}/10^4$	0.3	0.3	6.661 (394 K) 2.440 (444 K) 4.587 (671 K)	–
Viscosity of liquid	Pa s	10–1000	0.1–10	0.01 (388 K) 93 (460 K) 1.8 (588 K)	5.5 × 10^{-4} (240 K) [e] 3.9 × 10^{-4} (273 K) [e]
Viscosity of vapor	Pa s	–	–	–	1.2 × 10^{-5} (273 K) [e] 1.6 × 10^{-5} (373 K) [e]
Tensile strength at 300 K [h]	kPa	24	24	1.4	–
Compressive strength at 300 K [h]	kPa	207	207	21	–

Note: If a given value is followed by a number in brackets, that number denotes a composition, pressure or temperature on which the value is dependent.

[a] Shornikov *et al.* (1999).
[b] See Schaeffer and Fegley (2004).
[c] Peleg and Alcock (1974).
[d] After Touloukian and Ho (1970).
[e] International Critical Tables (ICT, 1929).
[f] Estimated by Schenk *et al.* (2004).
[g] Lunar mare and martian shield basalts (see references in Cattermole, 1996).
[h] After Nash *et al.* (1986).

two – orthorhombic (α) and monoclinic (β) sulphur – are thermodynamically stable. Both are composed of cycloocta (S_8) molecules. Material properties of S_8 are provided in Table 5.1. Other forms of solid sulphur, most notably allotropes containing cyclohexa (S_6) sulphur and cyclodeca (S_{10}) sulphur, are produced by the reaction of sulphur compounds (Meyer, 1976). S_6, S_{10}, and other allotropes containing more sulphur atoms are metastable, reverting to a more stable form. Monoclinic sulphur will convert to the orthorhombic form within a month if temperatures drop below 368 K. Other solid allotropes are produced by quenching of liquid sulphur.

The equilibrium composition of liquid sulphur is strongly dependent on temperature. Liquid sulphur consists of a complex mixture of allotropes. Meyer (1977) identified S_2, S_3, S_4, S_5, S_6, S_7, S_8, S_{12}, polycatena (long S_8 chains), S_∞ (polymers), and S_π, a mixture of allotropes. Above 432 K, linear polymers form. At this point, almost all properties of liquid sulphur suffer a discontinuity, a result of the polymerization process (Tobolsky and Eisenberg, 1959; Touro and Wiewiorowski, 1966b), where S_8 rings convert to S_8 chains, which then join to form polycatena.

The average number of catena (chains) per polychain increases from 1 at 393 K to a peak of 10^5 between 438 K and 450 K (MacKnight and Tobolsky, 1965). Above 450 K, the average chain length decreases. The equilibrium of the liquid is affected by impurities such as iodine, chlorine, carbon dioxide, hydrogen sulphide, and arsenic and by the other allotropes in the melt.

A third class of species, S_π, manifests itself mainly in the liquid below 432 K and consists of a complex mixture of S_6, a portion of mostly S_8 but with an average composition of $S_{9.2}$, and a component comprised of large rings of S_n, where $20 < n < 33$.

Sulphur vapor consists of polymers ranging from S_2 to S_{12} (Meyer, 1976), with the fraction of each constituent determined by temperature and pressure. The number of S atoms in the molecule decreases as vapor temperature increases. At 1000 K, S_2 is the most abundant allotrope (Rau *et al.*, 1973), becoming even more dominant at higher temperatures. If exsolving from liquid magma or in contact with molten rock, S_2 is the most likely initial allotropic form of sulphur (Reynolds *et al.*, 1980).

5.4.1 Viscosity

The viscosity of liquid sulphur undergoes dramatic changes as a function of temperature, as shown in Figure 5.3. After melting begins, viscosity decreases with increasing temperature to <0.01 Pa s at ≈430 K, after which viscosity increases very rapidly to a maximum of ≈93 Pa s at 463 K, the result of an increase in concentration and length of polycatena. The decrease in viscosity above 463 K is due to a decrease in the number of S_8 chains per polycatena molecule (Touro and Wiewiorowski, 1966b). Viscosity is profoundly reduced by the presence of impurities (organics,

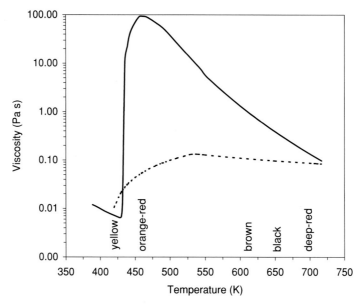

Figure 5.3 Viscosity of sulphur vs. temperature (after Tuller, 1954). The dotted line is for sulphur doped with 0.1% to 0.2% H_2S. H_2S and other impurities cap the ends of sulphur polycatena, preventing the formation of long catena and reducing viscosity (Rubero, 1964). Also shown are colors of pure sulphur liquid.

ammonia, hydrogen sulphide [H_2S]) that cap the ends of catena. The effect of H_2S on viscosity is also shown in Figure 5.3.

At temperatures close to its melting point and at temperatures greater than $\approx$523 K, sulphur is less viscous than most terrestrial lavas.

5.4.2 Other thermodynamic properties

Solid sulphur has a density of about 2000 kg m^{-3}, and liquid sulphur decreases in density from 1800 kg m^{-3} at 392 K to $\approx$1600 kg m^{-3} at 717 K. The effect of polymerization between 432 K and 460 K causes considerable variability of other physical properties, such as density and specific heat capacity (Table 5.2). The phase diagrams of sulphur and SO_2 are shown in Figure 5.4. Stable sulphur pools at a temperature of $\approx$500 K have been postulated to exist on the surface of Io (Lunine and Stevenson, 1985).

5.4.3 Color and temperature

The colors of various sulphur allotropes, especially those that retain their color on quenching, are of great significance when considering Io and the appearance of possible sulphur flows and other deposits on the surface. Sagan (1979) proposed that

Table 5.2 *Physical properties of liquid sulphur with temperature*

Temperature (K)	Viscosity (Pa s)	Viscosity, H_2S-doped S (Pa s)	Density (kg m^{-3})	Specific heat capacity (J kg^{-1} K^{-1})	Thermal conductivity (W m^{-1} K^{-1})
392	0.013	0.0127	1805.7	996.3	0.1326
400	0.011	0.0108	1798.8	1031.0	0.1349
410	0.009	0.0089	1790.1	1123.0	0.1377
420	0.007	0.0105	1781.4	1269.1	0.1405
430	0.006	0.0215	1773.5	1469.3	0.1434
440	26.3	0.0325	1771.2	1511.8	0.1462
450	59.2	0.0435	1771.6	1476.6	0.1490
460	91.9	0.0545	1765.4	1443.7	0.1518
465	82.0	0.0600	1762.3	1428.1	0.1532
470	73.3	0.0655	1759.2	1412.9	0.1546
480	59.0	0.0765	1753.0	1384.0	0.1575
490	48.1	0.0875	1746.8	1356.9	0.1603
500	32.5	0.0985	1740.6	1331.3	0.1631
525	13.0	0.1260	1725.1	1273.5	0.1702
550	5.7	0.1276	1709.6	1223.1	0.1772
575	2.7	0.1180	1694.1	1178.8	0.1843
600	1.3	0.1099	1678.6	1139.6	0.1913
625	0.7	0.1030	1663.1	1104.8	0.1984
650	0.4	0.0969	1647.6	1073.5	0.2054
675	0.22	0.0917	1632.1	1045.4	0.2125
700	0.13	0.0870	1616.6	1019.9	0.2195
717	0.10	0.0842	1606.1	1003.9	0.2243

Note: H_2S-doped S data from Rubero (1964).

color variations along lava flows and in the vicinity of volcanic vents were a result of the preservation, by quenching, of short-chain sulphur allotropes, and Pieri *et al.* (1984) used the color–temperature and viscosity–temperature relationships to explain possible flow features in the Ra Patera region in terms of allotrope concentration changes. However, a re-evaluation of *Voyager* imagery showed that the proposed color variations along lava flows were not actually present in the *Voyager* data (see Young, 1984; McEwen, 1988). Nevertheless, Io was the source of sulphur in the Io torus, so sulphur, in some form, was present on the surface or was a major plume constituent.

The color of elemental sulphur is as complex as its chemistry. Colors of pure solid allotropes are provided in Table 5.3 and colors of pure liquid sulphur in Figure 5.3. Pure liquid sulphur is bright yellow from its melting point up to 423 K. At higher temperatures, the color of liquid sulphur turns orange, then red, and, at temperatures in excess of 600 K, to brown, and almost to black in all but the purest sulphur. These latter changes may be due to impurities, an indication of the reactivity of sulphur

Table 5.3 *Color of solid sulphur allotropes*

Allotrope	Symbol	Color
Cycloduo, cycloquatro (quenched)	S_2, S_4	Red
Cyclohexa	S_6	Orange-red
Cyclohepta	S_7	Light yellow
Orthorhombic-α	S_8	Bright yellow
Monoclinic-β	S_8	Light yellow
Cycloenna	S_9	Deep yellow
Cyclodeca	S_{10}	Yellow-green
Cyclooctadeca	S_{18}	Lemon yellow
Cycloicosa	S_{20}	Pale yellow

From Meyer (1976).

above 600 K. Pure liquid sulphur changes to a deep, opaque red at the boiling point of 717 K.

The color of quenched solid sulphur is strongly dependent on the rate of cooling. A melt at 700 K quenched to 193 K will be bright yellow, but, if quenched in liquid nitrogen (64 K to 77 K), a bright red glass is produced. The surface of Io (90 K to 130 K) would quench sulphur in the range between yellow ($\approx$193 K) and red ($\approx$78 K).

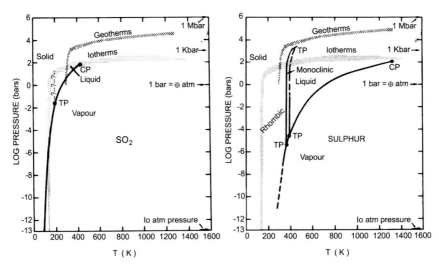

Figure 5.4 Pressure–temperature phase diagrams for sulphur dioxide and sulphur (Kieffer, 1982), with Iotherms from Smith *et al.* (1979a). Temperatures in the ionian crust probably do not increase with depth as quickly as shown here, in regions away from volcanic activity. CP is the critical point and TP the triple point. © 1982 The Arizona Board of Regents. Reprinted by permission of the University of Arizona Press.

5.4.4 Sulphur volcanism

Lava flows of elemental sulphur have been found at several terrestrial volcanoes. These flows are almost certainly "secondary" flows, which form from pre-existing deposits (e.g., Greeley *et al.*, 1990). The process begins with fumarolic activity that deposits elemental sulphur near the surface in poorly consolidated volcanic talus. Sulphur accumulates over time. Subsequent magmatic activity melts the sulphur, which then mobilizes to form a flow. In 1889, the Japanese volcano Siretoko-Iosan (literally "sulphur mountain") erupted a large amount of molten sulphur. In 1936, another event took place during which more than 80 000 tons of very pure sulphur were erupted (Watanabe, 1940). The highly fluid sulphur eventually filled a valley to a length of 1400 m, in places to a depth of 50 m. Much of the sulphur was dark brown in color and of low viscosity.

A similar event took place along the southwest rift of Mauna Loa, Hawai'i, in 1950 (Skinner, 1970). The resulting sulphur flow was yellow, about 27 m long and 14 m wide, with a thickness of 0.1 m to 0.45 m (Greeley *et al.*, 1984). A small tube, about 0.3 m wide, was found within the flow. The entire flow was emplaced as a single unit. The sulphur was over 99.9% pure, with a trace amount of calcium.

The bright yellow color of the Mauna Loa sulphur flow indicates an absence of red S_2 and S_4 allotropes, which would be present if sulphur had been heated to above 450 K. Because sulphur heated to no more than 420 K reverts to a lemon yellow, as noted by Cattermole (1996), the sulphur at these locations was erupted above its melting point but below 420 K. Other sulphur flows have been found at Volcan Azufre (Galapagos Islands), Ebeko (Kurile Islands), and Lastarria (Chile). Lakes of liquid sulphur up to 25 by 20 m in size were observed at Poàs (Costa Rica) by Oppenheimer and Stevenson (1989).

Industrial sulphur flows form when liquid sulphur is pumped into vats to solidify and cool. Studying such flows, Greeley *et al.* (1990) found that sulphur flows advance as both unconfined lobes and via channels. Flows can crust over, and, as a result, breakouts along flow margins are common. The very low thermal conductivity of sulphur means that heat loss from the interior of a flow, under an insulating crust, is low.

Such forms of secondary volcanism should be common on Io, given an abundance of sulphur compounds on the surface. Fink *et al.* (1983) studied the emplacement of sulphur lava flows in the ionian environment and concluded that, once mobilized, sulphur flows could conceivably flow for a considerable distance, with some additional morphological effects not encountered with basalts. The unique viscosity–temperature profile of sulphur means that zones of low viscosity would form within a laminar sulphur lava flow erupted above 560 K (Fink *et al.*, 1983), which might cause the outer flow layers and crust to slough off, leading to a unique morphology that could be used for identifying sulphur flows. This process was not apparent at the 1950 Mauna Loa sulphur flow because the sulphur was not hot enough for multiple zones with widely differing viscosities to develop.

5.5 Sulphur dioxide (SO_2)

Physical properties of SO_2 are listed in Table 5.1. *Voyager* IRIS detected gaseous SO_2 in the volcanic plume at Loki (Pearl *et al.*, 1979b). SO_2 plays a major role in Io's explosive volcanism as the driving force behind most plume activity (e.g., Smith *et al.*, 1979a) and also forms a tenuous atmosphere (Kumar, 1979). At the same time as the *Voyager* discovery, laboratory experiments identified SO_2 absorption features in Io's spectrum that were caused by SO_2 frost on Io's surface (Fanale *et al.*, 1979; Smythe *et al.*, 1979).

Galileo mapped SO_2 distribution over much of Io. SO_2 is ubiquitous on the surface of Io except in the location of hot spots, where extreme heat drives off local SO_2. This highlights the difficulty of identifying other substances on Io's surface: SO_2 covers almost everything to some degree. Hot spots are the exception but, unfortunately, these locations are almost always either sub-pixel, or available data lack the high spectral resolution necessary for detailed spectrographic analysis. Away from active volcanic centers, SO_2 is found in varying concentrations and grain sizes (Carlson *et al.*, 1997; Douté *et al.*, 2001, 2004).

Released from magma during volcanic activity, SO_2 can condense as frost on Io's surface. Deposits are buried under subsequent silicate and sulphurous deposits and can be remobilized through thermal exchange with hot silicates or hot sulphur, both on the surface and by intrusions (e.g., Keszthelyi *et al.*, 2004a). Recent crustal modeling infers that the upper few kilometers of Io's crust are rich in SO_2, although this layer may exhibit great local variability (Jaeger and Davies, 2006). Localized high concentrations are in line with initial *Voyager* observations of surface morphology (McCauley *et al.*, 1979).

SO_2 plays a role in the volcanic cycle on Io in very much the same way that water does on Earth. H_2O and CO_2, common volcanic volatiles on Earth, have not been detected on Io in significant concentrations. The high level of volcanic activity and low gravity of Io means that most or all water has been lost. CO_2 does not last significantly longer than water but may survive as a minor volcanic volatile (Kumar, 1979). The relatively heavy SO_2 emerges as the most robust volatile, a compound that can survive on Io's surface. Io has subsequently evolved a surface and lithosphere rich in SO_2. Magma–SO_2 interactions drive much of Io's volcanic plume activity (Kieffer, 1982; Johnson *et al.*, 1995). Exsolving SO_2 reduces the bulk density of magma and accelerates magma in conduits toward the surface. Magma rising through the crust to the surface interacts with crustal SO_2. Surface flows

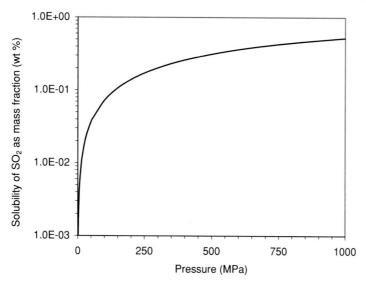

Figure 5.5 Theoretical solubility of SO_2 in basalt magma on Io as a function of pressure, after Leone and Wilson (2001) and Matson *et al.* (2006b). Basalt becomes saturated with a weight fraction of SO_2 of $\approx$0.2% at a pressure of about 300 MPa.

mobilize SO_2 in the same manner that terrestrial lava flows moving over permafrost mobilize water (Kieffer *et al.*, 2000; Milazzo *et al.*, 2001).

The Io thermal gradient in the upper 80% of the lithosphere is essentially zero (O'Reilly and Davies, 1981) because the rate at which internal heat is conducted to the surface is surpassed by the rate at which the surface is buried by cold volcanic deposits. This near-zero thermal gradient would result in sulphur's being solid in the lithosphere to depths approaching 25 km, except where locally mobilized by silicate activity. On the other hand, SO_2 (liquid at temperatures below the freezing point of sulphur) should be a low-viscosity liquid throughout most of the lithosphere (Keszthelyi *et al.*, 2004a).

5.5.1 Solubility of SO₂ in magma

Knowledge of the solubility of SO_2 in basalt is sparse. Using Mysen's (1977) data for an albite melt, Leone and Wilson (2001) developed the expression

$$n_{SO2} = 7.5 \times 10^{-12} P - 2.6 \times 10^{-21} P^2 + 3.2 \times 10^{-31} P^3 \tag{5.3}$$

to determine the equivalent saturation mass fraction of SO_2, n_{SO2}, in basalt. P is pressure in Pascals. At high pressure (above 300 MPa, a depth of $\approx$50 km on Io and $\approx$11 km on Earth), the magma saturates at a SO_2 fractional mass of $\approx$0.2%. During ascent, the fractional mass of dissolved SO_2 decreases rapidly with decreasing pressure and is largely exsolved by the time magma nears the surface. Figure 5.5

Magmas and volatiles

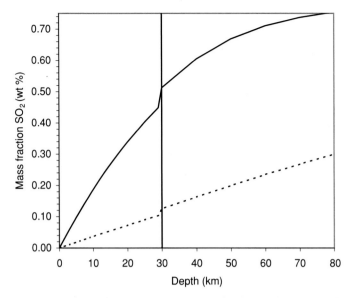

Figure 5.6 Theoretical solubility of SO_2, as a mass fraction, in basalt, as a function of depth for Earth (solid line) and Io (dashed line). The vertical solid line at 30 km represents a lithosphere–mantle boundary, where mean density changes from 2800 kg m^{-3} in the lithosphere to 3300 kg m^{-3} in the mantle. Earth's steeper pressure gradient means that SO_2 saturates basalt on Earth at a shallower depth than on Io.

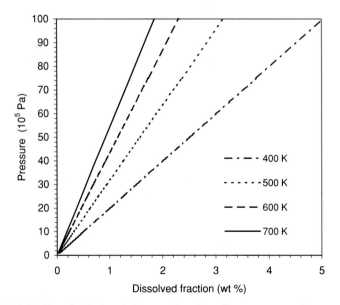

Figure 5.7 Solubility of SO_2 in sulphur as a function of temperature and pressure. After Touro and Wiewiorowski (1966a).

shows the theoretical SO$_2$ mass fraction in basalt for Io as a function of pressure and Figure 5.6 the theoretical solubility as a function of depth for Io and Earth.

5.5.2 Solubility of SO$_2$ in sulphur

The solubility of SO$_2$ in sulphur (Figure 5.7) assumes particular interest on Io, where both volatiles are being buried and recycled together. Touro and Wiewiorowski (1966a) determined that SO$_2$ solubility increased with increasing pressure and decreasing temperature, such that

$$\ln(c/P_b) = 931/T - 5.32, \tag{5.4}$$

where P_b is pressure in bars (10^5 Pa), T is temperature in kelvin, and c is concentration in weight %. Liquid sulphur rising from depth would start to exsolve SO$_2$ close to the surface, a process that is offset if the temperature of the sulphur decreases due to heat loss.

Section 3

Observing and modeling volcanic activity

6

Observations: thermal remote sensing
of volcanic activity

Many aspects of volcanic activity such as thermal emission, gas emission, and the changing shape of a volcano can be studied remotely (see Mouginis-Mark *et al.*, 2000). This chapter, however, limits discussions of remote sensing to techniques also available in the study of volcanism on Io.

6.1 Remote sensing of volcanic activity on Earth

Remote sensing has become an essential tool for terrestrial volcanologists since its origins in the data collected in the mid 1960s by the High-Resolution Infrared Radiometer (HRIR) on *Nimbus 1*. Those data were used to show that the Hawaiian volcano Kilauea had a higher infrared radiance than Mauna Loa, its then-inactive neighbor (Gawarecki *et al.*, 1965). More than four decades later, Earth-orbiting platforms are now being used to detect and monitor volcanic activity at different temporal, spatial, and spectral resolutions (Plate 3). Reviews of the development of spacecraft, orbits, sensor capabilities, and data analysis techniques up to the launch of the first Earth Observing System (EOS) spacecraft (*Earth Observing 1* [*EO-1*], *Terra* and *Aqua*) are summarized in a series of papers collected in the monograph *Remote Sensing of Volcanic Activity* (Mouginis-Mark *et al.*, 2000). The reader is specifically directed to Mouginis-Mark and Domergue-Schmidt (2000) for their comprehensive appraisal of the strengths and limitations of terrestrial satellite remote sensing capabilities. More recent EOS spacecraft observations of active volcanism are described by Ramsey and Flynn (2004) and references therein, and by Davies *et al.* (2006a).

The flight of the first Earth-orbiting high-spatial-resolution hyperspectral imager, Hyperion (Pearlman *et al.*, 2003), and the Advanced Land Imager (ALI) on *EO-1* (Ungar *et al.*, 2003); the Advanced Spaceborne Thermal Emission and Reflection Radiometer (ASTER) (Yamaguchi *et al.*, 1998), the high-spatial-resolution multi-spectral (visible and infrared) imager on *Terra*; and the Moderate-Resolution

93

Imaging Spectroradiometer (MODIS) on *Terra* and *Aqua* have yielded observations of volcanoes at spatial resolutions as high as 10 m per pixel (ALI), temporal coverage up to four times a day or better for high-latitude targets (MODIS), contemporaneous observations across a wide range of wavelengths (ASTER), and spectral resolutions of 10 nm (hyperion has 196 usable, discrete bands from 0.4 to 2.5 μm, covering visible and short infrared wavelengths). In the final years of the twentieth century and early years of the twenty-first, the proliferation of orbiting sensors has increased the pace of data acquisition dramatically, leading to the development of automated systems to process and mine the huge volumes of collected data for nuggets of high-value science content. Direct broadcast of satellite imaging data, for example, from MODIS, bypasses traditional routes of data transmission via a small number of ground stations and has been coupled with automatic data-processing applications to rapidly detect anomalous (above-background) thermal emission.

Two such detection systems are based at the University of Hawai'i. MODVOLC (Wright *et al.*, 2004b) processes daily MODIS data, and GOESvolc (Harris *et al.*, 2000a) processes *GOES* (*Geostationary Operational Environmental Satellite*) data from the Pacific Rim at lower spatial, but higher temporal (15-minute), resolution.

Sophisticated computer applications have been developed to correct infrared data for atmospheric effects, involving two fundamental elements: (1) a radiative transfer model capable of estimating the magnitude of atmospheric emission, absorption, and scattering; and (2) the acquisition of the necessary atmospheric parameters (e.g., temperature, water vapor, ozone, and aerosol profiles) at the time and location of the observation (Abrams, 2000). A widely used model is MODTRAN (Anderson *et al.*, 1993), which includes look-up tables for atmospheric structure as a function of season and location.

Currently, a trade-off exists between high-orbit sensors, which yield high-temporal-resolution coverage but at low spatial resolution, and low-orbiting spacecraft, which return high-spatial-resolution data at low temporal resolution. One solution has been to utilize high-orbit assets with global or hemispheric coverage, or low-orbit assets with wide fields of view and low spatial resolution, to trigger low-orbiting assets with high-spatial-resolution capability to obtain data. Such a capability has been developed at NASA's Jet Propulsion Laboratory, where an autonomous system re-tasks *EO-1* on receipt of detections of current or impending volcanic activity from MODVOLC alerts (Davies *et al.*, 2006b). ASTER is now also being retargeted on receipt of notifications of ash emissions from northern Pacific Rim volcanoes (Ramsey and Dehn, 2004).

Even more rapid responses, at best within a few hours of initial observation acquisition, have been obtained by the Autonomous Sciencecraft Experiment (ASE) on *EO-1*. ASE autonomously processes Hyperion data onboard *EO-1*, detects anomalous thermal emission, and re-tasks the spacecraft to obtain more observations. Results from onboard data analyses are downlinked within hours of acquisition for further processing on the ground, preceding the full Hyperion dataset by weeks (Chien *et al.*, 2005; Davies *et al.*, 2006a). Such rapid notifications of remote eruptions can be used for alerting authorities and as an aid in hazard and risk assessment.

The future holds the promise of low-orbiting spacecraft that will be capable of handling the huge volumes of data from hyperspectral instruments of high spatial resolution and a wide imaging swath, thereby enabling daily observation of volcanoes worldwide.

6.2 Remote sensing of volcanic activity on Io

Visible-wavelength observations, at higher spatial resolutions than possible with infrared instruments, reveal volcanic plumes against the blackness of space by both forward- and backscattered light and identify areas of volcanic activity by changes in surface albedo or color, and in doing so provide valuable context for observations obtained by other instruments. In sufficiently high-resolution images, the style of eruption (lava flows as opposed to a lava lake, for example) can be identified. Visible-wavelength observations show that the highest temperature and most intense hot spots glow brightly in nighttime or eclipse images and in rare cases can be seen in daylight.

Nevertheless, volcanism on Io and Earth is best observed and monitored at infrared wavelengths. As described in earlier chapters, data from ground-based telescopes of anomalous, above-background thermal emission have been collected over a time span of decades (e.g., Veeder *et al.*, 1994; Rathbun *et al.*, 2002). Spatial resolution has rapidly improved from the first observations of disc-integrated thermal emission from Io, progressing to multiple-wavelength data of the same, then multiple-wavelength observations from Earth orbit, to high-spatial-resolution, multiple-wavelength observations obtained using large optical interferometers and adaptive optics techniques (e.g., Macintosh *et al.*, 1997; Marchis *et al.*, 2001; Macintosh *et al.*, 2003), where individual volcanoes are resolved.

From spacecraft, observations of thermal emission have been obtained by instruments on *Voyager*, *Galileo*, and *Cassini*. Like observations of terrestrial volcanism, techniques have been developed to determine the area and extent of volcanic activity from the analysis of visible and infrared data: changes in volcano behavior are determined from changes in a sequence of observations and from models adapted or created to quantify and understand the processes taking place.

6.3 Remote sensing of thermal emission

Infrared remote sensing most directly detects the properties of the uppermost, microns-thick layer of the surface. From the magnitude and spectral distribution of thermal emission, the temperature and area distribution of the radiating surfaces can be determined. These values for lava surface temperature, most simply thought of as having a bi-modal distribution consisting of a cooling crust and a hotter "core" temperature component revealed through cracks in the surface crust of a dome or a'a flow (Crisp and Baloga, 1990a; Pieri *et al.*, 1990), can be determined by fitting combinations of Planck function blackbody curves to multi-spectral data obtained at appropriate visible and infrared wavelengths (e.g., Flynn and Mouginis-Mark, 1992; Flynn *et al.*, 1993). The radiative power output is used to determine several eruption parameters, not least of which is an estimation of volumetric eruption rate (e.g., Crisp and Baloga, 1990a, 1990b; Pieri *et al.*, 1990; Harris *et al.*, 1997a, 1998, 1999a; Davies *et al.*, 2000a; Harris *et al.*, 2000b; Davies *et al.*, 2001; Wright *et al.*, 2001; Harris and Neri, 2002; Davies, 2003b).

The determination of lava surface temperature from spacecraft data is indeed challenging. Silicate lavas on Earth are typically emplaced at temperatures of ≈ 1000 K to 1500 K (e.g., Calvari *et al.*, 1994; Hon *et al.*, 1994; Kilburn, 2000) and, as cooling is very rapid at these temperatures (Head and Wilson, 1986; Hon *et al.*, 1994) and the areas at the highest temperatures are generally very small (e.g., Flynn *et al.*, 1993; Davies, 1996), it is often very difficult to determine the temperature of the hottest components without *in situ* measurements. Nevertheless, great strides have been made with terrestrial data since the first detailed work on high-spatial-resolution thermal volcano monitoring (Francis and Rothery, 1987). These advances are summarized in Harris *et al.* (1998), Flynn *et al.* (2000), and Wright *et al.* (2001). Foremost in the development of techniques to quantify eruption processes from thermal emission was pioneering work on *Landsat* data, from the development of the "dual-band" method (Matson and Dozier, 1981) to identify a hot component in pixels, to the recognition of the importance of choosing the correct background temperature (Glaze *et al.*, 1989). To these advances are added the development of increasingly complex techniques to constrain temperature and area (Pieri *et al.*, 1990; Oppenheimer, 1991; Oppenheimer and Rothery, 1991; Flynn *et al.*, 1994; Harris *et al.*, 1998, 1999a). While these techniques were being developed for observations of terrestrial activity, similar techniques were also being developed for understanding observations of volcanism on Io (Carr, 1986; Johnson *et al.*, 1988; Blaney *et al.*, 1995; Davies, 1996; Howell, 1997).

Analyses of terrestrial eruption data are aided to no little extent by knowledge of the thermal and chemical properties of the lava being erupted. This knowledge often allows values to be set for some variables, such as the temperature at which lava is erupted.

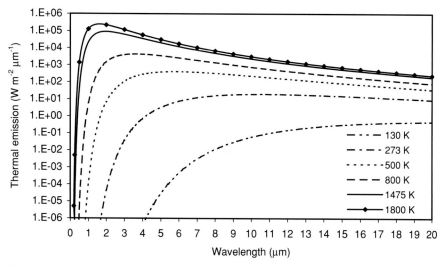

Figure 6.1 The effect of changing temperature on thermal emission at different wavelengths.

6.4 Blackbody thermal emission

Spectral radiance – the quantity of radiation emitted by a blackbody at temperature T at a given wavelength λ – is governed by Planck's Law:

$$F_{(\lambda, T)} = \frac{2hc^2}{\lambda^5 \left(e^{hc/\lambda pT} - 1\right)} \varepsilon, \tag{6.1}$$

where $F_{(\lambda, T)}$ is monochromatic thermal emission in W m^{-2} µm^{-1}, λ is in µm (10^{-6} m), T is in kelvin (K), h is Planck's constant (6.63×10^{-34} J s), c is the speed of light (3×10^8 m s^{-1}), p is Boltzmann's constant (1.38×10^{-23} J K^{-1}), and ε is emissivity. Assuming an emissivity ε of 1, Equation 6.1 is often shown as

$$F_{(\lambda, T)} = \frac{c_1}{\lambda^5 \left(e^{c_2/\lambda T} - 1\right)}, \tag{6.2}$$

where $c_1 = 3.7413 \times 10^8$ W µm^4 m^{-2} and $c_2 = 1.4388 \times 10^4$ µm K. The higher the temperature T (K) of the object, the more radiation is emitted at all wavelengths and the shorter the wavelength of maximum thermal emission λ_{max}, determined by Wien's Law, where

$$\lambda_{max} = \frac{2898}{T}. \tag{6.3}$$

Figure 6.1 shows the effect of temperature on blackbody thermal emission. The energy emitted per unit area is shown for non-volcanic background temperatures for Io ($\approx$130 K) and Earth ($\approx$273 K), for magma liquidus temperatures of basalt (1475 K) and ultramafic magma ($\approx$1800 K), and for cooling surfaces at 800 K and 500 K. This latter temperature is the estimated temperature of a postulated stable liquid sulphur lake on Io (Lunine and Stevenson, 1985). As temperature increases, thermal emission increases and the peak of thermal emission moves to shorter wavelengths. At 2 μm, a surface at 1800 K emits ≈2.5 times more energy than a surface at 1475 K, 150 times the energy of a surface at 800 K, and hundreds of thousands times more than a body at Io's non-volcanic peak daytime surface temperature of ≈130 K. Hence measurements of thermal emission at different wavelengths can be used to determine the temperature of the emitting surface.

If the spectral radiance can be measured at a given wavelength, then the brightness temperature of an object can be determined from re-arranging Equation 6.1 to yield

$$T_{(\lambda,K)} = (hc/\lambda p)(\ln[(2hc^2)/(\lambda^5 F) + 1])^{-1}. \qquad (6.4)$$

This relationship can be used to convert the spectral radiance measured at the spacecraft of a body at wavelength λ to brightness temperature, the temperature of a pixel-filling body. From Earth orbit, instruments sensitive in the short-wavelength infrared (SWIR) from ≈0.7–5 μm typically have spatial resolutions of ≈30–60 m pixel^{-1}, and hence pixel areas of 900 m^2 to 3600 m^2. In remote-sensing data of terrestrial activity, thermal anomalies rarely fill an entire pixel with a source at a single temperature. As described earlier, volcanic temperature and area distributions are more complex. Things are different on Io, where the larger scale of volcanic activity means that even in distant observations some hot spots may fill entire pixels (e.g., Loki Patera, with an emitting area of over 21 000 km^2, does show some isothermal pixels in 0.5-km pixel^{-1} NIMS data [Davies, 2003a]).

Laboratory and field work continues to produce refinements to remote-sensing data analysis. Perhaps one of the most pressing questions concerns the emissivity of new lava. Recent analyses of field data indicate that the emissivity of liquid basalt may be considerably less than the canonical value of 0.9 to 1. At short infrared wavelengths, emissivity may be as low as 0.55 at 1050°C (1323 K) to 0.85 at <500°C (<773 K) (Abtahi *et al.*, 2002). The manner in which emissivity changes as a function of temperature (itself a function of time) and wavelength has yet to be fully understood and incorporated into derivations of surface temperature distribution and of magma eruption temperature.

6.5 Multi-spectral observations

Observations at multiple wavelengths allow color temperatures – the emitting temperatures of the thermal sources – to be determined. In the simplest case, a pixel can be thought of as being filled with active thermal areas. For example, an a'a flow can be represented by components at two different temperatures: a high-temperature (the active areas of the flow and vent, and cracks in the flow revealing the molten interior) and the low-temperature crust. A pixel containing such a thermal distribution will have different pixel-integrated brightness temperatures at different wavelengths because emitted radiation varies with temperature as well as with wavelength (Equation 6.2).

The resulting thermal emission from the two-temperature scenario described above is determined from

$$F_{(\lambda, T)} = \left(\frac{c_1}{\lambda^5 \left(e^{c_2/\lambda T_h} - 1 \right)} \varepsilon \, A_h \right) + \left(\frac{c_1}{\lambda^5 \left(e^{c_2/\lambda T_w} - 1 \right)} \varepsilon \, A_w \right), \quad (6.5)$$

where T_h (K) is the temperature of the hot component with a total area of A_h (m^2) and T_w (K) is the temperature of the warm component with a total area of A_w (m^2), both components of emissivity ε. If the sum of the active areas is sub-pixel, imagers at longer wavelengths detect thermal emission from the passive, non-volcanic background that has been heated by the Sun or, in night observations, radiating heat absorbed during the day. On Earth, passive background (at $\approx$300 K) emission peaks at $\approx$12 μm; on Io, the passive background (at $\approx$100 K) emission peaks at $\approx$23 μm.

Equation 6.5 therefore becomes

$$F_{(\lambda, T)} = \frac{c_1 \varepsilon}{\lambda^5} \left(\frac{A_h}{\left(e^{c_2/\lambda T_h} - 1 \right)} + \frac{A_w}{\left(e^{c_2/\lambda T_w} - 1 \right)} + \frac{A_b}{\left(e^{c_2/\lambda T_b} - 1 \right)} \right), \quad (6.6)$$

where T_b is the background temperature (K), the same emissivity for all components is assumed, and the area of the pixel A_{pix} is the sum of the areas $A_{total} = A_h + A_w + A_b$. Figure 6.2 shows the resulting thermal emission spectrum for a 100 km $\times$ 100 km area on Io that contains sub-pixel hot and warm thermal anomalies.

Daylight observations include the re-emission of reflected incident sunlight (see Figure 6.3), the intensity of which may greatly exceed volcanic thermal emission. Reflected sunlight must be removed from the observed thermal emission spectrum before data analysis can be performed, a potentially problematic process if the albedo of the emitting surface is not known (often the case for Io). Daylight observations also contain emission from absorbed incident sunlight that warms the surface (see Veeder *et al.*, 1994). For practical reasons, therefore, nighttime data are always preferred for studying volcanic thermal emission.

6.6 The "dual-band" technique

Measurements of radiance at as few as two wavelengths can be used to constrain temperatures and areas. Using the "hot crack" and "cool crust" model, a thermal distribution consisting of two areas at different temperatures has four variables:

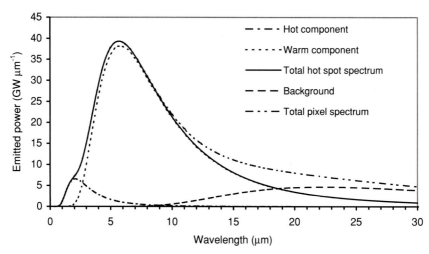

Figure 6.2 Nighttime spectrum of thermal emission from a 100-km × 100-km pixel on Io's surface containing three thermal sources: (a) a hot (1400 K) area of 0.1 km^2 (pixel area fraction $F_h = 0.00001$) (dash-dot-dash line); (b) a warm area (500 K) of 100 km^2 ($F_w = 0.01$) (dotted line); and the remainder of the pixel (9899.9 km^2) at a background temperature of 100 K ($F_b = 0.98999$) (dashed line). The total hot-spot spectrum is shown as a solid line, and the total pixel spectrum is a dash-dot-dot-dash line.

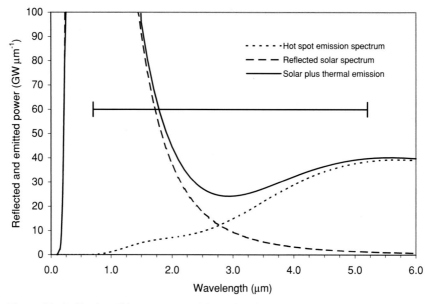

Figure 6.3 Stylized resulting spectrum of thermal emission from a hot spot on the surface of Io in daylight. Absorptions in the spectra are not shown. The reflected solar radiance is much larger than the volcanic thermal emission at short infrared wavelengths (<2 μm), complicating derivation of the highest surface temperatures present. Above 3 μm, the volcanically active (hot) surfaces dominate thermal emission. The NIMS wavelength range is also shown (horizontal bar).

A_h at T_h, and A_w at T_w. The dual-band method, first applied to *Landsat* data by Rothery *et al.* (1988), relies on determining at-sensor fluxes ($F_{s(\lambda_1)}$ and $F_{s(\lambda_2)}$) at two different wavelengths (λ_1 and λ_2). On Earth, at-sensor flux is dependent on atmospheric transmission (τ_x), upwelling (R_{xU}), and downwelling (R_{xR}) factors. The solution involves solving the simultaneous equations

$$F_{s(\lambda_1)} = \varepsilon_1 \tau_1 \left(A_h F_{\lambda_1, T_h} + (1 - A_h) F_{\lambda_1, T_w} \right) + (\tau_1 R_{1R} + R_{1U}) \qquad (6.7)$$

and

$$F_{s(\lambda_2)} = \varepsilon_2 \tau_2 \left(A_h F_{\lambda_2, T_h} + (1 - A_h) F_{\lambda_2, T_w} \right) + (\tau_2 R_{2R} + R_{2U}). \qquad (6.8)$$

On an airless body like Io, R_{xU} and R_{xR} are zero and τ_x is 1, so the equations simplify to

$$F_{s(\lambda_1)} = \varepsilon_1 \left(A_h F_{\lambda_1, T_h} + (1 - A_h) F_{\lambda_1, T_w} \right) \qquad (6.9)$$

and

$$F_{s(\lambda_2)} = \varepsilon_2 \left(A_h F_{\lambda_2, T_h} + (1 - A_h) F_{\lambda_2, T_w} \right). \qquad (6.10)$$

The equations are often solved by assuming a value for one of the unknowns and solving for the other values.

Lava composition – and resulting thermo-physical quantities such as specific heat capacity – are often known on Earth or can be inferred from knowledge of tectonic setting and aided by analyses from previous eruptions. For example, Hawaiian volcanoes typically erupt basalt at $\approx$1423 K, a temperature measured directly using thermocouples inserted into active pahoehoe flows. This temperature can be used to constrain the highest temperature component fits to relevant data (e.g., Flynn *et al.*, 2000). For Io, no such "hands-on" constraint can be applied. The only clues to composition are from temperature fits to remote-sensing data, as the ubiquitous presence of sulphur dioxide on Io, and the low resolution of the NIMS data (spatially for Prime Mission and GEM encounters [June, 1996–August, 1999] and spectrally during close fly-bys [October, 1999–January, 2002], unfortunately with a jammed NIMS spectrometer grating) make determining detailed surface composition from spectral analysis challenging. In the absence of any other evidence and with the knowledge that any observed surface has probably cooled to some extent, the derived "hot" temperature applies a minimum to the temperature of the molten material being erupted. Therefore, any derived temperature of 1200 K or more rules out sulphur volcanism as the primary magma, a measurement of temperature above 1600 K rules out basalt (assuming no exotic processes such as superheating are taking place), and so on.

6.7 Surface temperature distributions and effect on thermal emission

The example used in Section 6.5 is for a lava flow surface with two temperature components. Terrestrial lava flows generally acquire one of two surface morphologies:
pahoehoe or a'a. Generally, a'a flows have a much higher average surface temperature because the flow interior is continually exposed by fractures in the outer
crust. Pahoehoe flows, on the other hand, generally have lower average surface
temperatures because the lava interior is effectively insulated by a relatively continuous visco-elastic skin. The two-temperature component model is therefore only
an approximation of the actual temperature distribution. A range of temperatures is
present between the lava eruption temperature and the temperature of the coolest,
oldest surface. The observed integrated thermal emission spectrum is therefore
the sum of thermal emission from all of these areas. This phenomenon has been
empirically modeled for lava flows on Io to interpret infrared observations of thermal emission (Carr, 1986; Davies, 1996; Howell, 1997) and has been observed
in the field by Wright and Flynn (2003), who measured thermal emission from
Hawaiian pahoehoe flows using a radiometrically calibrated infrared camera with a
high dynamic range. Histograms of the frequency of temperatures present showed
a continuum of temperatures (Figure 6.4).

6.8 Hyperspectral observations

A hyperspectral imager by definition is capable of more than 100 measurements of
radiance in a single spectrum. The *Galileo* NIMS and *EO-1* Hyperion spectrometer
are two such instruments. The detailed spectra gathered by these instruments allow
determination of composition, especially when spectrally distinct materials are
present, and of sub-pixel thermal anomalies. Multiple-component blackbody models are well constrained by these data (Davies *et al.*, 2001; Wright *et al.*, 2004a). The
increased sampling and high dynamic range allow for more precise measurements
of thermal flux and, especially with NIMS, the detection of the sometimes subtle
differences in thermal emission caused by different eruption and lava emplacement
modes (see Chapter 8).

6.9 Analysis of hyperspectral thermal emission data

The spatial resolution of a spacecraft sensor, $R_{spatial}$, is given by

$$R_{spatial} = S\Omega , \qquad (6.11)$$

where S is range to target and Ω is the instrument field of view. For example,
Hyperion has $\Omega = 4.3 \times 10^{-5}$ radians, with range S of 7.05×10^5 m (705 km).

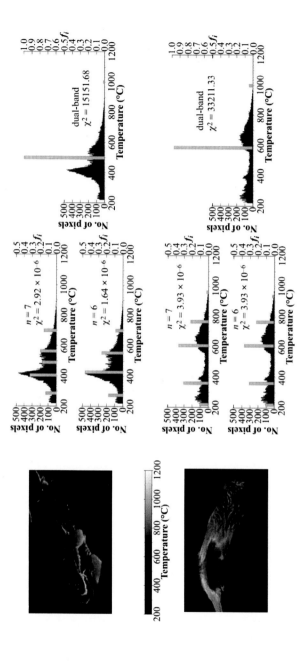

Figure 6.4a One of the latest technological developments to aid study of volcanic thermal emission is a small, easily transportable, calibrated infrared camera capable of measuring thermal emission from areas in excess of 1700 K without saturating. Shown are images obtained with such a camera of pahoehoe flows on the slopes of Kīlauea, Hawai'i. The histograms show temperature and area distributions. n is the number of temperatures used to fit the data; X^2 represents the residuals between data and model fit (Wright and Flynn, 2003).

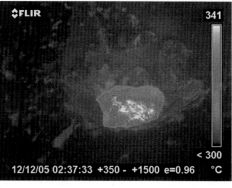
b

Figure 6.4b Infrared camera (FLIR Systems, Inc. *ThermaCAM* P65) image of the Mt. Erebus, Antarctica, lava lake on December 12, 2005. The lava lake is ≈40 m across. The infrared camera again reveals the surface temperature distribution in great detail. Figure by A. G. Davies.

From Equation 6.11, $R_{spatial}$ is 30 m. *Galileo* NIMS has $\Omega = 5 \times 10^{-4}$ radians, yielding $R_{spatial} = 350$ km at 700 000 km and 0.5 km at 1000 km. The area of a pixel, A_{pix}, is, therefore,

$$A_{pix} = (S\Omega)^2 \qquad (6.12)$$

or 900 m² for Hyperion. For NIMS, the area of a pixel is over 100 000 km² at a range of 700 000 km and 0.25 km² at 1000 km. A single NIMS pixel therefore often contains an active volcanic area in its entirety, with all of the emitting surfaces at different temperatures combining to form the observed thermal emission spectrum.

The power output F_λ (W μm⁻¹) from a pixel as a function of wavelength is derived from the measured radiance I_λ (W m⁻² str⁻¹ μm⁻¹), which for terrestrial data has been corrected for atmospheric absorption. Power output is given by

$$F_\lambda = I_\lambda \pi (S\Omega)^2 / \cos\theta, \qquad (6.13)$$

where θ is the emission angle in degrees.

The shape of the spectrum can be fitted with a blackbody curve or a combination of blackbody curves. Examples of one- and two-temperature fits to NIMS data are shown in Figure 6.5; see also Davies (2003b). Once a color temperature T_c (K) has been determined, the thermal output per unit area, $F_{(\lambda,Tc)}$ (W m⁻² μm⁻¹), is given by Equation 6.2. The area A of the sub-pixel thermal source is therefore the total energy detected at wavelength λ (F_λ) divided by $F_{(\lambda,Tc)}$, such that

$$A = F_\lambda / F_{(\lambda,Tc)}. \qquad (6.14)$$

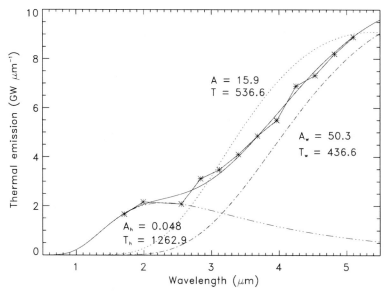

Figure 6.5 Nighttime NIMS data (stars) of thermal emission from Prometheus on Io. Like many hot spots on Io seen by NIMS, Prometheus shows an increase in thermal emission as wavelength increases. Fits to the data indicate the peak of thermal emission lies at 6.6 μm. A two-temperature fit (solid line) closely matches the data. The presence of the hot area (T_h > 1200 K) indicates exposure of erupting lava. The hot area (A_h) spectrum is the sum of thermal emission from molten lava exposed at the active vent, breakouts where new lava flow lobes are forming, and molten magma seen through cracks in the cool crust. The larger area (A_w) corresponds to areas of solid, cooling crust on lava flows. A single-temperature fit (A and T) is also shown. Note how a single-temperature model underestimates temperature and total emitting area. Dotted line = single-temperature fit. Dash-dot-dash line = warm component of two-temperature fit. Dash-dot-dot-dot-dash line = hot component of two-temperature fit. From Davies (2003b).

An example: NIMS data

The processing of NIMS volcanic thermal emission data is described in Chapter 3. Steps in data reduction include the removal of radiation-induced data artifacts, reflections from spacecraft booms in certain observation geometries, and the separation of full NIMS spectra into the spectra obtained in individual grating positions. Consider the NIMS G1INNSPEC01 Io observation of June 28, 1996 (Plate 7a and Figure 3.2). This observation was obtained at a range S of 700 750 km (7.0075×10^{10} cm) and NIMS measured a radiance I_λ of 2.2 μW cm^{-2} str^{-1} μm^{-1} at 5 μm from the active volcano Prometheus. The emission angle θ was 19 degrees. A single-source temperature T_c of 536 K was fitted to a thermal emission spectrum of Prometheus (Figure 6.5) obtained in the NIMS grating position with the highest value of integrated thermal emission (Davies, 2003b). The temperature fit was achieved using an algorithm that minimizes residuals between model output values and the data

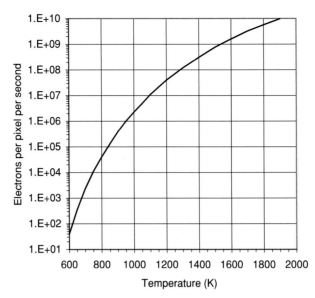

Figure 6.6 Relationship between electron count and brightness temperature for observations obtained through the SSI clear (CLR) filter. To determine temperature, the total signal (electrons/pixel/second) must be divided by the fraction of the pixel filled by the hot source. Alternatively, an assumed temperature can be used to determine fractional area occupied by the thermal source. From McEwen *et al.* (1997).

to derive the most probable model value (best fit). A number of methods for doing this are described by Press *et al.* (1992).

Equations 6.11 and 6.12 produce a spatial resolution of 350 km pixel^{-1} and a pixel area of 112 000 km^2. Equation 6.13 yields a 5-μm power output F_λ (corrected for emission angle) of 8.97 GW μm^{-1}. At 536 K, Equation 6.2 yields a 5-μm power output per unit area, $F_{(\lambda, Tc)}$, of 564 W m^{-2} μm^{-1}; and from Equation 6.14, the area A of the emitting surface is found to be 15.9 × 10^6 m^2.

The same exercise is performed for a two-temperature blackbody fit to the data, as shown in Figure 6.5. Like many hot spots on Io seen by NIMS, Prometheus shows an increase in thermal emission as wavelength increases. Fits to the data indicate the peak of thermal emission lies at ≈6.6 μm. The presence of a hot area with T_h > 1200 K indicates the ongoing exposure of molten silicate lava. The hot component spectrum is the sum of a range of areas at different high temperatures: molten lava exposed at the active vent, breakouts where new lava flow lobes are forming, and molten lava seen through cracks in the cool crust. The larger area (A_w) corresponds to areas of solid, cooling crust with an average temperature (T_w) of ≈440 K. Note that in this example the single-component best-fit color temperature, although fitting the data at some wavelengths, underestimates the temperature of the hot area, overestimates the temperature of the warm component, and underestimates the total thermal emission. Two- or multi-component blackbody models generally fit the data more closely than single-temperature models.

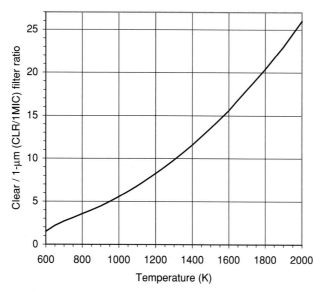

Figure 6.7 Relationship between thermal emission and the ratio of intensity through two filters, yielding a minimum color temperature. This example is for the *Galileo* SSI clear (CLR) and 1-μm (1MIC) filters, based on pre-*Galileo*-launch filter responses. Reproduced with permission from McEwen *et al.* (1998b), *Science*, **281**, 87–90, © 1998 AAAS.

6.10 Analysis of SSI thermal emission data

Visible-wavelength imagers (such as SSI) can detect thermal emission from the newest, hottest areas of lava flows. Data are digitally stored and the energy measured by the detector can be related to the temperature of the emitting surface, once the response characteristics of the filter and detector are factored in (Figure 6.6). A single-filter SSI observation (a single data point) yielded a brightness temperature (McEwen *et al.*, 1997), and dual-filter observations yielded color temperatures (Figure 6.7) (McEwen *et al.*, 1998b). SSI had a minimum brightness temperature detection limit of ≈700 K. Combining SSI and NIMS data collected almost contemporaneously provides a high degree of constraint on likely magma liquidus temperature, which itself imposes a constraint on likely magma composition (Davies *et al.*, 1997, 2001).

7

Models of effusive eruption processes

Having identified the location of ongoing or recent volcanic activity and quantified the thermal emission, many questions remain. Is the eruption emplacing flows? How thick are they? How far will they flow? Models of the physical processes taking place can at least constrain the answers to these questions, quantifying the eruption parameters (volumetric flux, areal coverage rate), constraining eruption behavior (total volume erupted, time taken, style of eruption), and allowing comparison with other eruptions, both on Io and Earth.

Volcanology has been transformed over the past four decades by the consideration of the processes taking place from the perspective of applied physics, leading to the development and application of mathematical models of eruption mechanics and flow emplacement. Models are compared with remote and field observations and laboratory studies and then retained and refined – or discarded if found to be unrealistic. The laws of physics being universal, the resulting physical models derived on Earth can be used on other Solar System bodies so long as local conditions are taken into account.

The need for process modeling arose from data collected by the first generations of planetary missions during the 1960s and 1970s, when the importance of large-scale volcanism throughout the inner Solar System was realized, as was the primary role played by basalt (described in the massive *Basaltic Volcanism on the Terrestrial Planets*, published by the Lunar and Planetary Science Institute [BVSP, 1981]).

On Earth and Io, visible and infrared data reveal the presence of active volcanism. Explosive eruptions are identified by active plumes, lava fountains, and the emplacement of a wide variety of pyroclastic deposits. Effusive activity results in the quiescent emplacement of lava flows and changes in the level of activity within lava lakes.

Because almost all data returned from *Voyager* and *Galileo* are at resolutions greater than 1 km pixel^{-1}, many features that are useful to the understanding of effusive eruption mechanisms on Earth (active vent size, measurements of channel width, levee width and depth, and so on) are not resolvable except in rare cases. Techniques have been developed to interpret the low-resolution Io data, obtained not only from spacecraft but also from Earth-based telescopes.

Visible-wavelength imagery reveals the areal extent of new flows. Infrared data yield the level of activity and the location of the active flow areas. From a series of observations, the duration of activity can be constrained, and a persistent hot spot can be monitored for changes in the level of activity. From time-series data, rates of areal coverage are calculated and, where flow thickness or levee depth can be measured or inferred, the average volumetric eruption rate can be derived.

Identifying and quantifying the processes by which lava cools on the surface of Io yield the relationship between temperature and surface age (Carr, 1986; Davies, 1996; Howell, 1997). When combined with observed distributions of temperatures and areas, this temperature–age relationship is key in determining the style of eruption taking place, enabling additional process-specific modeling of eruptions.

Using a model of lava flow emplacement and estimates of volumetric eruption rate, magma thermo-physical properties can be constrained (e.g., Schenk *et al.*, 1997; Williams *et al.*, 2001a, 2001b; Schenk *et al.*, 2004). These values are then used to model the ascent and eruption of magma from the lithosphere or deeper, a process dependent on models of lithospheric structure and depth of magma generation (e.g., Leone and Wilson, 2001; Wilson and Head, 2001; Davies *et al.*, 2006c).

This chapter describes models that have been used to interpret *Galileo* data, so as to understand, from the surface expression of activity, the ascent of magma from its region of origin to the surface via a conduit; the effusive emplacement of lava on the surface; and how lava cools and solidifies. By modeling all of these processes, the layers of a volcano can be peeled back to constrain the emplacement mechanism, eruption style, magma rheology, crustal structure, and the depth where the magma originates; and the role volcanism plays on Io as a mechanism for removing heat from the interior can thus be quantified.

7.1 Cooling of lava on Earth and Io

Watching the emplacement of small pahoehoe flow lobes on the slopes of the Hawaiian volcano Kilauea, the observer is struck by the rate at which the toes of basalt lava, typically 10–20 cm high, cool. Erupting at a temperature of $\approx$1423 K ($\approx$1150 °C), the lava is a bright orange color when first exposed, but in a few seconds the color changes to bright red, which fades to black. As heat is lost, a glassy surface crust forms. The crust initially deforms plastically but rapidly becomes rigid and brittle. The lava crystallizes inward from the cooling surfaces until the

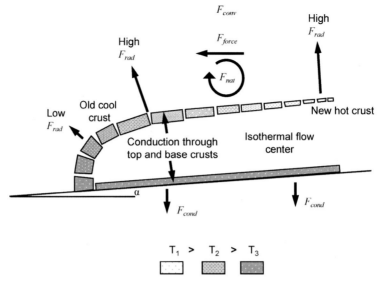

Figure 7.1 Mechanisms of heat loss from a lava flow. Heat is lost from a lava flow on Earth, shown here flowing down a slope of α radians, by conduction into the substrate F_{cond}, by atmospheric convection (F_{conv}, which is the larger value of natural [F_{nat}] and forced [F_{force}] convection terms), and by radiation F_{rad} from all exposed surfaces. The highest values of F_{rad} are from young, hot surfaces and from hot lava exposed by cracks in the crust. On Io, F_{conv} is zero.

lava body is completely solid. After about a half hour, it is possible to walk on the new lava surfaces, although the amount of radiated heat is intense. The atmosphere immediately above a flow or just downwind can be heated to temperatures in excess of 400 K, hot enough to singe hair or worse, yet the area of heating is localized to the immediate vicinity of recently emplaced flows or skylights. Just upwind of these locations, the air is cool, at ambient temperature.

The mechanisms of heat loss from a lava flow are shown in Figure 7.1. A lava body loses heat via conduction into the ground over which the flow passes (F_{cond}) by radiation from all exposed surfaces (F_{rad}) and, on planetary bodies with a tangible atmosphere, by convective heat transport (F_{conv}), which may be natural (F_{nat}) or forced (F_{force}) in the presence of a wind. Total heat loss F_{tot} at any given time for a body with a significant atmosphere is, therefore,

$$F_{tot} = F_{rad} + F_{cond} + F_{conv},$$ (7.1)

where the greater of F_{nat} and F_{force} is used for F_{conv}. For airless bodies such as Io and the Moon,

$$F_{tot} = F_{rad} + F_{cond}.$$ (7.2)

7.1.1 Heat loss by radiation (F_{rad})

Radiative heat loss from a flow surface is controlled primarily by the absolute surface temperature T_{surf} and the emissivity ε of the lava. The heat loss by radiation, F_{rad} (W m^{-2}), from exposed flow surfaces at temperature T_{surf} (K) is given by the Stefan-Boltzmann equation for blackbody radiation

$$F_{rad} = \varepsilon\sigma\left(T_{surf}^4 - T_{env}^4\right), \tag{7.3}$$

where σ is the Stefan-Boltzmann constant (5.67 × 10^{-8} W m^{-2} K^{-4}) and T_{env} is the effective temperature (K) of the environment into which heat is being lost. For $T_{surf} = 700$ K and a terrestrial ambient temperature $T_{env} = 300$ K, $F_{rad} = 13$ kW m^{-2}. The dependence on the fourth power of temperature means that cooling is at its most rapid in the earliest stages after eruption.

7.1.2 Heat loss by conduction (F_{cond})

Heat loss from the upper surface of the flow is sustained by the ability of the thickening crust to conduct heat from the hot interior to the cooling surface. Heat loss from the base of a lava flow is by conduction to the underlying substrate, which is heated up. If enough heat is supplied to the substrate, it may be thermally eroded. Typical values of thermal and physical properties of basalt and ultramafic lavas and planetary environment constants are provided in Table 7.1. Silicate lavas are typically characterized by low thermal conductivities and high specific heat capacities. Heat is therefore lost only gradually through the thickening crust. On Hawai'i, drilling into the Kilauea Iki, Makaopuhi, 'Alo'i, and 'Alae lava ponds after the cessation of volcanic activity (Peck et al., 1966; Wright and Okamura, 1977; Peck et al., 1979) allowed the depth of the 1273 K (1000°C) isotherm to be charted with time (Wright et al., 1968). The interior of the 12-m-deep 'Alae lava pond stayed molten for more than 400 days (Peck et al., 1966).

Field and theoretical studies of lava flows show that the interface temperature between the flow and the underlying ground rapidly reaches approximately the half-way point between the relatively low original ground temperature and the high lava melt temperature (e.g., Turcotte and Schubert, 1986). With lava flows, a basal crust forms, although this may be buckled and torn by the movement of lava above. If the flow is turbulent, the constant renewal of molten material at the lava–ground interface may lead to thermal erosion of the substrate once it has warmed to its melting temperature (e.g., Head and Wilson, 1981; Jarvis, 1995; Williams et al., 2001c). Modeling the temperature profiles in the crusts that form at the top and base of a body of lava shows that, on Io, heat loss by conduction into the ground beneath a lava flow (assuming the substrate is of a composition similar to the lava) is ≈20%

Table 7.1 *Physical constants and model input parameters and values*

	Units	Symbol	Io	Earth
Environment temperature	K	T_{env}	3	300
Gravity	m s^{-2}	g	1.79	9.81
Mantle density	kg m^{-3}	ρ_{mant}	e.g., 3000	4600 (average)
Lithosphere density	kg m^{-3}	ρ_{litho}	e.g., 2200	2870 (average)
Lava heat loss mechanisms		F_{rad}, F_{cond}, F_{conv} (F_{nat}, F_{force})	Radiation (F_{rad}), Conduction (F_{cond})	Radiation (F_{rad}), Conduction (F_{cond}), Convection ($F_{nat,force}$)

	Units	Symbol	Basalt	Ultramafic[a]
Liquidus (eruption) temperature	K	T_{erupt}	1400–1600	1800–1900
Latent heat of fusion	J kg^{-1}	L	3–4 $\times 10^5$	8–10 $\times 10^5$
Specific heat capacity	J kg^{-1} K^{-1}	c_p	1200–1500	1500
Emissivity	W m^{-2} K^{-4}	ε	0.9	0.9
Thermal conductivity	W m^{-1} K^{-1}	k	0.9	1.0
Thermal diffusivity	m^2 s^{-1}	κ	Various	Various
Liquid density	kg m^{-3}	ρ_{lava}, ρ_{magma}	2600	2800
Solid density	kg m^{-3}	ρ_{solid}	2800	2845
Viscosity	Pa s	η	Variable (see Ch. 5)	Variable (see Ch. 5)
Temperature at flow stop[b]	K	T_{stop}	1073	1498
Temperature loss at flow stop[b]	K	ΔT	402	402
Lava flow solidification fraction[b]		Δf	0.045	0.045
Lava lake solidification fraction[b]		Δf	0.03–0.045	0.03–0.045
Lava lake temperature range[b]	K	ΔT_{magma}	150–200	150–200

[a] Type magma taken to be Commondale komatiite. Values from Williams *et al.* (2001a).
[b] Harris *et al.* (1998, 1999a).

of the heat lost by radiation while a lava flow is still partially molten (Davies *et al.*, 2005). Eventually, all heat conducted into the ground is conducted back up to the surface, where it is lost by radiation to space.

7.1.3 Heat loss by convection (F_{nat} or F_{force})

Convection is heat transfer by movement of macroscopic particles of a fluid (such as the terrestrial atmosphere) with different temperatures. Natural – or free – convection results from density differences caused by temperature gradients, as, for example, when air above a lava flow is heated. With forced convection, movement in the air is caused by outside means, such as wind. The heat lost by convection varies as a function of flow surface temperature, decreasing as surface temperature falls.

F_{force}, forced convective heat loss per unit surface area of the flow or lava lake, is given by

$$F_{force} = W f (T_{surf} - T_{env}) \, \rho_{atmos} \, c_{p,atmos}, \tag{7.4}$$

where W is wind speed (m s^{-1}), f is a friction factor equal to 0.0036 (Greeley and Iverson, 1987; Keszthelyi and Denlinger, 1996), T_{surf} is the temperature of the lava surface, T_{env} is the temperature of the atmosphere, ρ_{atmos} is the atmospheric density, and $c_{p,atmos}$ is the atmosphere specific heat capacity at constant pressure. ρ_{atmos}, and $c_{p,atmos}$ are evaluated at the mean temperature of the silicate–atmosphere interface $(T_{surf} + T_{env})/2$.

In the absence of any wind or other such "outside" agent, natural convection takes over from forced convection. Head and Wilson (1986) showed that heat loss due to natural convection (F_{nat}) is given by

$$F_{nat} = A \left(\frac{c_{p,atmos} \, \rho_{atmos}^2 \, g \, \beta (T_{surf} - T_{env})}{\eta_{atmos} \, k_{atmos}} \right)^{1/3} (T_{surf} - T_{env}) \, k_{atmos}, \tag{7.5}$$

where k_{atmos} is the thermal conductivity of the atmosphere, η_{atmos} is the viscosity of the atmosphere, g is the acceleration due to gravity, β is the volume expansion coefficient of the gas (i.e., air), and A is a dimensionless factor reflecting the physical geometry of the system. For a heated plate facing upward (a representation of a lava flow) and for turbulent air movement, $A = 0.14$ (McAdams, 1954). Equation 7.5 also shows that heat loss due to natural convection is independent of any length scale factors.

7.1.4 Effect of planetary environment and atmosphere

Head and Wilson (1986) calculated the heat losses due to each type of heat transfer for flows on Earth and Venus to determine the heat loss per second for different lava flow surface temperatures. The total heat loss per unit area from the upper flow surface F_{tot} was the sum of the radiative heat loss F_{rad} and F_{conv}, where F_{conv} was whichever of the two convection heat losses (F_{force} and F_{nat}) was largest. Heat loss from a terrestrial lava flow surface is shown in Figure 7.2. Radiative heat loss exceeds convective heat loss until the surface temperature of the flow decreases below 398 K.

7.2 Modeling lava solidification and cooling

It is useful to quantify the relationship between surface temperature, surface age, and crust thickness on lava flows because it allows further constraint of eruption style and quantification of eruption parameters from measurement of surface temperature.

7.2.1 The importance of latent heat

As lava solidifies, the latent heat of fusion is liberated. This latent heat is a large quantity of energy per unit mass (see Table 7.1). Settle (1979) noted the buffering effect this had on lava: the liberated latent heat has to be removed before further cooling takes place. The total amount of heat energy H (J) per kilogram released by cooling, solidifying magma is given by

$$H = L + c_p(T_1 - T_2), \tag{7.6}$$

where c_p is specific heat capacity, T_1 is the initial temperature, T_2 is the final temperature, and L is the latent heat of fusion. For basalt, $c_p = 1500\,\mathrm{J\,kg^{-1}\,K^{-1}}$, $L = 4 \times 10^5\,\mathrm{J\,kg^{-1}}$, and a terrestrial cooling range can be $T_1 = 1470\,\mathrm{K}$ to $T_2 = 300\,\mathrm{K}$.

From Equation 7.6, 1 kg of basalt yields $H = 2.2 \times 10^6$ J, when fully solidified and cooled. Of this heat, about 20% is accounted for by latent heat release. On Io, a colder environment, an additional 3×10^5 J is liberated in cooling from 300 K down to 100 K.

7.2.2 Cooling before lava solidification

The thickness of the crust on a lava flow or lava lake is the depth at which the solidus temperature is reached. As discussed earlier, Settle (1979) stressed the importance of latent heat release in retarding the cooling of the interior of lava flows. The following methodology was developed by Head and Wilson (1986) for determining the cooling of a lava flow taking latent heat release into account. First, however, a cautionary note: from a mechanical point of view, a lava flow may be taken to be "solid" when some mass fraction (>50%) of crystals is reached and the flow can no longer move. The models described here deal with heat content, and it is convenient to refer

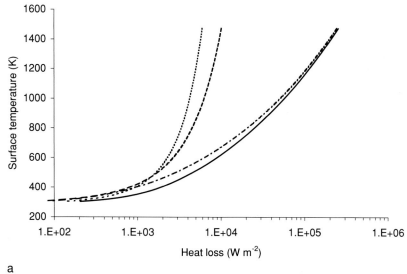

a

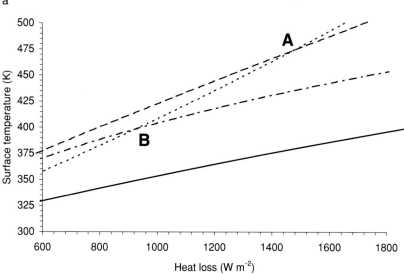

b

Figure 7.2 Figure 7.2a shows the heat loss per unit area from a terrestrial lava surface as a function of surface temperature and process of heat removal. Solid line = total heat loss (F_{tot}). Dotted line = heat loss by forced convection (F_{force}). Dashed line = heat loss by natural convection (F_{nat}). Dash-dot-dash line = heat loss by radiation (F_{rad}). In Figure 7.2b, Point A marks the temperature (475 K, when $F_{force} = F_{nat} = 1480\,W\,m^{-2}$) above which $F_{nat} > F_{force}$. Point B marks the temperature (398 K, when $F_{force} = F_{rad} = 920\,W\,m^{-2}$) above which $F_{rad} > F_{force}$.

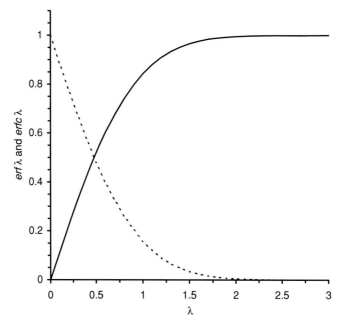

Figure 7.3 The error function *erf* (solid line) and complementary error function *erfc* (dotted line).

to a flow as being "solid" when 100% crystallization is attained, that is, when all of the latent heat has been exchanged. The thermal models address the chemical or mineralogical solidus, rather than the point at which the flow is mechanically "solid."

From Head and Wilson (1986), the thickness C (m) of the lava crust at time t (s) is given by

$$C = 2\lambda\sqrt{\kappa t}, \tag{7.7}$$

where κ is thermal diffusivity (m^2 s^{-1}) of the lava and λ is a dimensionless quantity such that

$$\frac{L\sqrt{\pi}}{c_p(T_{erupt} - T_{surf})} = \frac{\exp(-\lambda^2)}{\lambda\,erf(\lambda)}, \tag{7.8}$$

where T_{erupt} is the magma liquidus temperature (K), T_{surf} is the surface temperature (K) of the flow at time t, c_p is specific heat capacity (J kg^{-1} K^{-1}), L is latent heat of fusion (J kg^{-1}), and *erf* is the error function. Figure 7.3 shows the *erf* and *erfc* (complementary *erf*) functions. The heat flow through the surface of the lava flow is equal to the temperature gradient at the surface multiplied by the thermal conductivity of the lava, and must be equal to the total heat loss F_{tot} (W m^{-2}) from the upper surface such that

$$F_{tot} = \frac{k(T_{erupt} - T_{surf})}{\sqrt{\pi \kappa t}\ erf(\lambda)}, \qquad (7.9)$$

where k is thermal conductivity (W m^{-1} K^{-1}).

The method of solution is as follows. A surface temperature T_{surf} and values for T_{erupt}, c_p, and L are selected and the left-hand side of Equation 7.8 solved. The right-hand side of Equation 7.8 is then solved iteratively to find λ, which is used with the value of T_{surf} in Equation 7.9 to find time t, the time from commencement of cooling. The derived values of t and λ are then used in Equation 7.7 to find the crustal thickness at time t for surface temperature T_{surf}. For the crust that forms at the base of the lava flow, in the case where the lava and substrate are of the same composition, the relationship is

$$\frac{L\sqrt{\pi}}{c_{host}(T_e - T_{host})} = \frac{\exp(-\lambda^2)}{\lambda(1 + erf(\lambda))}, \qquad (7.10)$$

where $c_{p,host}$ is the specific heat capacity of the host and magma and T_{host} is the temperature of the solid material below the lava flow or surrounding an intrusion (i.e., "host" in the context of receiving the heat from the magma body). The value of λ derived from Equation 7.10 for time t is used in Equation 7.7 to determine crust thickness.

Figure 7.4 shows the development of top and base crusts for a basalt flow on Io using the thermo-physical values in Table 7.1. This example assumes a substrate of the same composition as the lava. On the ionian surface, although initially the crust forms more quickly at the base, quenched as it is against the cold (105 K) substrate, after less than 20 minutes the upper crust is thicker than the base crust and remains so until complete solidification. The solidification time for any thickness of flow, and the thickness of the top and base crusts, can be found from Figure 7.4.

Figure 7.5 shows the surface cooling curves for basalt flows with different thermal conductivities on Io and Earth. Regardless of composition, cooling is initially very rapid. For the case in which thermal conductivity $k = 1$ W m^{-1} K^{-1}, in less than 10 s lava has cooled from the eruption temperature (1475 K for basalt, $\approx$1870 K for an ultramafic magma such as the Commondale komatiite) to 1200 K. After a few minutes, surface temperatures fall to less than 1000 K. After an hour, the surface is cooler than 750 K. The presence of atmospheric cooling components F_{force} and F_{nat}, not present on Io, means that flow surfaces initially cool faster on Earth. Surface temperatures have to decrease below $\approx$375 K before the efficiency of radiative cooling on Io catches up and overtakes cooling via all processes on

Models of effusive eruption processes

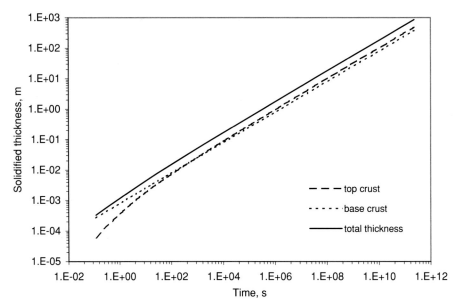

Figure 7.4 Development of top and base crusts on a basalt lava flow on the surface of Io (Davies *et al.*, 2005). Reprinted by permission of Elsevier.

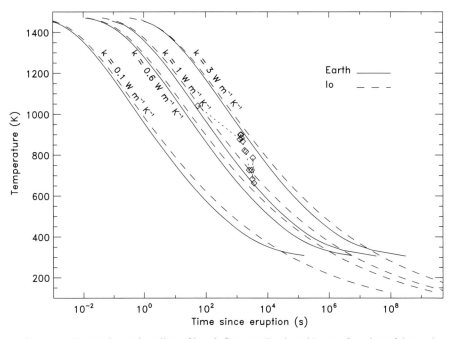

Figure 7.5 Comparison of cooling of basalt flows on Earth and Io as a function of thermal conductivity: the lower the thermal conductivity, the faster the surface cools because heat is removed faster than it can be conducted to the surface. Also shown (diamonds) is the cooling of a Hawaiian pahoehoe flow as measured by Flynn and Mouginis-Mark (1992). From Davies (1996). Reprinted by permission of Elsevier.

Earth. The higher environment temperature on Earth reduces the rate of cooling as flow surface temperature approaches ambient temperature.

7.2.3 Cooling after solidification

The model described above can be used for thick lava bodies and deep magma pools for a considerable length of time. On Io it takes more than 70 years for an upper crust 50 m thick to form. However, with a lava body of finite thickness, the time will come when top and base crusts meet and the source of latent heat is removed. For subsequent cooling, another model is required. One such model that has been applied to post-solidification lava flow cooling (Davies *et al.*, 2005) utilizes a finite-difference technique called the Schmidt graphical method (e.g., Lydersen, 1979).

The starting point is the temperature profile in the lava body at the point of solidification. Having selected a flow thickness, the top and base crustal thicknesses and time since onset of cooling at time of solidification are found from Figure 7.4 for a flow on Io. Equations 7.7 through 7.10 are used to determine λ for both the top and base crusts at the total solidification time. Temperature T_x in the upper crust as a function of depth x is given by

$$T_x = \left[\left(\frac{T_{erupt} - T_{surf}}{erf(\lambda)} \right) erf\left(\frac{x}{2\sqrt{\kappa t}} \right) \right] + T_{surf}. \qquad (7.11)$$

The temperature in the base crust at a distance y from the total solidification plane within the lava flow is given by

$$T_y = \left[\left(\frac{T_{erupt} - T_{host}}{1 + erf(\lambda)} \right) \left(1 + erf\left[\frac{y}{2\sqrt{\kappa t}} \right] \right) \right] + T_{host}, \qquad (7.12)$$

where T_{host} is the original temperature of the crust underlying the flow, taken to be 105 K. From the temperature profile constructed using Equations 7.11 and 7.12, the aforementioned Schmidt graphical method is used to model subsequent cooling. Here, the differential forms of the Fourier equation for the conduction of heat,

$$\frac{\partial T}{\partial t} = \kappa \frac{d^2 T}{dx^2}, \qquad (7.13)$$

where κ is the thermal diffusivity (m^2 s^{-1}), are replaced with the finite differences, which, when inserted into Equation 7.13, yield

$$T_{x,(t+\Delta t)} - T_{x,t} = \kappa \frac{\Delta t}{\Delta x^2} (T_{(x+\Delta x),t} - 2T_{x,t} + T_{(x-\Delta x),t}), \qquad (7.14)$$

where Δt (the time increment) and Δx (the distance increment) are independent variables, and Δt can be chosen to give

$$\kappa \frac{\Delta t}{\Delta x^2} = \frac{1}{2}. \tag{7.15}$$

Equation 7.14 then reduces to

$$T_{x,(t+\Delta t)} = \frac{1}{2}(T_{(x+\Delta x),t} + T_{(x-\Delta x),t}). \tag{7.16}$$

The right-hand side of Equation 7.16 contains only temperatures at time t, and the left-hand side gives the temperature at time $t + \Delta t$. As an example, for a time interval Δt of 2000 s, and using a value of $\kappa = 7 \times 10^{-7}\,\mathrm{m^2\,s^{-1}}$, Δx is found from Equation 7.15 to be 0.052915 m.

With a constant environment temperature T_{env} and a constant heat transfer coefficient h given by

$$h = \frac{F_{rad}}{\sigma\,T_{surf}^4/(T_{surf} - T_{env})}, \tag{7.17}$$

where σ is the Stefan-Boltzmann constant, the heat flow F_{rad} through the surface of the slab per unit area is

$$F_{rad} = \frac{T_{surf} - T_{env}}{k/h}, \tag{7.18}$$

where k is the thermal conductivity of the lava (Table 7.1). Figure 7.6 shows how the temperature profile at a given time can be used to generate the profile after the next time increment.

The temperature curve extended from the surface of the lava slab intercepts the temperature of the environment T_{env} line at a distance k/h from the surface. To include the surface temperature in the calculations, the first interval is located at a distance $\Delta x/2$ outside the surface. Successive applications of Equation 7.16 allow heat to be radiated from the surface. As shown in Figure 7.7, initial values are set up using

$$k/h = \frac{T_{surf} - T_{env}}{\tan \theta}, \tag{7.19}$$

where θ is shown graphically in Figure 7.7, and the temperature $\Delta x/2$ from the surface is

$$AA(1) = \left[\tan \theta \left(\frac{k}{h} - \frac{\Delta x}{2}\right)\right] + T_{env}. \tag{7.20}$$

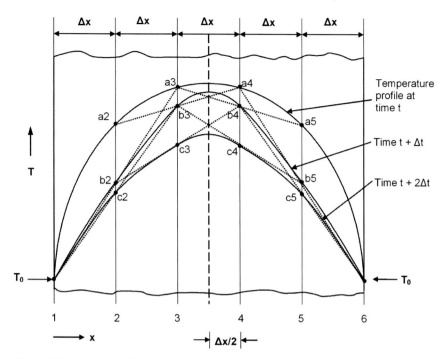

Figure 7.6 Derivation of successive cooling profiles in a cooling body using the Schmidt graphical method. In this example, the surface temperature on either side of the slab is fixed at T_0, which equals T_{surf}. The starting profile, T_0-a2-a3-a4-a5-T_0 at time t is used to generate a profile at time $t + \Delta t$. This profile is T_0-b2-b3-b4-b5-T_0. Point b2 is the average of T_0 and a3. Point b3 is the average of a2 and a4. Point b4 is the average of a3 and a5. Profile c, at time $t + 2\Delta t$, is similarly constructed from profile b. The relationship between the width of each element Δx and time increment Δt is dependent on thermal diffusivity, and shown in Equation 7.15. From Davies et al. (2005). Reprinted by permission of Elsevier.

The surface temperature T_{surf} is known at the initiation of the process, and for subsequent time steps is found from

$$T_{surf} = \frac{AA(2) + AA(1)}{2}. \tag{7.21}$$

Heat flow at the surface is found by multiplying the temperature gradient at the surface by the thermal conductivity of the silicate crust. As the flow surface cools, the heat flow – and, therefore, the heat transfer coefficient – change. These quantities are recalculated at each time step.

This is an approximation of a complex process. The model described above assumes constant thermal properties throughout the forming crust, which is not the case because, for example, porosity decreases with depth. Figure 7.8 shows the effect of flow thickness on observed surface temperature for 1-m- and 10-m-thick basalt flows and compares the cooling of basalt, ultramafic lava, and a sulphur

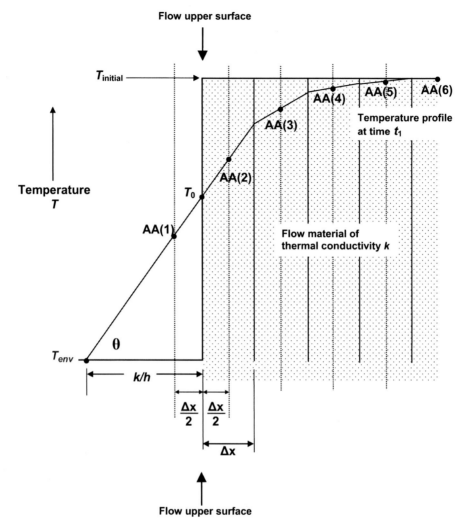

Figure 7.7 Cooling of slab with defined coefficient of heat transfer *h* at surface. The new surface temperature T_0, which equals T_{surf}, is the result of finding the balance between the transfer of heat to the surface and heat loss by radiation, itself a function of surface and environment temperature (T_{env}). Equations 7.17–7.19 are used to calculate a new value of *h* at every time interval (Δt). From Davies *et al.* (2005). Reprinted by permission of Elsevier.

flow. Table 7.2 shows the solidification times for different flow thicknesses and compositions.

7.2.4 Use of the age–temperature relationship

The treatment of heat loss from a lava flow as derived by Head and Wilson (1986) and Davies *et al.* (2005) provides a reasonable approximation to the lava cooling

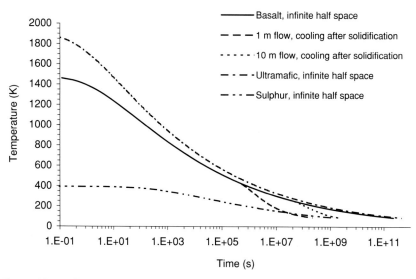

Figure 7.8 Cooling of basalt, ultramafic, and sulphur flows on the surface of Io. When total solidification is reached, surface cooling is more rapid. Surface cooling curves for 1-m- and 10-m-thick basalt flows after flow solidification are shown.

process. The identification of a color temperature from remote-sensing data now constrains the age of the emitting surface, if lava composition is known or assumed. This relationship between surface temperature and surface age assumes a greater importance when studying Io's volcanism. On Earth, heat transfer processes mean that lava flow surfaces reach ambient temperatures much faster than on Io. Infrared observations after only a few days may no longer detect the lava flow surfaces. The slower rate of cooling and a much lower background temperature on Io means that thermal anomalies persist for weeks to months, often for years, allowing a more robust study of post-eruption cooling and how other processes modify the cooling process (e.g., how albedo and emissivity change as the lava surfaces become cool enough to be coated with first sulphur and then sulphur dioxide).

7.2.5 Integrated thermal emission models

Having determined how lava cools on the surface of Io, the evolution of thermal emission as a function of wavelength can be modeled. In the simplest case, assuming a steady eruption volumetric flux, a lava flow covers ground at a steady rate. A crust forms on the surface, and the distal end of the flow cools as heat is lost and a thickening crust forms, as shown in Figure 7.1. Heat is lost through cracks in the crust; from breakouts where new, young (and therefore hot) flow lobes form; and at the hot vent where lava is being first exposed. If lava tubes have formed, incandescent lava can commonly be seen through skylights in the tube roof.

Table 7.2 *Solidification times for different lava flows on Io*[a]

Lava type	Flow thickness (m)	Time to total solidification, t_s	Base thickness (m)	Top thickness (m)	Surface temp. T_s at time t_s (K)
Basalt	0.1	61 min	0.048	0.052	733
Basalt	0.2	4 hrs	0.094	0.106	635
Basalt	0.5	1 day	0.23	0.27	521
Basalt	1	3.8 days	0.46	0.54	447
Basalt	2	15 days	0.90	1.10	382
Basalt	5	92 days	2.25	2.75	309
Basalt	8	228 days	3.57	4.43	277
Basalt	10	1 yr	4.45	5.55	237
Basalt	20	4 yrs	8.8	11.2	222
Basalt	50	25 yrs	22.1	27.9	178
Basalt	75	53 yrs	33	42	162
Ultramafic	0.1	1.25 hrs	0.046	0.054	805
Ultramafic	0.2	4.8 hrs	0.09	0.11	693
Ultramafic	0.5	29 hrs	0.22	0.28	565
Ultramafic	1	5 days	0.44	0.56	483
Ultramafic	5	112 days	2.13	2.87	331
Ultramafic	10	448 days	4.20	5.80	280
Ultramafic	20	5 yrs	8.4	11.6	237
Ultramafic	50	30 yrs	21	29	190
Ultramafic	75	67 yrs	31	44	172
Sulphur	0.1	3 hrs	0.049	0.051	296
Sulphur	0.2	11 hrs	0.098	0.102	267
Sulphur	0.5	2.6 days	0.238	0.262	227
Sulphur	1	10 days	0.46	0.54	199
Sulphur	5	233 days	2.24	2.76	142
Sulphur	10	2.5 yrs	4.41	5.59	122
Sulphur	20	9.9 yrs	8.7	11.3	104

After Davies *et al.* (2005).
[a] For basalt and ultramafic lava, thermal conductivity $k = 1\,\mathrm{W\,m^{-1}\,K^{-1}}$ is used; temperature of host (T_{host}) = 105 K; emissivity $\varepsilon = 1$.

The thermal emission observed in multi-wavelength, low-spatial-resolution data is the sum of thermal emission from all radiating surfaces. Models of cooling lava on Io have been developed (Carr, 1986; Davies, 1996; Howell, 1997; Keszthelyi and McEwen, 1997b) that determine how thermal emission changes with evolving temperature and area distributions.

The Io flow cooling model of Davies (1996) uses the cooling curve in Figure 7.5 to calculate how the integrated thermal emission spectrum from an expanding flow changes with time. The resulting model spectra are then used to fit data (e.g., Davies, 1996; Davies *et al.*, 2000a). In the simplest case, the following assumptions are made:

1. A single eruption source is present.
2. The rate of areal increase (dA/dt) is constant.
3. The heat from levees is ignored, as these cool quickly and would not be easily detected by NIMS.
4. The lava is of basaltic composition.

Total thermal output F_{rad} at a selected time is given by

$$F_{rad} = \sum_{i=1}^{i=n_t} A_i \, \sigma \, \varepsilon \, \left[T_i^4 (1 - c_f) + T_{erupt}^4 \, c_f \right], \qquad (7.22)$$

where c_f is the fraction of the surface occupied by cracks (the "crack fraction"); n_t is the number of temperature bins of size 1 K from T_{erupt} to the lowest flow surface temperature present T_{low} (K); T_{erupt} is the crack temperature, set to the lava eruption temperature (K); σ is the Stefan-Boltzmann constant; ε is emissivity (taken to be 1); and T_i and A_i are bin temperature (surface temperature) (K) and area (m^2) at time t (s). The model output thermal emission curves are fitted to multi-wavelength data, yielding a surface age distribution and total area. These results are then used to determine the average areal coverage rates for different eruptions.

Thermal emission at a given wavelength as a function of time ($F_{\lambda, t}$) is therefore

$$F_{\lambda, t} = \sum_{i=1}^{i=n_t} \frac{A_i \, \varepsilon \, c_1}{\lambda^5} \left[\frac{(1 - c_f)}{(e^{c_2/\lambda T_i} - 1)} + \frac{c_f}{(e^{c_2/\lambda T_{erupt}} - 1)} \right], \qquad (7.23)$$

where constant $c_1 = 3.7413 \times 10^8$ W μm^4 m^{-2}, and constant $c_2 = 1.4388 \times 10^4$ μm K. The model generates a series of thermal emission spectra at each non-uniform time step taken to be the time needed for surface temperature to decrease from temperature T (K) at time t_T (s) to temperature T-n at time t_{T-n}, with the decrease in temperature $n = 1$ K. Initially, cooling is very rapid, and the corresponding areas at these temperatures are very small. Model spectra examples are shown in Figure 7.9.

As time progresses, larger and cooler areas form as the distal parts of the flow cool and the peak of the integrated thermal emission spectrum moves to longer wavelengths (Figure 7.10). The model, when fitted to observational data, generates a spectrum consisting of many individual blackbodies; the summed area of the individual areas; and the age of the oldest exposed surface. Figure 7.11 shows the relationship between temperature, age, and area of the model fitted to data of an Io "outburst" eruption (Marchis *et al.*, 2002). Additional model components are added for cases where complex thermal spectra are observed that are poorly fitted using a single component (e.g., Davies *et al.*, 2001).

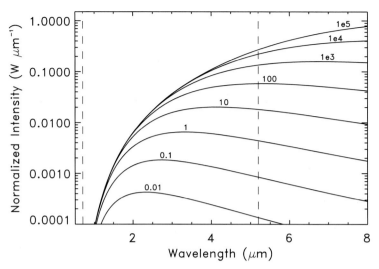

Figure 7.9 The normalized model spectral signature of an active flow as a function of time. The spectrum is created from integrating over all flow surface areas and evolving temperature range. As the distal ends of the flow cool, the surface temperature distribution changes along with the thermal emission spectrum. Each curve shows time in hours. The NIMS wavelength range lies between the vertical dashed lines. From Davies *et al.* (2000a). Reprinted by permission of Elsevier.

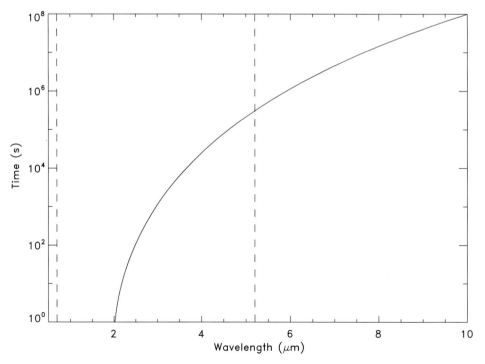

Figure 7.10 The change in wavelength of maximum emission with time, from the Io flow cooling model. The NIMS wavelength range lies between the vertical dashed lines. From Davies *et al.* (2000a). Reproduced by permission of Elsevier.

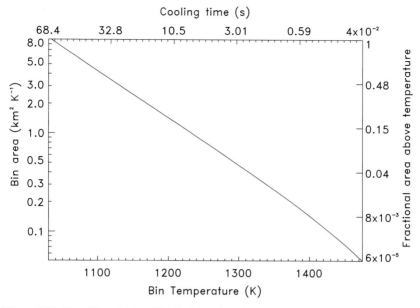

Figure 7.11 Fit of the Davies (1996) thermal emission model to ground-based, multi-spectral data of a titanic thermal outburst from Surt observed using adaptive optics on February 22, 2001 (Marchis *et al.*, 2002). The figure shows the area at a selected temperature and the time for each parcel of incandescent lava to cool to a given temperature. The total thermal emission is the sum of thermal emissions in each 1 K temperature bin, and the total hot-spot area is the sum of areas in each 1 K bin; for instance, there is 8.9 km² emitting at 1034 K, 8.8 km² at 1035 K, 8.7 km² at 1036 K, and so on, to 0.05 km² at 1475 K. The integrated thermal spectrum adds up to what Marchis *et al.* observed as one large, hot sub-pixel feature. Areas hotter than 1034 K sum to 804 km². After Kargel *et al.* (2003a).

In summary, the cooling model just described yields

1. the temperature range present (T_{erupt} at $t = 0$ to T_{low} at $t = t_{low}$);
2. the area at each temperature (A_{erupt} to A_{low}), using a 1 K bin size; and
3. the total emitting area A_{total} (the sum of areas from A_{erupt} to A_{low}).

Average areal coverage rate is therefore the total area divided by the longest cooling time, or A_{total}/t_{low}.

7.3 Volumetric rates (Q_F and Q_E)

7.3.1 Determination from thermal emission

The text above described how the inferred temperature range can be used to determine an average areal coverage rate. How, then, can we determine the third dimension – the thickness of the flow – and hence the volume of lava erupted? An individual eruption has an *eruption rate* (Q_E), which is the discharge rate averaged over all or the major part of the eruption, derived by determining the total volume erupted divided by the duration of the eruption; and an *effusion rate* (Q_F), the instantaneous volumetric flux at any given time, such as might be derived from an *in situ* measurement of flow velocity and a knowledge of flow cross-sectional area (Wadge, 1981). Volumetric eruption rate Q_E can be determined from high-resolution images if observations bracket the eruption duration and flow thickness can be measured (e.g., derived from the lengths of shadows cast by the edge of the flow). Application of a model of varying effusion rate (Wadge, 1981) quantifies the effusion rate as a function of time and determines the maximum value of Q_F.

Determining Q_F from infrared observations relies on understanding the style of volcanic activity taking place. The distribution of energy across visible and infrared wavelengths varies as a function of eruption style and, hence, the mode of flow emplacement. The spectrum of active pahoehoe flows, where most thermal emission comes from cooling crust, is very different from that of a lava fountain. The shape of the thermal emission spectrum can therefore be used to constrain the eruption mode using infrared observations of volcanic activity.

Estimates of volumetric fluxes derived from thermal emission measurements were first advocated by Pieri and Baloga (1986) and Crisp and Baloga (1990a). More sophisticated approaches have been developed, including the addition of terms for heat loss and gain caused by various processes and testing against field data (e.g., Keszthelyi and Denlinger, 1996; Harris *et al.*, 1997a, 1997b, 1998, 1999a, 2000b).

7.3.2 Lava flows and calculation of Q_F

Investigations of heat loss from new lava flows and lava lakes on Earth have yielded the relationship between thermal emission (a measurable quantity) and effusion rate, with input values constrained using a vast body of previous research based on *in situ* investigations. Such intimate knowledge of thermo-physical properties does not accompany *Galileo* NIMS data, but compositional parameters for Io's lavas can be constrained from available temperature, spectral, and geomorphological analyses.

Harris *et al.* (1998) showed how the instantaneous effusion rate Q_F (m^3 s^{-1}) of an active pahoehoe flow is given by

$$Q_F = F_{active}/\rho_{lava}(c_p \, \Delta T + L \, \Delta f), \tag{7.24}$$

where F_{active} is the total thermal flux from the active part of a flow, the net rate at which heat is being transferred out of the system; ρ_{lava} is lava density; c_p is specific heat capacity; ΔT is the temperature range from liquidus temperature to the temperature at which the flow comes to a halt; L is the silicate latent heat of fusion; and Δf is the change in crystallization fraction over ΔT. Values for basalt and ultramafic lavas are given in Table 7.1. Determining F_{active} is often problematic, depending on such factors as whether flows are moving along tubes and the degree of flow interaction with ground and seawater, and even whether it is raining or had recently rained (Keszthelyi, 1995; Harris *et al.*, 1998). Another complication, noted by Wright *et al.* (2001), is the importance of distinguishing active flows from inactive flows that still have hot or warm surfaces contributing to the observed thermal emission and that may have been emplaced at an entirely different effusion rate. Such a distinction is not possible with Io data, so a different approach is adopted that uses F_{tot} rather than F_{active} (Davies, 2003b).

For persistently active volcanoes emplacing lava flows, a two-component or multi-component blackbody fit to NIMS data yields F_{rad}, and F_{tot} is calculated from Equation 7.2 (again, $F_{cond} = 20\%$ of F_{rad}). Davies (2003b) showed that the minimum effusion rate $Q_{F(NIMS)}$ (m^3 s^{-1}) is given by

$$Q_{F(NIMS)} = \frac{F_{tot}}{\rho_{lava}\left(L + c_p\left[T_{erupt} - T_{NIMS}\right]\right)}, \qquad (7.25)$$

where T_{erupt} is the lava eruption temperature (K) and T_{NIMS} is the NIMS low-temperature detection threshold of 220 K for hot spots that are sub-pixel (Smythe *et al.*, 1995).

7.3.3 Lava lakes and Q_F

Once the total thermal flux from an active lava lake is known, the mass flux M (kg s^{-1}) needed to maintain that heat loss can be estimated using the methodology of Harris *et al.* (1999a), such that

$$M = (F_{rad} + F_{conv})/(L\Delta f + c_p\Delta T_{magma}), \qquad (7.26)$$

where F_{rad} is the radiative thermal emission, F_{conv} is the convective thermal emission, L is the latent heat of fusion, Δf is the crystallized mass fraction, c_p is the specific heat capacity of the magma, and ΔT_{magma} is the temperature through which the magma cools before recirculating into the lake. Values of Δf, c_p, and ΔT_{magma} are derived from the study of lava lakes at Mt. Erebus (Ross Island, Antarctica), Erta'Ale (Ethiopia), and Pu'u 'O'o (Kilauea, Hawai'i) by Harris *et al.* (1999a), and are given in Table 7.1. As before, on Io F_{conv} is zero.

7.3.4 Discharge rate variability

The rate at which magma is discharged varies substantially during many eruptions. As demonstrated by Wadge (1981) using several well-documented eruptions, Q_F rapidly reaches a maximum value after a relatively short period of waxing flow and then decreases during the longer waning stages of the eruption. The point at which maximum Q_F is reached divides waxing and waning eruption stages. An example of waxing and waning Q_F is shown in Figure 7.12.

For exponential decay, the relationship between the peak value $Q_{F(max)}$ and that at some subsequent time t $(Q_{F,t})$ is given by

$$\ln\left(\frac{Q_{F,t}}{Q_{F(max)}}\right) = bt \quad \text{or} \quad Q_{F,t} = Q_{F(max)}e^{bt}, \tag{7.27}$$

where t is time and b is the eruption-specific decay constant. For example, for an eruption where $Q_{F(max)} = 10^3$ m^3 s^{-1}, $Q_{F,t} = 0.01$ m^3 s^{-1}, $t = 250$ days, and $b = -0.023496$. The value of Q_F can be found for any value of t, yielding 889 m^3 s^{-1} after 10 days, 347 m^3 s^{-1} after 50 days, 107 m^3 s^{-1} after 100 days, and 10 m^3 s^{-1} after 200 days.

7.3.5 Factors influencing waxing flow

Initial effusion rates during the period of waxing flow are low. Wadge (1981) attributed the low initial rate to the loss of energy from the magma during the propagation of the conduit to the surface by forced intrusion. As the magma advanced, mechanical energy was lost, the magma cooled against the relatively cool country rock, and viscosity therefore increased. As the walls of the conduit or fissure reached thermal equilibrium with the magma, this cooling effect decreased and one inhibition on magma ascent was removed. Wadge also noted that, theoretically, viscous dissipation within the magma could lead to an increase in waxing flow (Fujii and Uyeda, 1974; Hardee and Larson, 1977). In viscous dissipation, frictional heating in the magma exceeds the ability of country rock to conduct away the heat: the magma becomes superheated, lowering viscosity and increasing flow rate. Although frictional heating has not definitively been proven to take place on Earth, it may be an important process on Io, possibly causing superheating of magma by hundreds of kelvin (Keszthelyi et al., 2005a).

7.3.6 Factors influencing waning flow

Waning flow is usually the dominant stage of a basaltic eruption (Wadge, 1981). As the eruption proceeds, the energy stored within the volcano is gradually released.

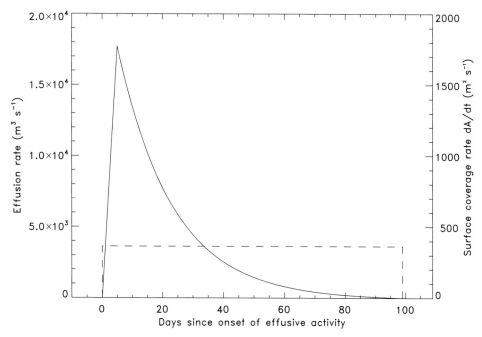

Figure 7.12 An example of the variation in effusion rate (Q_F) showing waxing flow over five days followed by a longer period of waning flow (94 days). The maximum value of Q_F is $1.8 \times 10^4\ \mathrm{m^3\,s^{-1}}$. Waning flow decays exponentially. The areas under both curves are the same and represent the total volume erupted, in this case 31 km^3. The eruption rate (Q_E, the dashed line) is 3630 m^3 s^{-1}, but only correlates with the actual effusion rate at two times. The y-axis on the right-hand side shows the implied rate of areal coverage if emplaced flows are 10 m thick. The area under the time-dA/dt curve is the total area covered by the flows, 3100 km^2.

This process includes the release of elastic strain energy within the rocks containing the magma storage reservoir. Energy is transferred through the pressure of the reservoir rocks to the magma. The measured deformations of Kilauea (e.g., Dzurisin *et al.*, 1980; Cervelli and Miklius, 2003) are consistent with an expanding and contracting reservoir system. Theoretically, the decompression of magma in a reservoir of fixed size, after an influx of new magma, is also exponential (Blake, 1981).

7.3.7 *Areal coverage rate*

Varying effusion rate translates into a varying surface coverage rate. Figure 7.12 therefore also illustrates how the areal coverage rate changes with time, as the eruption waxes and wanes, for an assumed flow thickness of 10 m. At the peak of the eruption (max Q_F), ground is covered at a rate of $3.2 \times 10^3\ \mathrm{m^2\,s^{-1}}$. With data on the final volume erupted and the duration of the eruption, the Wadge curve can be used to iteratively determine the effusion rate profile. This technique is particularly useful because it can be applied after the eruption.

7.4 Models of lava emplacement

Considerable advances have been made over the past 30 years in modeling the geometric and rheological behaviors of lava and their effect on the emplacement of lava flows. Reviews of this work can be found in McBirney and Murase (1984), Pinkerton and Wilson (1994), and Rowland *et al.* (2004). Two main strategies have shaped the modeling of lava flows. The first applies the results of Hulme (1974), who demonstrated that many flow parameters, such as flow width and thickness, are determined by the rheological properties of the lava, specifically viscosity and yield strength at the time of emplacement. Hulme (1974) modeled lava as a Bingham body (where stress in the lava has to exceed the lava yield strength before the lava can flow) and estimated lava yield strength from channel dimensions. Use of these relationships relies on knowing the surface gravity and the slope of the ground on which the lava is emplaced (see also Pinkerton and Wilson, 1994). Once rheological properties are selected, the dimensions of flows are used to estimate other properties, such as flow velocity.

The second strategy for determining eruption parameters studies the properties of flowing lava (e.g., Baloga and Pieri, 1986; Dragoni, 1989; Crisp and Baloga, 1990a). These techniques focus on calculating the effects on flow rheology of heat loss as the flow advances, and determining in turn the subsequent effects on flow movement. As noted by Rowland *et al.* (2004), "the refinement of both the lava-rheology and thermal-budget techniques has taken the form of developing ever more complicated mathematical expressions that take into account more and more details of flow processes and lava properties."

As models become more sophisticated, the reliance on the accuracy of input variables also increases. Both approaches rely on precise knowledge of the rheology of the lava erupted. This is a problem for Io, about which no precise information exists. Nevertheless, "generic" silicate thermal and rheological properties can be used, as a best estimate, to match model output to observations (see Table 7.1). Basalt is often chosen as the material erupted (see Chapter 9). Flows of basaltic composition are abundant on Earth and are common as well to all terrestrial planets (BVSP, 1981). The apparent low relief of flows seen in *Voyager* data, coupled with post-*Voyager* detections of high eruption temperatures indicative of silicate volcanism (Johnson *et al.*, 1988; Veeder *et al.*, 1994), point to a relatively low-viscosity (≤ 1000 Pa s) silicate lava composition.

For Hulme's lava-rheology approach, the equations in the following section are considered appropriate for unconfined, high-effusion-rate, short-duration flows (Pinkerton and Wilson, 1994).

The Hulme model

In the Hulme model, rheologically controlled levees form that confine the moving, active part of the flow. Initial levee widths w_b can be calculated (Hulme, 1974) from

$$w_b = \frac{\tau}{2g\rho\,\alpha^2},$$
(7.28)

where α is the slope in radians, τ is magma yield strength, g is acceleration due to gravity, and ρ is the magma density. For any given rheology, levee widths are purely a function of slope. The flow-channel width w_c can be calculated (Wilson and Head, 1983) from

$$w_c = \left(\frac{(24Q_F\,\eta)^4 g\,\rho}{\tau^5\alpha^6}\right)^{1/11},$$
(7.29)

where $w_c/2w_b > 1$, or

$$w_c = \left(\frac{24Q_F\,\eta}{\tau\alpha^2}\right)^{1/3},$$
(7.30)

where $0 < w_c/2w_b < 1$, where η is the Bingham viscosity of the lava. The total width of channel and levees is w_t, where $w_t = w_c + 2w_b$. The centerline thickness of the flow d_c (m) is given by Hulme (1974) as

$$d_c = \sqrt{\frac{\tau\,w_t}{g\rho}}.$$
(7.31)

The flow average velocity v (m s^{-1}) is given by

$$v = \frac{Q_F}{A},$$
(7.32)

where A is the cross-sectional area of the flow (m^2) and Q_F is the effusion rate (m^3 s^{-1}). As a result of conductive cooling, the flow will cease to move when its dimensionless Gratz number Gz reaches some critical number, Gz_{crit}. For basalt flows, this appears to be ≈ 300 (Pinkerton and Sparks, 1976), and the value appears to be similar for other lava compositions (Pinkerton and Wilson, 1994). The Gratz number is calculated (Head and Wilson, 1986) using

$$Gz = \frac{d_c^2}{\kappa t},$$
(7.33)

where κ is the thermal diffusivity of basalt and t (s) is the time since leaving the vent. The maximum flow length l (m) (Head and Wilson, 1986) is given by

$$l = \frac{d_c^2 v}{\kappa G z_{crit}}.$$ (7.34)

Using these equations, the flow advance rates for isothermal Bingham liquids can be calculated. The flow advance rate yields rate of area increase of the flow, so long as the mass eruption rate and slope do not change.

This model assumes that the flow is laminar: the dimensional relationships in Equations 7.28 to 7.34 are not applicable to turbulent flow. A flow will be laminar if the dimensionless Reynolds number (Re), given by $d_c v \rho / \eta$, is less than ≈ 2000. On slopes up to six degrees (similar to Hawaiian shield volcano regional slopes), flows resulting from mass eruption rates as large as 10^6 m^3 s^{-1}, with a viscosity of 1500 Pa s, do not exceed the Reynolds-number limit for laminar flow.

The lower gravity on Io (Earth gravity/Io gravity = 5.5) means that, for the same lava and slope, ionian flows are 2.6 times thicker than their terrestrial counterparts and move more slowly by the same factor, channels are narrower by $\approx 20\%$, and levees are wider by a factor of 5.5. If magmas on Io have lower viscosities and yield strengths, then channels are wider and flows are shallower and move faster. The calculated rate of areal increase of a basalt flow on Io as a function of slope and mass eruption rate is shown in Figure 7.13. Channel width as a function of mass eruption rate and slope is shown in Figure 7.14. Measurement of a lava channel width or levee width can be used to estimate magma yield strength and mass eruption rate.

7.5 Supply to the surface: conduit geometry

Magma rises as a result of positive buoyancy between the crust and the lower-density liquid. Magma typically moves up through the crust as linear dikes (Rubin, 1993). If the dike reaches the surface, initial effusion takes place along the resulting fissure.

The standard fluid dynamic expressions (e.g., Knudson and Katz, 1979) for the velocity V of a fluid in a fissure of width w that is much longer than it is wide (so that end-effects can be neglected) are

$$V_l = \frac{w^2 g \Delta \rho}{12 \eta},$$ (7.35)

when the motion is laminar and

$$V_t = \sqrt{\frac{w g \Delta \rho}{f \rho}},$$ (7.36)

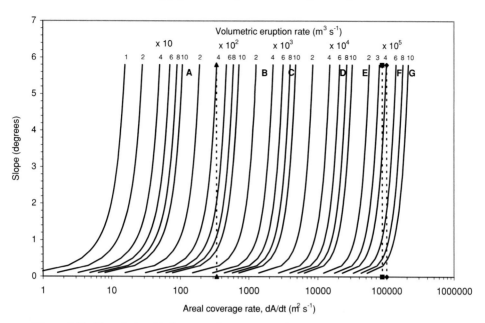

Figure 7.13 The relationship between slope, volumetric eruption rate (Q_E), and areal coverage rate (dA/dt) for basalt-composition lava on Io using Equations 7.28 to 7.32 (Hulme, 1974; Wilson and Head, 1983). Eruption rates are also shown for some terrestrial, ionian, and lunar eruptions: A = Kilauea Iki, Hawai'i, 1959; B = Loki, Io, from *Voyager* data (Carr, 1986); C = Laki, Iceland (1783–4, peak Q_F); D = Mare Imbrium, Phase II; E = Jan. 1990 Io outburst in Loki region (Blaney et al., 1995); F = Jan. 1990 Io outburst in Loki region (Davies, 1996); G = Mare Imbrium, Phase III. Estimates of areal coverage rates are shown: triangles = Loki in 1979 (Howell, 1997); squares = 1990 outburst (Davies, 1996); diamonds = 1990 outburst (Blaney et al., 1995). Base image from Davies (1996). Reprinted by permission of Elsevier.

when the motion is turbulent. Here g is the acceleration due to gravity (9.81 m s^{-2} for Earth, 1.79 m s^{-2} for Io); $\Delta\rho$ is the density difference between magma and crust, a positive buoyancy of typically $\approx$400 kg m^{-3}; η is the magma viscosity (typically 30 Pa s for basalt); the magma density ρ is 2600 kg m^{-3}; and f is a dimensionless friction coefficient of order 0.01. These numbers yield a pressure gradient dP/dz of 540 Pa m^{-1} for Io. In each case the Reynolds number for the motion, Re, is

$$Re = \frac{2wV\rho}{\eta}. \tag{7.37}$$

The decision as to whether V_l or V_t is relevant in a given case is made on the basis that the corresponding Reynolds number should be self-consistent: if the motion is laminar, the value of Re obtained using V_l should be less than $\approx$2000; and if the motion is turbulent, the value of Re found using V_t should be greater than $\approx$2000. In practice, the smaller of V_l and V_t is always the appropriate velocity to use for V.

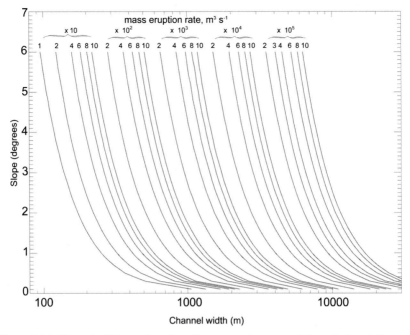

Figure 7.14 Channel width as a function of mass eruption rate and slope for basalt flows on Io, as determined using Hulme (1974) and Wilson and Head (1983). From Davies (1996). Reprinted by permission of Elsevier.

The analogous equations for magma rise in a conduit that is circular in section with diameter d are

$$V_l = \frac{d^2 g \Delta \rho}{32 \eta} \qquad (7.38)$$

and

$$V_t = \sqrt{\frac{d g \Delta \rho}{2 f \rho}} \qquad (7.39)$$

and

$$Re = \frac{d V \rho}{\eta}. \qquad (7.40)$$

Although many eruptive fissures become localized with time into near-circular vents due to excessive cooling of magma in the narrow ends of the fissures, it is not considered likely that magma systems have a circular section to great depth, but the results are included for completeness.

Once the magma rise speed V has been established, the effusion rate Q_F is given by

$$Q_F = (\pi/4)\, d^2\, V \tag{7.41}$$

in the case of a circular vent and

$$Q_F = w^2\, RV \tag{7.42}$$

in the fissure case, where R is the ratio of fissure length to width. When the ratio approaches unity, the width and length become equal, end-effects are no longer negligible, the conduit is likely to be approximately circular, and Equations 7.38, 7.39, 7.40, and 7.41 are used instead of Equations 7.35, 7.36, 7.37, and 7.42.

Equations 7.35 to 7.42 can be used to calculate Q_F for a range of vent widths, shapes, and aspect ratios. The relationship between conduit geometry and volumetric effusion rate for magma on Io and Earth is illustrated in Figure 7.15a for a circular conduit and in Figure 7.15b for a dike, which intersects the surface as a linear fissure. The reliance of flow velocity on planetary gravity means that rise velocities on Io are smaller than on Earth by a factor of 2.3, all other factors being equal. The same factor allows a systematic increase in dike width on Io and, consequently, higher effusion rates by a factor of more than five.

7.6 Crustal structure controls on ascent of magma on Io

7.6.1 *Ascent of magma*

The processes affecting the ascent of magma on the Earth and Moon were modeled by Wilson and Head (1981), and more recently for Io by Leone and Wilson (2001) and Davies *et al.* (2006c), from which the following treatment is derived. Magmas reach the surface of a planetary body either directly from primary sources (partial melt zones in the mantle) or from reservoirs closer to the surface. A melt is positively buoyant in the mantle but not necessarily positively buoyant throughout the lithosphere. To rise to the surface from the mantle, the magma only has to have a net positive buoyancy, balancing the mass of the magma column with the integrated mass of crust and mantle through which the magma ascends.

This process is demonstrated by considering the following example, taken from Davies *et al.* (2006c). The overburden pressure R_{source} at the level in Io's mantle where magma originates is given by

$$R_{source} = g(\rho_{litho}\, D_{litho} + \rho_{mantle}\, D_{mantle}), \tag{7.43}$$

where ρ_{litho} is the average density of lithosphere, 2200 kg m^{-3}; ρ_{mantle} is the density of mantle, 3000 kg m^{-3}; D_{litho} is the thickness of the lithosphere; and D_{mantle}

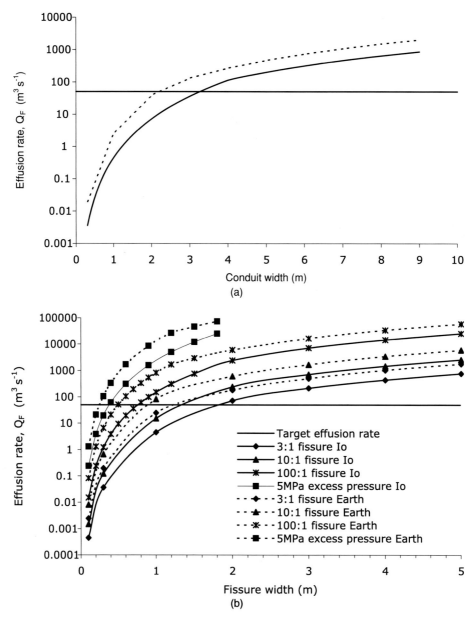

Figure 7.15 (a) Effusion rate on Io (solid line) and Earth (dashed line) as a function of circular conduit geometry. The horizontal line is for an effusion rate (Q_F) of 50 m³ s⁻¹. (b) Effusion rate on Io and Earth as a function of dike geometry. For a target rate of 50 m³ s⁻¹ (the horizontal line), a conduit on Io is systematically larger than its terrestrial counterpart, a result of a lower gravity on Io. On either planet, nevertheless, vast quantities of magma can be erupted from long fissures in a short period of time. From Davies *et al.* (2006c). Reprinted by permission of Elsevier.

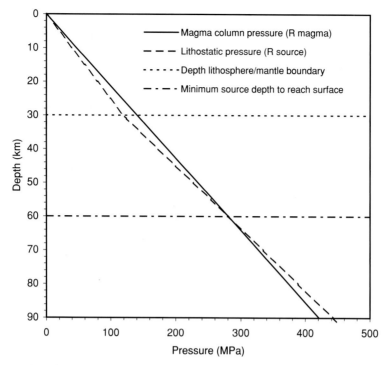

Figure 7.16 Rise of magma on Io. Magma will reach the surface directly from its source region if the overburden pressure R_{source} (dashed line) exceeds the weight of the column of magma, R_{magma} (solid line). For the example density values given in the text and a lithosphere 30 km thick (dotted line), magma originating from greater than 60 km will reach the surface directly from the source region (Davies et al., 2006c). Reprinted by permission of Elsevier.

is the depth below the lithosphere–mantle boundary where magma originates. If D_{litho} is 30 km and D_{mantle} is 20 km so that D_{source} is 50 km, then $R_{source} = 227\,\text{MPa}$.

The pressure R_{magma} needed to raise a column of basalt to the surface from depth D_{source}, ignoring compressibility effects and magma bulk density reduction due to exsolution of volatiles close to the surface, is given by

$$R_{magma} = g\rho_{magma}\,D_{source}, \qquad (7.44)$$

where ρ_{magma} is the density of magma, given above as 2600 kg m^{-3} for basalt. From Equation 7.44, $R_{magma} = 234\,\text{MPa}$. With R_{source} of 227 MPa, there is not quite enough lithostatic (overburden) pressure to raise magma to the surface directly from the source region. There is, therefore, a depth below which R_{source} is always greater than R_{magma}. All other factors being equal, magma originating from below this level (in this case, 60 km, where $D_{source} = D_{magma} = 281\,\text{MPa}$) will always reach the surface (Figure 7.16). Magma originating from shallower depths will stall, accumulating in a zone centered on the neutral buoyancy level to form the nucleus of a magma chamber (Davies *et al.*, 2006c).

7.6.2 Lithospheric magma reservoirs

For a neutrally buoyant magma chamber with the magma chamber roof at a depth D_{roof} below the surface, excess pressure R_{excess} is needed, first, to fracture the magma reservoir and, second, to lift magma to the surface. The absolute pressure at the magma chamber roof is then $R_{excess} + g\, \rho_{litho}\, D_{roof}$, and if magma is to reach the surface this must be equal to the static weight of the magma column of height D_{roof}, that is,

$$D_{roof} = \frac{R_{excess}}{g(\rho_{magma} - \rho_{litho})}. \qquad (7.45)$$

The value of R_{excess} sufficient to fracture the magma chamber under a weak terrestrial basalt volcano is typically ≈ 7 MPa (Parfitt, 1991). Taking values that bracket this value, 5 MPa and 10 MPa, and using $\rho_{magma} = 2600\ \mathrm{kg\,m^{-3}}$ and $\rho_{litho} = 2200\ \mathrm{kg\,m^{-3}}$, the depths of magma column supported by these pressures are 6.9 km and 13.9 km, respectively (Davies *et al.*, 2006c). Neutrally buoyant magma chambers at deeper levels require greater excess pressures to allow magma to reach the surface. However, greater excess pressures are not likely because they would cause a weak volcano reservoir to fail and magma to escape vertically and laterally to form dikes and sills. Reservoirs in the ionian lithosphere are considered "weak" because the lithosphere is composed of layers of lava flows and pyroclastic material. It is quite possible that the dikes might not reach the surface; there is probably a considerable amount of intrusive activity taking place on Io that is caused by the stalling of magma columns in a weak, relatively low-density crust (Davies *et al.*, 2006c).

If a magma chamber is too close to the surface, the release of pressure during an eruption may result in the roof failing in compression, and failure of the roof in tension may occur during pressurization of the magma chamber, in either case forming a caldera. A magma chamber of volume 1100 km³ and a diameter of ≈ 40 km has a vertical extent of order ≈ 1 km. To ensure normal stability of its roof, the depth of the roof of the reservoir would have to be at least ≈ 2 km to 3 km (Rowan and Clayton, 1993), making the depth of its center ≈ 2 km to 4 km. Alternatively, if the reservoir had a more ellipsoidal shape, its diameter would need to be ≈ 6.5 km and its center located at a depth of ≈ 8 km to 10 km to place the roof at least 2 km to 3 km below the surface.

7.6.3 The effect of volatiles on crustal structure

On Io, volcanic resurfacing on a global scale, mostly through plume fallout, results in a mixture of sulphurous volatiles and silicate pyroclastic debris. It is expected that this mixture leads to a relatively low mean crustal density close to the surface. The Io thermal gradient in the upper 80% of the lithosphere is essentially zero (O'Reilly and Davies, 1981) because the rate at which internal heat is conducted to the surface is surpassed by the rate at which the surface is buried by cold volcanic deposits. This zero thermal gradient would result in sulphur being solid through most of the lithosphere, to depths approaching 25 km, although SO_2 is liquid throughout most of the lithosphere (Keszthelyi *et al.*, 2004a). However, there will be many pockets of liquid sulphur in the lithosphere at shallower depths, the result of thermal exchange with silicate intrusions, especially stalled magma columns that subsequently form sills, laccoliths, and magma chambers. At a temperature of ≈432 K, the viscosity of sulphur reaches a minimum of 0.006 Pa s, at which point sulphur is a mobile liquid. At much lower temperatures (<200 K), SO_2 becomes a low-viscosity liquid. These volatiles will be driven off from the vicinity of the silicate intrusions to be replaced with silicates. Bulk crustal density would increase.

In detail, therefore, the temperature and density structure of the Io lithosphere may be very complex. Further work is needed to refine our understanding of the rise of magma through the ionian lithosphere (e.g., Jaeger and Davies, 2006).

8

Thermal evolution of volcanic eruptions

Chapters 6 and 7 describe models that are used to quantify thermal emission from volcanic processes and the subsequent use of those results to quantify the movement of magma from the interior of a planet to the surface. This chapter demonstrates how different types of eruptions generate different thermal emission spectra, characteristic "thermal signatures" of eruption style, and how those spectra change with time. Determination of eruption style is important because it allows application of the appropriate model to determine effusion rate. Observed temporal behavior allows eruption mechanism to be constrained.

Because most ionian volcanic thermal-emission data are at resolutions rendering the entire emitting area sub-pixel, it is necessary to understand how eruption style affects the shape of the integrated thermal emission spectrum. Such analyses are dependent on, first, adequate spectral coverage and, second, temporal coverage. Spectral resolution need not be particularly high so long as data at appropriate wavelengths are available. Two wavelengths (2 μm and 5 μm are particularly effective) can be used to constrain the silicate volcanism mode because the resulting 2-μm to 5-μm ratio (2:5-μm ratio) is very sensitive to the changing surface temperatures of lava surfaces of ages ranging from seconds to months, sometimes to years (Davies and Keszthelyi, 2005). Table 8.1 shows the timescales of different volcanic processes and the instruments best suited to observe them. NIMS data, with up to 17 data points from 0.7 to 5.2 μm (for a spectrum obtained in a single grating position), have been used to chart thermal emission variability for a variety of eruption styles (Davies, 2001; Davies *et al.*, 2001, 2006c).

Temporal coverage is of equal importance to spectral coverage because the cooling behavior of lava is influenced by the eruption mode and continuing activity. Rare high-spatial-resolution observations can definitively ascertain eruption mode by revealing individual lava flows or imaging a lava fountain, but single high-spatial-resolution observations of a target at night (and sometimes in day imagery)

Table 8.1 *Timescales of volcanic processes*

Time scale	Process	Best seen by
Seconds	Cooling of pyroclasts	SSI, ASTER
	Cooling of lava surfaces to 1000 K	
Hours	Lava fountaining	SSI, NIMS, ASTER,
	Transportation of magma in conduits	MODIS
Hours–days	Cooling of lava flow surfaces to ambient temperature (Earth)	ASTER, MODIS
Weeks–months	Cooling of thin flow surfaces to 200 K (Io)	NIMS, IRIS
	Growth of lava domes (Earth)	
Months–years	Cooling of thick flow surfaces to 200 K (Io)	NIMS, PPR, IRIS
	Growth of lava domes (Earth)	
Years–100s of years	Subsequent cooling of lava flow surfaces (Io)	PPR, IRIS
	Cooling of shallow intrusions (possibly detected at the surface – see Section 17.5)	
1000 years – 10^6 yrs	Cooling of deep plutons	
1–5 Myr	Lifetime of large magmatic systems – Yellowstone (Earth) – Loki Patera? (Io)	

are not as illuminating as one would imagine in the quest to identify eruption mode. Temporal behavior is more reliable.

Ultimately, the true test of a model is that it must not only explain what is observed but must also successfully predict subsequent activity. The *Galileo* dataset allowed for such model testing.

8.1 Effusive activity: landforms and thermal emission evolution

Different styles of volcanic activity (here taken to be the mechanisms by which lava is emplaced on the surface of a planet) exhibit different thermal emission spectra and spectral evolutions. Table 8.2 shows the expected thermal variation with time for different styles of eruption on Earth and Io. Eruption styles are classified as pahoehoe lava flow fields, channeled flows and inflated sheet flows, lava lakes, lava fountains, and lava domes.

8.1.1 Pahoehoe lava flow fields

Pahoehoe flows (Plate 15d) are emplaced in a relatively low-energy environment with laminar flows producing overlapping flow lobes. The most-often quoted example of this style of activity is the ongoing (1983–2006) Pu'u 'O'o-Kupaianaha

Table 8.2 *Thermal signature and evolution of different eruption styles*

Eruption type	Subtype	Location	Thermal characteristics	Short-timeframe change	Long-timeframe change
Lava fountains	Can feed a variety of different flow types (open-channel is most common: see image below)	Fixed	Intense short-wavelength thermal emission. Very high flux densities. Highest color temperature.	Very short duration ($\approx$hours?)	Not seen unless event repeats: emission rapidly decays with time.
Lava lake (1)	Stage 1 (active overturning)	Fixed	Very high flux density, most emission at short wavelengths (<3 μm). High color temperature.	Continues as long as crust disruption persists: progresses to Stage 2	Persistent hot spot, cyclic activity seen.
Lava lake (2)	Stage 2 (some rifting)	Fixed	Medium flux density. Medium color temperature.	Progresses to Stage 3	Persistent hot spot, part of cycle of activity.
Lava lake (3)	Stage 3 (quiescent, stable crust)	Fixed	Low flux density. Low color temperature.	Quiescent: cooling persists until return to Stage 1	Persistent hot spot, part of cycle of activity.

Flow type	Location	Flux density	Thermal source behavior	Thermal emission
Ponded flow (stagnant flow)	Fixed	High to low flux density. High to low color temperature.	Fixed location, cooling with time like a lava flow	Thermal emission follows predictable cooling curve to extinction or renewal.
Channeled flows, sheet flows	Wandering	Higher flux density than insulated flows. Initial high color temperature.	Thermal source increases in size	Flows eventually stop and cool.
Insulated and tube-fed flows (pahoehoe flows)	Wandering, persistent thermal source	Low flux density. Low color temperature.	Source increases in area in downslope direction	Source migrates and increases in size with time. Lava tubes form.
Lava domes	Fixed, gradual increase in size	Low flux densities. Very low color temperature.	Small incandescent areas: occasional explosive activity	Weeks–months. Not identified on Io.

From Davies and Keszthelyi (2007).
Key: Flux density = average thermal emission per unit area.

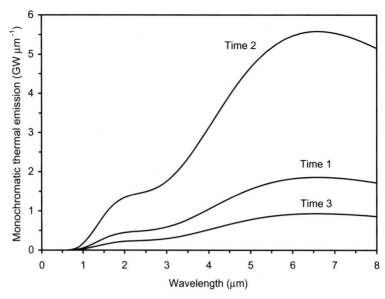

Figure 8.1 Model thermal emission from insulated pahoehoe-type flows. Thermal emission at Time 1 is from an area of 10 km^2 at 430 K and a hot component at 1400 K, with an additional area of 0.01 km^2. Time 2 is a period when the emitting areas have increased by a factor of three. Time 3 is a period when the emitting areas have declined to half those at Time 1. For all cases, emissivity $\varepsilon = 0.95$. The ratio of thermal emission at two different wavelengths at any given time remains the same. From Davies and Keszthelyi (2007).

eruption of Kilauea, Hawai'i, where relatively low-viscosity, low-gas-content basalt forms predominantly pahoehoe flows. In Hyperion hyperspectral data (Plate 3d), these active flows appear as clusters of hot pixels that change position over time as the location of surface activity moves. Activity takes place in eruption phases that may persist for months or longer.

Figure 8.1 shows how the thermal emission from an ongoing eruption of insulated pahoehoe flows varies as a function of active flow area. As seen by NIMS, thermal emission from an eruption of this style can be fitted with a two-temperature distribution, one for the cool crust on insulated, pahoehoe-like flows and the second for a much smaller area at higher temperature: cracks in the cool crust that expose molten lava beneath, the active vent, and newly exposed lava at breakouts (Davies *et al.*, 1997). A characteristic of this eruption type is that the shape of the integrated thermal emission spectrum, which can be expressed as the ratio of monochromatic thermal emission between any two wavelengths between 0.7 and 5.2 μm (the *Galileo* NIMS wavelength range), does not change even when effusion rate increases. Although increased effusion rate leads to an increased areal coverage rate, thermal emission at all wavelengths increases proportionately so long as the style of emplacement remains constant (Davies *et al.*, 2006c).

8.1.2 Channeled flows and inflated pahoehoe sheet flows

Channeled flows (Plate 15c) are higher-effusion-rate eruptions where a lava flow channel develops. The thermal emission per unit area emitted from these events is greater than that of pahoehoe flows because the rapidly forming surface crust is disrupted through shear forces at the edges of the channel, revealing more hot material. These eruptions form a wandering, linear thermal source, controlled by local topography. Such events are relatively short-lived, typically on a scale of days to a week or two, and thermal emission decreases rapidly with time after the end of the eruption. Carr (1986) developed a model for channeled flows that incorporated several flow components, including flow levees, to model *Voyager* IRIS data. The integrated thermal emission spectrum from channeled flows looks broadly similar to that from insulated flows, except that two-temperature fits to channeled flow spectra yield slightly higher temperatures and/or a larger hot fraction.

Inflated pahoehoe sheet flows are generated from high-volume basalt eruptions. Consider the example from Chapter 7 of a 99-day eruption with a peak effusion rate of 3.2×10^4 m^3 s^{-1} that emplaced 10-m-thick flows. Knowing the area covered by flows every day of the eruption allows the thermal emission from each surface element to be calculated using the age–temperature relationship in Figure 7.8. The evolution of the integrated thermal emission for this eruption is shown in Figure 8.2.

8.1.3 Lava lakes

Lava lakes (Plate 15b) are volcanic features of particular interest. They are geographically fixed, persistent hot spots. Strictly speaking, an active lava lake is directly connected to a deeper magma source. For example, the lava lake at the summit of Mt. Erebus, on Ross Island, Antarctica, is at the top of a conduit leading to a magma chamber. Magma circulates through the volcanic system.

Thermal emission from a lava lake varies over time, as the crust is disrupted and molten lava is exposed. The crust can be disrupted in several ways: pulses of lava entering the lake cause convection, leading to crust movement and breakup; gas exsolved from the magma bursts through the crust; and cold, dense crust founders. Generally, a sequence of observations over time should reveal whether a suspected lava lake is of an active nature (persistent and regularly overturning) or is an inactive lake, with no deep magma source. Such an inactive lake is formed by a ponded flow, which will solidify over time. The thermal emission trends diverge with time, with active lava lakes showing persistent thermal emission, whereas a lava pond continues to cool until it is no longer detectable.

A particularly well-studied terrestrial active lava lake was the Kupaianaha lake (on Kilauea, Hawai'i; Plate 15b). Intensive observations over its lifetime

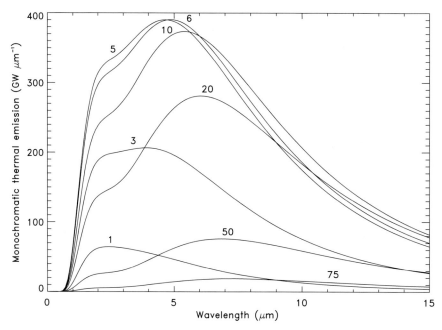

Figure 8.2 Thermal emission evolution from large inflating flows on Io using the example in Chapter 7 (Figure 7.12) for a 99-day eruption, during which 31 km³ of lava is erupted. Peak effusion rate is reached on Day 5, after which effusion rate decreases exponentially. Using the cooling model for basalt on Io (Chapter 7), the integrated thermal emission spectrum is determined for each day, with a 2% crack fraction only on the area currently being emplaced. Peak thermal emission (4×10^{12} W) is reached on Day 6. The small numbers are time in days. After Davies *et al.* (2006d).

(1986–1990) revealed three stages of surface activity (Flynn *et al.*, 1993). Stage 1 was characterized by active lava fountains and rifting, which resulted in the highest thermal emission from the lake. Stage 2 was marked by rifting events between plates of crust. Times of quiescence, when a thickening crust covered the surface of the lake, were designated as Stage 3. More than 99% of the time, the lava lake was quiescent. Examples of thermal emission for the three stages of activity in a hypothetical ionian lava lake are shown in Figure 8.3.

Episodic or periodic overturning of a lava lake leads to a unique and distinctive temporal signature, where thermal emission regularly increases and decreases. The replacement and cooling of lava lake surface crusts, and the resulting thermal emission, have been modeled (Davies, 1996; Rathbun and Spencer, 2006; Matson *et al.*, 2006b). Figure 8.4 shows the variability of thermal emission from a Loki-Patera-sized lava lake (area 2.15×10^4 km²) resurfaced over a period of 196 days and quiescently cooling for another 344 days; the ≈540-day resurfacing cycle was discovered by Rathbun *et al.* (2002).

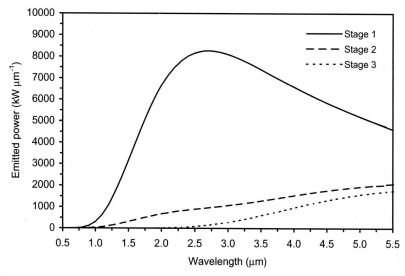

Figure 8.3 Thermal emission in the NIMS wavelength range from a hypothetical lava lake on Io exhibiting the three stages of surface activity observed by Flynn *et al.* (1993) at Kupaianaha, Hawai'i. The lava lake is 100 m in diameter. Stage 1: crust temperature at 450 K, 5% of area ($\approx$400 m^2) at 1100 K due to lava fountaining. Stage 2: crust temperature at 450 K, 0.5% of lake ($\approx$40 m^2) at 1100 K due to rifting. Stage 3: quiescent lake surface at 450 K with no crack fraction. From Davies and Keszthelyi (2007).

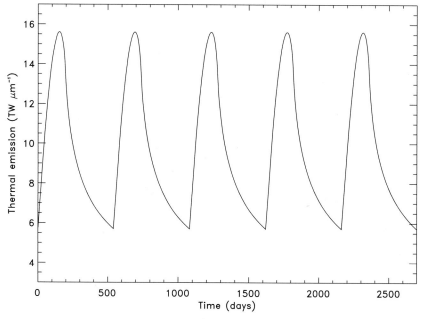

Figure 8.4 The periodic overturn of a lava lake generates a distinctive thermal signature, shown here for a lava lake with area 2.15×10^4 km^2, the area of the low-albedo floor of Loki Patera on Io. The resurfacing cycle is as follows: replacement of the crust exposes hot lava and thermal emission increases as the resurfacing wave propagates across the patera. This takes $\approx$200 days, and is followed by $\approx$340 days of cooling before the cycle repeats. Thermal output is derived using the model of Matson *et al.* (2006b). From Davies and Keszthelyi (2007).

8.1.4 Lava (fire) fountains

Lava fountains (also known as fire fountains) occur when gas-rich, low-viscosity magma reaches the surface. Expanding volatiles accelerate the lava to form fountains that on Earth may exceed 1 km in height. Exceedingly high short-wavelength thermal emission often results in the saturation of visible and SWIR detectors on orbiting platforms. Lava-fountain events are relatively rare and short-lived, typically lasting only a few hours. Fountaining ends when the most gas-rich lava has been erupted and the strain inherent in the magmatic system has begun to decrease. On Io, the absence of a thick atmosphere means that a lower magma gas content than on Earth is needed for the formation of lava fountains. Lava fountains may be more common on Io than on Earth (Davies, 1996). Lava fountains feed flows that spread over the surface. If fountain activity takes place within a caldera (e.g., Kilauea Iki, in 1959) or flows fall into a pit or crater (Plate 16b), a lava lake or pond forms. After the fountain phase ceases, the effusion rate may still climb before the waning phase of the eruption begins (see Chapter 7). After the eruption ceases, the emplaced flows cool in the same fashion as channeled flows or inflated pahoehoe flows described previously, depending on emplacement mechanism and emplaced flow thickness. Figure 8.5 shows the evolving thermal emission from a thermal model fit to an eruption on Io in 1990 that was most likely a lava-fountain episode (Davies, 1996).

8.1.5 Domes

Volcanic domes are rounded, steep-sided mounds built by magma, such as dacite or rhyolite, which is too viscous to move far from the vent before cooling and crystallizing. Domes can generate deadly pyroclastic flows when the sides of a dome, or erupting lava flows on a dome, collapse down a steep slope to form an avalanche of hot lava fragments and gas. The high viscosity of the lava (10^5–10^9 Pa s) traps exsolving gas, leading to explosive activity when pressure exceeds the strength of the dome. Such features have not been detected on Io. The relatively low thermal emission from such features may be one reason why domes have not been detected, but a more likely reason is that such high-silica-content magmas may not exist on Io.

8.2 Flux density as a function of eruption style

Having quantified thermal emission and derived the size of the emitting areas, the flux density – the average thermal emission per unit area – can be calculated. Usefully, different eruption styles yield different flux densities. Eruptions where the eruption style is more vigorous – where large areas at high temperatures are

Table 8.3 *Flux densities for a selection of volcanoes and eruption styles*

Terrestrial volcano	Style of activity	Flux density (kW m^{-2})	Reference
Krafla 1984 (Iceland)	Open channel flow	≈0.5–1.5	Keszthelyi *et al.* (2001b)
Mt. Etna 1992 (Italy)	Open channel flow	1.2–3.5	Keszthelyi *et al.* (2001b)
Kilauea (USA)	Insulated flows	0.3–0.7	Keszthelyi *et al.* (2001b)
Kupaianaha, Hawai'i, Stage 1	Lava lake	≈22	Davies *et al.* (2001)
Kupaianaha, Hawai'i, Stage 2	Lava lake	5.3	Davies *et al.* (2001)
Kupaianaha, Hawai'i, Stage 3	Lava lake	4.9	Davies *et al.* (2001)

Ionian volcano	Observation	Flux density (kW m^{-2})	Reference
Loki hot spot 1979	*Voyager 1* IRIS	0.3	Davies *et al.* (2001)
Pele hot spot 1979	*Voyager 1* IRIS	6–10	Davies *et al.* (2001)

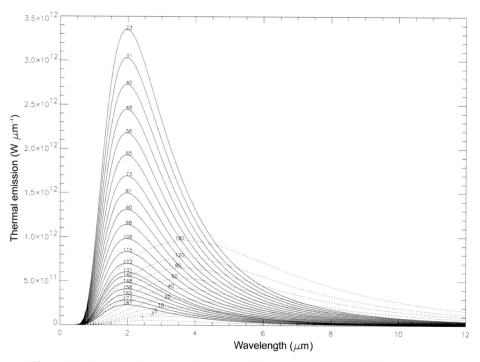

Figure 8.5 The thermal emission from diminishing lava fountains (solid lines) and lava flows that are increasing in size (dotted lines). The integrated thermal emission and change with time matches observations of an Io outburst eruption in 1990 that was recorded by ground-based observers (Veeder *et al.*, 1994). Time is given in minutes. The total peak integrated thermal emission is 1.1×10^{13} W, 11% of Io's total thermal emission. From Davies (1996). Reprinted by permission of Elsevier.

exposed – tend to have high flux densities. Styles of emplacement dominated by cool crust have lower flux densities. A rule of thumb is the shorter the wavelength of peak thermal emission, the greater the flux density. Table 8.3 shows flux densities for several terrestrial eruption styles as well as for the Io hot spots Loki and Pele, as derived from *Voyager* IRIS data (Davies *et al.*, 2001).

8.3 Summary

Different styles of effusive and explosive volcanic activity generate different temperature and area distributions. The resulting integrated thermal emission spectrum, and the way in which this spectrum changes with time (the "thermal signature" of an eruption), can be used to constrain mode of activity.

The better the temporal coverage, the better the chance of recognizing the eruption style, allowing the use of appropriate models for quantifying the eruption process.

Persistent activity with little change in spectral shape is characteristic of insulated flows. Eruptions producing open channel flows and voluminous inflationary sheet flows may begin with lava fountains and a rapid increase in emitting area. Areal coverage rates drop as time progresses, the eruption comes to an end, and thermal emission decays as the emplaced flows cool.

A lava lake may undergo periodic resurfacing, demonstrating a systematic variability in thermal emission with recognizable thermal flux densities.

Section 4

Galileo at Io: the volcanic bestiary

9

The view from *Galileo*

Galileo's arrival at Jupiter opened a new era of exploration of Io's volcanism. Plates 1 and 6 show Io as seen by *Galileo*. A global mosaic is shown in Plate 4, created using the best *Galileo* and *Voyager* image data (Becker and Geissler, 2005). Appendix 2 contains maps of Io with feature names.

After the disappointment of the loss of imaging during Jupiter orbit insertion, the first observations of Io by the Solid State Imaging experiment (SSI) and Near-Infrared Mapping Spectrometer (NIMS) were finally obtained in late June, 1996. The closest approach to Io during this encounter (Orbit G1) was 696 000 km, not as near as *Voyager 1* got to Io (20 570 km) but closer than *Voyager 2* (1 129 900 km). Eclipse observations by SSI showed high-temperature hot spots glowing in the darkness (McEwen *et al.*, 1997). NIMS obtained spectra indicating the presence of silicate-temperature volcanism (Davies *et al.*, 1997).

The first sunlit images of Io showed a surface that was, if anything, even more colorful than that seen by *Voyager*. Io's surface was dominated by black, red, yellow, white, and green hues, representing different mixtures of silicate and sulphur com-pounds. Over the next 3 years, Io was periodically observed by *Galileo*, mostly at long range (hundreds of thousands of kilometers) but culminating with a sequence of fly-bys as close as 182 km (Table 3.2). During the primary *Galileo* mission, SSI observations were obtained primarily at resolutions of between 10 and 20 km pixel^{-1}. Images obtained by SSI showed hundreds of paterae, many with dark floors, set into smooth plains; lava flow fields of different sizes; active plumes and plume deposits; and mountains, many much higher than any on Earth (see Table 4.1). In high-spectral-resolution observations, NIMS detected the absorption features of sulphur dioxide everywhere, although in different concentrations and implied grain sizes. Plumes were laying down deposits as they did during *Voyager* fly-bys, and in some of the same locations. NIMS and SSI data were analyzed, temperatures and areas of volcanic thermal sources derived, and areal coverage rates and volumetric fluxes quantified (e.g., Davies *et al.*, 1997; McEwen *et al.*, 1997, 1998a, 1998b; Lopes-Gautier *et al.*, 1999; Davies *et al.*, 2000a).

During the close fly-bys in the final phase of the *Galileo* mission, small areas of Io were observed at unprecedented resolution (≥ 6 m pixel^{-1}). These data allowed mapping of temperature distributions and identification of style of activity, and the quantification of thermal and eruption fluxes across a wide range of eruption sizes and styles, with data from SSI (Geissler *et al.*, 1999; McEwen *et al.*, 2000b; Keszthelyi *et al.*, 2001a; Geissler *et al.*, 2004a; Turtle *et al.*, 2004); NIMS (Lopes-Gautier *et al.*, 1999; Davies *et al.*, 2000a; Lopes-Gautier *et al.*, 2000; Davies *et al.*, 2001; Lopes *et al.*, 2001; Davies, 2003a, 2003b; Lopes *et al.*, 2004; Davies *et al.*, 2006c); and PPR (Spencer *et al.*, 2000b; Rathbun *et al.*, 2004).

9.1 Surface changes: *Voyager* to *Galileo*

Because extensive changes were observed in the 4 months between the *Voyager 1* and *Voyager 2* fly-bys, great changes on Io's surface were expected when *Galileo* data were compared with *Voyager* images. The first surprise was that some of the surface changes seen by *Voyager 2* had reverted to their appearance as seen by *Voyager 1* (McEwen *et al.*, 1998a). For example, the Surt plume deposit that appeared between *Voyager* encounters was no longer in evidence at the beginning of the *Galileo* mission. About 87% of Io's surface was unchanged from 1979 to 1999 (Geissler *et al.*, 1999), as illustrated in Figure 9.1. Nevertheless, about a dozen large-scale (>300 km) changes on Io were seen, most notably at Ra Patera, where the beautiful, sinuous flows seen by *Voyager* had been buried under new deposits from an active plume (Figure 9.2). This eruption created the large albedo change observed by the Hubble Space Telescope (Spencer *et al.*, 1997a; see Chapter 3).

The large, red, elliptical plume deposit around Pele was still in evidence, appearing much as it did when imaged by *Voyager 2*. The Prometheus plume was also still active, but the position of the plume deposits had shifted some 80 km to the west, and new flows were seen on the surface (McEwen *et al.*, 1998a). Loki Patera appeared unchanged from how it appeared to *Voyager 1*. This unchanged appearance was puzzling because the nearby Loki feature was a source of plumes during the *Voyager* encounters and its appearance changed markedly between *Voyagers 1* and *2*. Loki Patera, repeatedly observed from Earth, was also the most thermally powerful of Io's volcanoes. Yet continual volcanic activity had not altered its appearance after more than 20 years (Figure 9.3).

An exhaustive examination of the entire set of hundreds of *Galileo* SSI images was carried out to identify and chart changes on Io's surface (McEwen *et al.*, 1998a; Geissler *et al.*, 1999, 2004a). The task was complicated by photometric effects caused by varying illumination and viewing geometry, but this problem was

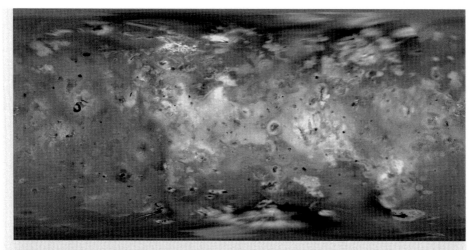

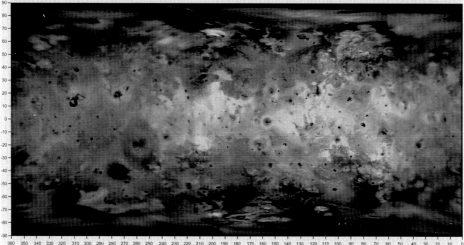

Figure 9.1 Io as seen by *Voyager* (top) and *Galileo* (bottom). These images are global mosaics obtained using a green filter. For such an active satellite, there was surprisingly little change on regional or global scales from 1979 to 1999. The absence of large-scale change is consistent with observations indicating long-term stability of surface albedo units (Morrison *et al.*, 1979). Courtesy of NASA.

overcome by comparing images taken under similar illumination and viewing angles to distinguish actual surface changes from photometric variation (Geissler, 2003).

Three broad categories of surface change were identified from SSI data: volcanic plume deposits, patera color or albedo changes, and seepages of SO_2 (Geissler *et al.*, 2004a). About 80 significant surface changes were identified during the course of the *Galileo* mission, ranging in scale from subtle changes in surface color or albedo on the floors of individual paterae to massive plume deposits, some exceeding 1000 km in diameter. One of the most notable events took place at Pillan in June, 1997, when

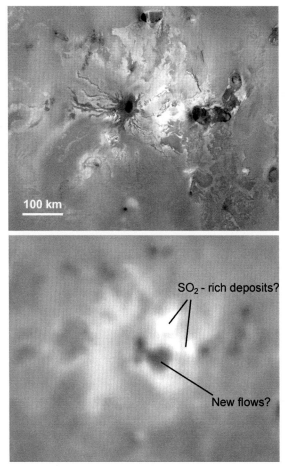

Figure 9.2 Changes at Ra Patera from *Voyager* (top, March, 1979) to *Galileo* (bottom, June, 1996). An eruption observed by the Hubble Space Telescope covered the flows observed by *Voyager* with new deposits. When *Galileo* observed this region, a plume was still depositing material on the surface (McEwen *et al.*, 1998a). Courtesy of NASA.

a black pyroclastic deposit more than 400 km across (Plate 9b) was laid down during a powerful eruption (see Keszthelyi *et al.*, 2001a). The Pillan 1997 eruption was one of the most closely studied of the eruptions observed by *Galileo* and is described in detail in Chapter 11. Most of the volcanic hot spots detected by SSI and NIMS during the *Galileo* mission did not exhibit surface changes on a scale much greater than $\approx$10 km in extent (Geissler, 2003), implying that, although much of Io's heat is lost through its volcanic centers, the bulk of resurfacing during the *Galileo* era was from plume fallout and emplacement of other pyroclastic material (Davies *et al.*, 2000a; Geissler *et al.*, 2004a). That said, large flow fields covering many thousands of square kilometers were seen, representing significant areal coverage by effusive

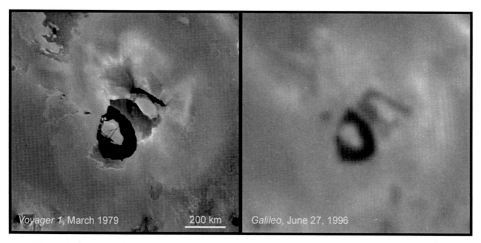

Figure 9.3 Although Loki Patera is the most powerful volcano on Io, little change took place between *Voyager* and *Galileo*, indicating that the process of resurfacing of Loki Patera does not involve explosive volcanic activity and the ejection of large amounts of material that would color the surface. Courtesy of NASA.

volcanism. At Prometheus, for example, flows covering more than 6700 km^2 were emplaced in the years between *Voyager* and *Galileo* (Davies, 2003b). The flows at Amirani, more than 300 km long, made up the longest active flow field in the Solar System (Keszthelyi *et al.*, 2001a), and *Voyager*-era flows at Lei-Kung Fluctus, still visible, covered more than 1.25×10^5 km^2. Areally extensive, voluminous flows clearly played an important role in resurfacing Io.

9.2 Color and composition

The different colored deposits and units of different albedo on Io's surface can be assigned likely compositions based on their color, spectral properties, and temperature (Table 9.1). Coverage by units of different color is shown in Figure 9.4. Sulphur and its compounds are the most likely candidates for most of the variegated colors of Io, a conclusion reached by *Voyager* investigators. Even after *Galileo*'s investigation of Io, only sulphur dioxide has been identified spectroscopically on the surface (Fanale *et al.*, 1979; Smythe *et al.*, 1979; Carlson *et al.*, 1997; Douté *et al.*, 2001). The loss of spectral resolution when the NIMS grating stuck effectively ended detailed investigation of possible silicate deposits at high spatial resolution during close fly-bys, which probably would have revealed at least some mineral constituents. Nevertheless, the shape of Io's full-disk reflectance spectrum (Figure 9.5) has features that cannot be explained by sulphur dioxide alone, and sulphur is the obvious candidate. Io is the source of the Io sulphur torus, and the presence of elemental sulphur would explain deep absorptions in the ultraviolet. Sulphur

Table 9.1 *Main surface unit colors and inferred compositions*

Effusive activity

Color	Composition	Magma	Example
Black	Silicates	Primary	Pillan, Prometheus, Amirani
Red	Short-chain sulphur	Secondary	Tupan Patera, Culann, Zamama
Yellow	Sulphur	Secondary	Emakong Patera
White	SO_2	Primary and secondary	Balder Patera

Explosive activity

Color	Composition	Example
Black	Silicates	Pillan, Tvashtar, Pele
Red	Short-chain sulphur	Pele, Culann, Dazhbog
White	SO_2	Thor

has been conclusively identified in the Pele volcanic plume (Spencer *et al.*, 2000a). Surface deposits of sulphur may contain impurities (e.g., hydrogen sulphide, which has an extraordinary effect on sulphur's viscosity/temperature profile; see Chapter 5) consistent with impurities found in terrestrial sulphur deposits. Deposits of impure sulphur have spectral properties that are at least consistent with much of Io's surface (Kargel *et al.*, 1999). Small amounts of impurities could explain many of the observed color variations, the subtle hues that tinge surface units with green, pink, and yellow (Geissler *et al.*, 1999; Kargel *et al.*, 1999).

9.2.1 Dark units

Dark, low-albedo areas are thought to be dominated by silicates, which form a wide range of flow and plume deposits but which cover less than 2% of Io's surface (Geissler *et al.*, 1999, 2004a). The floors of many paterae have relatively low albedoes and, because excess thermal emission is observed in these locations, low albedo is associated with volcanic activity. Black pyroclastic materials are deposited around some eruption sites and clearly are a major constituent of some plumes. Dark silicate lava flows range in emplacement style from the relatively quiescent eruptions at Prometheus to the voluminous flows of Pillan, which were probably fed by lava fountains and which might have been turbulently emplaced. Lava lakes at Pele and, probably, at Loki Patera are silicate in nature, based on their low albedo, derived surface temperatures, and spectral qualities. The spectral reflectances of these dark materials are consistent with silicates containing the magnesium-rich

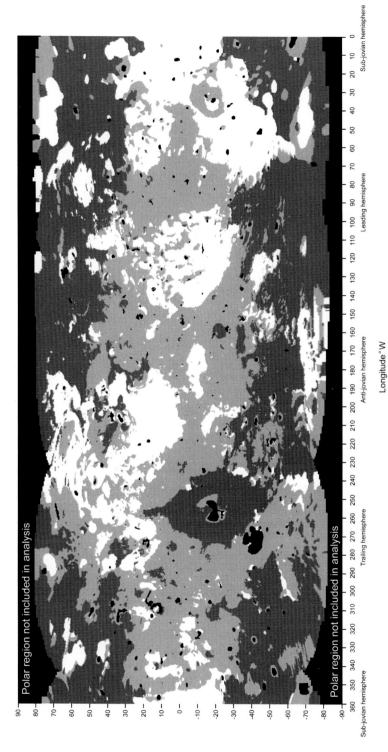

Figure 9.4 A composite global map of Io, produced by Geissler *et al.* (1999), showing main color units in 1999. The map is centered on the anti-jovian point, 180°W. About 39% of the surface consists of yellow material (light gray), with a slight concentration on the trailing hemisphere. About 31% is red material (dark gray), mostly confined to latitudes above ±30°. About 26% are white units (white). Slightly >1% is dark material (black). About 3% was unclassified, mostly at the poles (Geissler *et al.*, 1999). The plot does not show large deposits laid down late in the *Galileo* mission (e.g., at Tvashtar, Dazhbog, and Surt). Reprinted by permission of Elsevier.

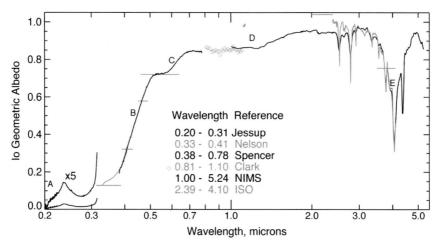

Figure 9.5 The Io reflectance spectrum. Spectral identifications are A = SO_2 gas (fine structure below 0.23 μm); B = S_8 or S_nO; C = S_4; D = unknown; E = Cl_2SO_2? Other features
are caused by SO_2 frost. The NIMS data (0.7 μm to 5.2 μm) especially show multiple
SO_2 absorption bands, with the deepest at 4.3 μm (Carlson *et al.*, 1997). References: Jessup *et al.* (2002); Nelson and Hapke (1978); Spencer *et al.* (1995a); Clark and McCord
(1980); Schmitt and Rodriguez (2003). From *Jupiter*, eds. Bagenal, F. *et al.*, reproduced by
permission of Cambridge University Press, © 2004.

mineral orthopyroxene and also with coatings containing varying concentrations
of sulphur (Geissler *et al.*, 1999; Turtle *et al.*, 2004). Orthopyroxene is common in
terrestrial mafic and ultramafic rocks.

9.2.2 Red and orange polar units

Red and orange plains cover a third of Io's surface, particularly at latitudes above
± 30° and toward the poles, and in isolated locations at lower latitudes. The origin of
these reddish-orange materials is enigmatic. In polar regions, short-chain allotropes
are thought to form through photo-dissociation of yellow cyclooctal sulphur (S_8) by
charged particle bombardment (Johnson and Pilcher, 1977; Nash and Fanale, 1977).

9.2.3 Yellow units

Yellow deposits cover ≈40% of Io's surface, especially in equatorial regions and
particularly on the trailing hemisphere. Yellow deposits contain significant amounts
of elemental sulphur, probably in the form of the stable (and yellow) S_8 allotrope.

9.2.4 Red units

Distinct from the red polar units, reddish deposits are seen in the vicinity of several
volcanically active centers, most notably in plume deposits at Pele and Culann
Patera. The interpretation is that these deposits are short-chain sulphur allotropes

(S_2, S_3, and S_4) formed, for example, by heating of cyclooctal (S_8) sulphur by silicate magma. Accordingly, the sources of the red plumes have been equated to vents for silicate lavas (McEwen *et al.*, 1998a). This interpretation is supported by the direct observation of sulphur in plumes (Spencer *et al.*, 2000a). These short-chain sulphur allotropes are short-lived on the surface of Io and soon revert to yellow S_8, although it is not known precisely how long this process takes. The change of these deposits to a pale yellow accounts in part for the relatively few changes observed on Io's surface between *Voyager* and *Galileo* (Geissler *et al.*, 1999).

Another possibility is that these deposits may be solid sulphuryl chloride (Cl_2SO_3) or sulphur dichloride (Cl_2S) that condensed from Cl-rich volcanic gases (Schmitt and Rodriguez, 2003). These compounds have not been detected in the Pele plume, but the detection of chlorine in the Io torus (Küppers and Schneider, 2000) indicates some source on Io, possibly NaCl (Lellouch *et al.*, 2003).

9.2.5 *White and gray units*

White and gray units make up 27% of Io's surface, primarily in equatorial regions and in localized deposits at higher latitudes. Small white areas are often found adjacent to yellow and dark deposits, forming auras around these units. In keeping with previous analyses (e.g., McEwen *et al.*, 1988), the large white areas are rich in relatively uncontaminated SO_2 of coarse-to-moderate grain size, as determined by NIMS (Carlson *et al.*, 1997). These units may be relatively young (Douté *et al.*, 2001, 2004). Gray units may be contaminated with silicate material.

9.2.6 *Green deposits*

Some deposits are distinctly greenish in color. In SSI and NIMS data they exhibit a negative spectral slope from 0.7 to 1.5 μm (Geissler *et al.*, 1999; Lopes *et al.*, 2001). The green deposits appear on dark, low-albedo areas and may be caused by the alteration of sulphur compounds deposited on recently emplaced silicate lava. The floor of Shamash Patera (152°W, 32°N) turned green while an eruption was depositing bright SO_2 deposits toward the south (Geissler *et al.*, 2004a). The greenish-colored materials on the surface cannot be explained by sulphur allotropes alone. Cyclodeca sulphur (S_{10}) is yellow-green (Meyer, 1976), but uncontaminated sulphur specimens generally have flat or positive spectral slopes in the visible to 1-μm range. The green deposits may be contaminated with iron (Kargel *et al.*, 1999) or, alternatively, these deposits may be silicates, perhaps olivine or pyroxene.

9.2.7 *Other compounds*

Various compounds have been proposed to explain absorption features in Io's reflectance spectrum (Figure 9.5). These compounds include the aforementioned

Cl_2SO_3 (Schmitt and Rodriguez, 2003), SO_3 (Nelson and Smythe, 1986), and H_2S (Nash and Howell, 1989); H_2S greatly reduces the viscosity of liquid sulphur (see Chapter 5).

9.2.8 Changes during Galileo

Over time, as the number of *Galileo* images grew ever larger and the evolution of new plume deposits and lava flows could be studied, reasons for the apparent absence of change emerged. Dark plume deposits appear to be ephemeral, fading away over a period of a few years. This fading was most notable at Pillan, where black, possibly silicate-rich, pyroclastic deposits were rapidly buried by other plume deposits rich in sulphur and SO_2 (see Plates 9b–c). Some of these newer deposits themselves are short-lived. Bright red deposits (see Plates 9 and 12) that are likely to be short-chain sulphur allotropes (S_3 and S_4) are associated with centers of high-temperature silicate activity and appear to either revert to yellow cyclooctal (S_8) sulphur with time in an ionian environment, or are buried under ubiquitous SO_2 plume deposits. Both processes probably act to produce a trend back to a stable "background" distribution of color units. Sulphur may be mobilized from surface frosts or may be exsolving from the lava itself. If exsolving from the lava, the sulphur may either be an integral component of the magma or incorporated into the magma through mechanical and thermal erosion of conduit walls during magma ascent.

Proposed bright yellow sulphur flows (see Chapter 14) most likely result from the thermal remobilizing of fumarolic sulphur deposits, a process observed on Earth (described in Chapter 5).

9.2.9 Colored lava flows on Earth and Io

Terrestrially, the colors and spectral qualities of silicate flows are changed by chemical alteration of the silicates, a process that can take place both while the flows are active and during the months and years after emplacement (Rothery *et al.*, 1996). On lava flows on Hawai'i and Mt. Etna, the formation of white and yellow deposits rich in sulphur and sulphur compounds was noted. Similar processes almost certainly affect flows on Io, with allowances for differences in available compounds. On Io, there is almost certainly a near-complete absence of water, a compound that plays an important role terrestrially by combining with sulphur and SO_2 to form sulphurous and sulphuric acids. From a field study carried out primarily on basalt volcanoes, Rothery *et al.* (1996) found that pale coatings of amorphous silica, sulphur, and sulphur-bearing compounds could form on basalt flows. Although some of these coatings are unlikely to form on ionian lava flows because of the absence of sufficient water, the formation of sulphurous acid (H_2SO_3) on Io through interaction of cryogenic SO_2 or SO_2/H_2O mixtures and high-energy proton irradiation has been proposed (Voegele *et al.*, 2004). If H_2SO_3 condenses on the surface, reaction with flows would take place. Endogenic sulphur-rich compounds would also condense on the lava surfaces. For Io, magmas are already thought to be rich in sulphur compared to terrestrial lavas because of the assimilation of crustal sulphur during ascent (Davies, 1988). Once flow surfaces on Io have cooled sufficiently, they are coated not only by fallout from plumes but also by degassing from the flows themselves after emplacement. The resulting albedo variations can be used for relative dating of flow surfaces (McEwen, 1988; Keszthelyi *et al.*, 2001a; Davies *et al.*, 2005).

9.3 Discovery of widespread silicate volcanism

The major discovery of *Galileo*'s initial encounters – and one that would forever change the perception of Io's volcanism – was that of multiple high-temperature hot spots in SSI eclipse observations (McEwen *et al.*, 1997). Prior to *Galileo*'s arrival, high-temperature silicate volcanism was thought to be a relatively rare event on Io, and it was not expected that SSI would be widely useful for direct observation of hot spots. Accordingly, the first SSI Io observations in June, 1996 (Orbit G1) were obtained of Io in eclipse with the intent of looking for Io's atmosphere and searching for plumes on the limb. To the delight of *Galileo* scientists, in clear-filter eclipse images obtained at resolutions of effectively ≈ 50 km pixel^{-1} (as a result of image smear during long exposures), SSI detected the glow from multiple hot spots (Plate 6f). These hot spots must have had brightness temperatures of at least 700 K to be detected by SSI (McEwen *et al.*, 1997). The temperature could be higher if the area were smaller, yielding the same thermal emission as seen through a single filter (Figure 6.6): a range of temperatures and areas could explain the observed single-filter intensity. Because it was unlikely that an isothermal area at 700 K would fill a pixel, and postulated stable sulphur lakes on Io's surface would not exhibit temperatures greatly in excess of ≈ 500 K (Lunine and Stevenson, 1985), these hot-spot detections were evidence of silicate volcanism at multiple locations on Io's surface.

During the same orbit, NIMS detected multiple sources of above-background short-infrared-wavelength thermal emission (Lopes-Gautier *et al.*, 1997). The single usable NIMS observation from this orbit (designated G1INNSPEC01: see Plate 7a) was obtained with the highest NIMS spectral resolution (408 wavelengths between 0.7 and 5.2 µm) and contained 16 hot spots, 14 in darkness (Lopes-Gautier *et al.*, 1997). All of the hot-spot spectra showed an increase in emission toward longer wavelengths (see examples in Figure 3.2) and were remarkably similar in shape to spectra of active silicate pahoehoe lava flows on Earth (e.g., Flynn and Mouginis-Mark, 1992).

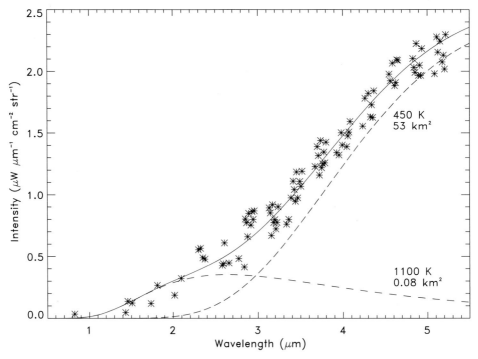

Figure 9.6 The Zamama thermal emission data, obtained on June 28, 1996 (Orbit G1), by NIMS, were best-fitted with two temperature components. Data are shown here in NIMS intensity units. A larger, cool component (53 km² at 450 K) was interpreted as the cooling crust on either silicate lava flows or a lava lake. A smaller, hotter component, with a total area of 0.08 km² at 1100 K, was attributed to red-hot lava exposed in cracks in the cool crust, at breakouts of new lava, and at the vent itself (Davies *et al.*, 1997). Later SSI images showed new flows were being emplaced at Zamama (Keszthelyi *et al.*, 2001a). From Davies *et al.* (1997).

Single-temperature blackbody fits (Lopes-Gautier *et al.*, 1997; Davies, 2003b) to the NIMS data produced temperatures around 500 K, although one fit – for Tupan Patera – was just above 600 K. These temperatures were too low to conclusively differentiate between silicate and sulphur compositions, being at the top end of the liquid sulphur range (up to 717 K at pressures found a few meters below the surface of Io; see Kieffer [1982], also Figure 5.4). However, the presence of thermal emission at wavelengths shorter than 2 μm, which was not well fitted by a single blackbody curve, indicated that higher temperature areas were present (Davies *et al.*, 1997).

During the June, 1996 (Orbit G1) encounter, only one hot spot (originally named "South Volund" because it lay just south of the *Voyager*-era Volund plume site, but later officially named "Zamama") was seen in darkness by both SSI and NIMS. By fitting a two-temperature blackbody thermal emission model to the NIMS Zamama data (Figure 9.6; see also Table 9.2a, b), a higher-temperature component of 1100 K

Table 9.2a *Two-temperature fits to NIMS G1INNSPEC01 data*

	Maximum intensity spectra[a]				Maximum temperature	
Volcano	T_{hot} (gp)[b] (K)	Area T_{hot} (km²)	T_{warm} (K)	Area T_{warm} (km²)	Max. T_{hot} (gp) (K)	Area T_{hot} (km²)
Monan	1007 (23)	0.191	427	33.2	1158 (5)	0.0401
Amirani	1022 (20)	0.348	420	173.1	1207 (16)	0.0950
Maui	1182 (4)	0.034	471	32.6	1295 (5)	0.0187
Tupan	974 (22)	0.270	449	31.1	1137 (9)	0.0808
Prometheus	1263 (14)	0.052	437	53.6	1479 (16)	0.0274
Culann	993 (8)	0.209	424	69.2	1230 (6)	0.0592
Zamama[c]	1100[d]	0.080	450	53	n/a	n/a

From Davies (2003b).

Table 9.2b *Single-temperature fits to NIMS G1INNSPEC01 data*

	"Maximum-intensity" fit (gp in brackets)		Maximum-temperature fit (gp in brackets)		Average of all suitable grating position spectra		
Volcano	T_{hot} (gp) (K)	Area at T_{hot} (km²)	Maximum T_{hot} (gp) (K)	Area at max T_{hot} (km²)	Average T_{hot} (K)	Average area (km²)	No. of spectra fitted
Hi'iaka	458 (17)	47.5	503 (1)	10.6	459 ± 20	18.3 ± 6.6	20
Zal	485 (20)	35.6	543 (8)	8.4	492 ± 22	9.1 ± 3.5	18
Gish Bar	457 (22)	24.0	497 (23)	11.4	477 ± 14	10.8 ± 2.4	18
Sigurd	449 (20)	36.0	470 (10)	13.6	436 ± 20	29.8 ± 12.7	16
Altjirra	369 (18)	351.8	405 (13)	130.8	382 ± 12	155.0 ± 50.4	19
Malik	586 (20)	7.7	641 (5)	3.7	604 ± 22	4.2 ± 0.7	15
Arinna Fluctus[e]	434 (4)	25.6	524 (3)	4.1	447 ± 34	14.0 ± 9.1	16

From Davies (2003b).
[a] The grating position spectrum with the most energy under the curve, as used in Davies *et al.* (2000b).
[b] gp = grating position. The largest NIMS product is a 408-wavelength observation, which consists of 24 spectra of 17 different wavelengths, each spectrum obtained in a different spectrometer grating position.
[c] Named South Volund in Davies *et al.* (1997).
[d] Fit to whole NIMS spectrum from Davies *et al.* (1997).
[e] Named 9606W in Lopes-Gautier *et al.* (1997).
Notes: Table 9.2a shows the results of two-temperature fits to the spectra that yield, first, the greatest integrated thermal output (taken to be the grating position spectrum most representative of the hot spot) and, second, the spectrum that yields the highest temperature.

Table 9.2b shows the single-temperature and area derivations for other volcanoes where the addition of a second temperature component was not statistically justified (Davies, 2003b). Fits are shown for the "maximum-intensity" spectrum, the spectrum yielding the highest temperature, and the average temperature from fits to all spectra.

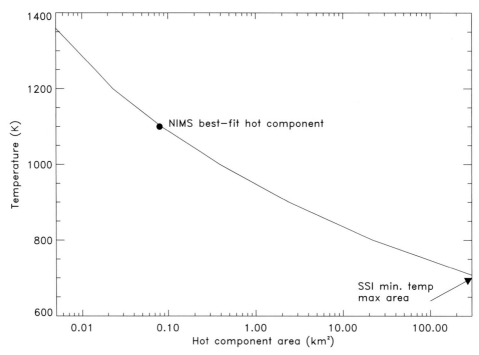

Figure 9.7 Zamama was observed by both SSI and NIMS on June 28, 1996. SSI measured the thermal emission from Zamama with one filter. The curve represents the possible combinations of temperature and area that fit the SSI clear-filter flux for Zamama (see Figure 6.6). The NIMS best-fit hot component falls exactly on this curve. This analysis was the first confirmation of active silicate-composition lavas from *Galileo* data (Davies *et al.*, 1997).

was derived, with an area of ≈ 0.08 km^2. The cooler component was at a temperature of 450 K and covered an area of ≈ 53 km^2 (Davies *et al.*, 1997). The total area of the emitting surfaces made up only a tiny fraction of a NIMS pixel area (122 000 km^2 for this observation). Comparing instrument data, the NIMS hot area and temperature exactly matched the measured SSI clear-filter flux (Figure 6.6) for an area at 1100 K (Figure 9.7). SSI did not detect the warm component seen by NIMS because it was cooler than 700 K. Two-temperature model fits to other spectra in this NIMS observation also showed high-temperature components indicative of silicate volcanism at Prometheus, Amirani, Culann, and other locations. SSI would continue to detect hot spots glowing in the dark – sometimes even in daylight. Most NIMS hot-spot data obtained in darkness would yield similar high-temperature areas from fits of two-, three-, and multiple-component blackbody models (Davies *et al.*, 1997; McEwen *et al.*, 1998b; Davies *et al.*, 2000a, 2001; Lopes *et al.*, 2001; Davies, 2003b). It rapidly became apparent that not only were the large volcanic eruptions observed from Earth silicate in composition but also that most of Io's volcanic thermal activity was consistent with silicate volcanism.

The spectral signature of Zamama, with intensity increasing with wavelength (the "thermal ramp"), was similar to the thermal signature of active, insulated lava flows or quiescent, active lava lakes or ponds (Chapter 8). This was the spectral shape seen most often by NIMS throughout the *Galileo* mission. The interpretation of the Zamama G1 data was that silicates had erupted onto the surface and created lava flows, or perhaps a lava lake was gently overturning (Davies *et al.*, 1997). The newly exposed lava was cooling rapidly, forming an insulated crust, but the hot molten interior was seen through cracks in the crust. High temperatures would also be seen at active vents, through skylights, and at the active margins of flows. Other temperature derivations from NIMS June, 1996 (G1) data are shown in Table 9.2. Later SSI observations would reveal that Zamama, Prometheus, and Amirani were indeed being resurfaced by lava flows (McEwen *et al.*, 1998a; Keszthelyi *et al.*, 2001a; Williams *et al.*, 2004; see Chapters 12 and 14).

9.4 The rise (and fall?) of ultra-high-temperature volcanism

As the quality of data acquired by *Galileo* improved, and increasingly sophisticated techniques for constraining magma eruption temperature were developed and applied, so temperature estimates of lavas erupting on Io increased. The progression is shown in Table 9.3.

After the discovery of silicate dominance of volcanism on Io, even more intriguing discoveries were to follow. Dual-filter SSI eclipse images allowed color temperature to be derived at several hot spots, and it was found that single-temperature fits in some cases yielded temperatures in excess of 1500 K (McEwen *et al.*, 1998b). Single-temperature fits can underestimate the hottest temperatures of emitting surfaces, often by hundreds of kelvins. The implication drawn from these temperature fits was that some magmas on Io may have liquidus temperatures in excess of 1600 K, approaching or exceeding the accepted range for basaltic lavas. A likely candidate for these very-high-temperature lavas was magnesium-rich (ultramafic) lava, akin to terrestrial komatiites (Williams *et al.*, 2001a). These lavas were widespread on Earth billions of years ago, but now only highly metamorphosed remnants remain. Many questions exist as to the mechanisms leading to the generation, eruption, and emplacement of ultramafic lavas on Earth. It seemed that Io was providing a look at processes that took place on an early Earth (Matson *et al.*, 1998).

Although at most locations wide error bars meant that basaltic volcanism could not be ruled out, three locations initially showed unambiguous temperatures in the ultramafic range: Pillan, Pele, and Tvashtar Paterae. The most dramatic example was seen at Pillan in 1997, where SSI data indicated temperatures of 1500 K and possibly higher (McEwen *et al.*, 1998b). Fits to a combined NIMS and SSI dataset derived a minimum temperature of >1870 K for the magmas at Pillan

Table 9.3 *Lava eruption temperature derivations: the rise (and fall?) of ultramafic volcanism on Io*

Temperature (K)	Location	Instrument	Reference	Notes
600	Disc-integrated	Ground-based	Sinton (1980b)	Vapor eruption proposed
654	Pele	IRIS	Pearl and Sinton (1982)	Modeled as silicate eruption (Carr, 1986)
900	≈35°W	Ground-based	Johnson et al. (1988)	Aug. 1986 outburst
1200	Loki region	Ground-based	Veeder et al. (1994)	Jan. 1990 outburst
1225	≈35°W	Ground-based	Blaney et al. (1995)	Aug. 1986 outburst
1200–1600	Loki region	Ground-based	Davies (1996)	Jan. 1990 outburst
1100+	Zamama	SSI+NIMS	Davies et al. (1997)	Combined NIMS-SSI dataset
1400+	Various locations	Ground-based	Stansberry et al. (1997)	
1500+	Various locations	Ground-based	Spencer et al. (1997c)	
1400–1600?	Pele	SSI+NIMS	Davies et al. (2001)	Combined NIMS-SSI dataset
1450–1600+	Pele	SSI, ISS	Radebaugh et al. (2004)	High-resolution SSI observation
1800–1825	Pillan 1997	SSI+NIMS	McEwen et al. (1998b)	Using SSI CLR:1MIC ratio and NIMS 3-temperature fit to NIMS spectrum of Pele and Pillan
1870+	Pillan 1997	SSI+NIMS	Davies et al. (2001)	Combined NIMS-SSI dataset, having isolated Pillan in NIMS data
1760 ± 210	Pele	NIMS	Lopes et al. (2001)	High-resolution NIMS observation. 1.8- to 3.8-μm data were saturated
2000+	Tvashtar	SSI	Milazzo et al. (2005)	Realization that model output was unreasonable
1600+	Pillan, Pele, Tvashtar	SSI, NIMS	Keszthelyi et al. (2005b)	Re-assessment of models used and instrument response
1400–1500?	Pillan, Pele, Tvashtar	SSI, ISS	Keszthelyi et al. (2006)	New model of lava-fountain thermal emission, not including 200 K superheating

Key:
NIMS = *Galileo* Near-Infrared Mapping Spectrometer
SSI = *Galileo* Solid State Imaging experiment
ISS = *Cassini* Imaging Sub-System
IRIS = *Voyager* Infrared Radiometer Interferometer and Spectrometer

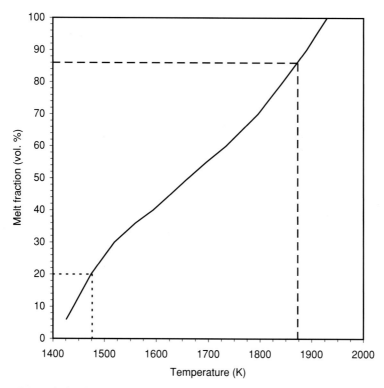

Figure 9.8 Melt fraction as a function of temperature, assuming an initial LL chrondrite composition (after Keszthelyi *et al.*, 2005a). Magmas at 1870 K (dashed line) require a melt fraction of more than 80%. A melt fraction of 20% yields magma at a liquidus temperature of ≈1470 K (dotted line).

(Davies *et al.*, 2001). Ultramafic temperatures (>1700 K) at Pele were derived from a sampling of spectra from a high-resolution NIMS observation obtained in February, 2000 (Lopes *et al.*, 2001). Finally, initial data analysis of SSI data at Tvashtar Paterae yielded temperatures higher than 1900 K (Milazzo *et al.*, 2005).

These very high temperatures have proven difficult to explain (Kargel *et al.*, 2003a; Keszthelyi *et al.*, 2004a, 2005a, 2005b, 2006a). Specifically, the estimates of lava temperatures at or higher than 1870 K require that the interior of Io either have an extremely unusual chemistry or be almost completely (>80%) molten. Neither of these possibilities fits comfortably within the current understanding of the evolution of the jovian system (e.g., Consolmagno, 1981; Lewis, 1982; Ross and Schubert, 1985; Ojakangas and Stevenson, 1986; Moore, 2001). If Io has a broadly chondritic composition (chrondrites are considered the basic building blocks of the Solar System), then recent tidal heating models suggest that no more than ≈20% partial melting of the interior should take place (e.g., Moore, 2001; Monnereau and Dubuffet, 2002). This alone suggests a maximum magma temperature of ≈1500 K (Figure 9.8), consistent with basalt lava.

Keszthelyi *et al.* (2005a, 2005b, 2006a) approached the re-evaluation of these high temperatures by (a) considering other processes that might superheat magma before eruption at the surface, such as viscous or frictional heating during ascent; (b) refining the uncertainties in the observations; and (c) taking a close look at the models and methodologies used for data reduction and analysis for Pillan, Pele, and Tvashtar Paterae.

9.4.1 Re-evaluation: viscous heating

Could magmas be superheated during ascent? Viscous heating of magma coupled with melt production at elevated pressure could produce significant superheating. These processes are negligible on Earth, but Io has a unique stress distribution in the lithosphere, the result of strong lateral compression caused by rapid resurfacing and burial (Jaeger *et al.*, 2003). Keszthelyi *et al.* (2005b) estimated that superheating of rapidly ascending magma from a depth of 30 km was likely in excess of 150 K. This superheating is consistent with a relatively deep origin of magmas erupted in outburst-type eruptions, and with lava fountains of the type implied at Pillan in 1997 and seen at Tvashtar Paterae in 2000.

9.4.2 Re-evaluation: derivation of temperature

Extracting a magma temperature from Pillan Orbit C9 data depended on the fluxes derived through the SSI clear (CLR) and 1-μm (1MIC) filters. The derivation of those fluxes and their use in determining temperature at Pillan are described by Davies *et al.* (2001) but can be summarized as follows:

1. The Pillan hot spot was completely saturated in the SSI observation, but a CLR:1MIC ratio was still estimated by (a) using the brightness of a streak caused by a serendipitous sequence error that slewed the scan platform in the middle of the observation and (b) gauging the brightness of the saturated area from the amount of bleeding in the detector.
2. NIMS data of the Pillan eruption were fitted with a lava cooling model (Davies, 1996). As described in Chapter 6, model output consisted of a range of areas at different temperatures ranging from a very small area at the magma liquidus temperature to increasingly larger areas at lower temperatures.
3. Using available SSI filter responses, the NIMS-derived temperatures and areas were used to synthesize the SSI CLR:1MIC response (see Figure 6.7). NIMS data were fitted with the cooling model output for increasing initial magma liquidus temperatures until the CLR:1MIC ratio was reached, at a temperature of 1870 K (Davies *et al.*, 2001).

The first problem with this methodology was that SSI filter response changed during the *Galileo* mission. Using a set of filter responses obtained closer to Orbit C9 than the set of pre-launch filter responses used by Davies *et al.* (2001), a new

CLR:1MIC ratio yielded a temperature in excess of 2000 K, a *higher* magma temperature than originally reported – entering the realm of unfeasibility.

The second problem was that the SSI C9 Pillan dual-filter observation utilized a slew; the flux from Pillan was determined from the intensity of the streak, not the saturated image at either end. A constant rate of slew was assumed, but it is more likely that the slew rate was not constant, starting slowly, accelerating to a peak, and then slowing down (L. Keszthelyi, pers. comm., 2005).

Additionally, the glow from the aurora, where charged particles ionize gas molecules in the tenuous atmosphere above the Pillan vent, may have contributed to the flux observed by SSI through the clear filter. This contribution would increase the estimate of CLR flux, increasing the CLR:1MIC ratio and leading to derivation of erroneously high magma temperatures. The full magnitude of these effects will probably never be known.

9.4.3 A new model for Pillan and Tvashtar Paterae

An unfeasible magma temperature of >2000 K at Pillan derived using the Davies (1996) model is the result of the same problem identified by Milazzo *et al.* (2005) in their analysis of the Tvashtar Paterae SSI data. Cooling models where heat loss is buffered by release of latent heat may not be applicable for deriving temperatures from lava fountains. The Davies (1996) and Keszthelyi and McEwen (1997b) models balance heat flow from a surface with the conduction of heat through a solid crust to the emitting surface from the molten interior. Heat loss is buffered by the release of latent heat, which may not necessarily be the case when working with clasts in lava fountains, which can freeze into glass without release of the latent heat of fusion.

A model of thermal emission from clasts in lava fountains was therefore developed (Keszthelyi *et al.*, 2006). A good fit between SSI color temperature, from CLR:1MIC ratios, and eruption temperature is provided by the simple linear relationship

$$T_E = (1.29T_c - 165), \tag{9.1}$$

where T_E is the eruption temperature in degrees Celsius and T_c is the color temperature derived from the SSI CLR:1MIC ratio, expressed in °C. T_E is converted to kelvin by adding 273.

Applying the model to the SSI data reduces the eruption temperature to 1650 K at Pillan and ≈1570 K at Tvashtar Paterae. These temperatures are still indicative of ultramafic composition; however, when superheating is taken into account, the unheated magmas might be at temperatures in the range ≈1400 K to 1500 K. Ultramafic magmas on Io are not ruled out. At the time of writing, however, available data and analyses do not conclusively support them.

9.4.4 Re-evaluation: Pele

Pele is a very different volcano from Pillan and Tvashtar. Pele is a persistent hot spot and does not have the huge lava fountains seen at Pillan and Tvashtar Paterae. Ultramafic temperatures were reported at Pele by Lopes *et al.* (2001) from high-resolution NIMS data. Lopes *et al.* (2001) fitted a two-temperature thermal model to a selection of a dozen or so spectra, yielding a median temperature of 1760 K ± 210 K for the hot component. The large error bars reflect the uncertainty from the small number of unsaturated data points in the spectra from this dataset. Although these results allow ultramafic temperatures, they are also consistent with the lower temperatures (1400–1600 K) derived from lower-spatial-resolution SSI and NIMS data (Davies *et al.*, 2001).

Analysis of *Galileo* SSI and *Cassini* Imaging Sub-System (ISS) data yielded temperatures for portions of Pele of 1605 K ± 220 K and 1420 K ± 100 K for different SSI filters, and 1500 K ± 80 K for ISS (Radebaugh *et al.*, 2004), straddling basaltic and ultramafic temperature ranges. On the other hand, because lavas at liquidus temperatures cool very rapidly, any temperature derived from remote-sensing data could underestimate actual eruption temperature by 100 K or more.

9.4.5 Re-evaluation: conclusions

Magma temperatures at Pillan may be as low as 1600 K (Keszthelyi *et al.*, 2006a), which is still in the range of ultramafic compositions, but, taking into account viscous heating of ≈200 K, mafic (basaltic) magmas can be erupted at this temperature. Temperatures derived from *Galileo* SSI and NIMS data at Pele and Tvashtar Paterae are also explainable using basaltic compositions, although ultramafic compositions are not necessarily ruled out. Decreasing the temperature of magmas from 1870 K to 1500 K decreases the amount of melting necessary at the top of the mantle from >80% to 20%. A melt fraction of 80% is difficult to maintain and would result in a liquid layer where tidal heating would be dissipated, effectively ending volcanism. It is possible that Io has entered or is entering such a state. In that case, Io's volcanism would decrease as tidal heat would be, to an increasing extent, dissipated in a deep global magma ocean. Eventually, heat loss would cause the ocean to solidify, tidal dissipation would cause heating and re-melting, and volcanism would begin again.

There is now consistency among observations, data analysis, and models of Io's interior composition and evolution (Keszthelyi *et al.*, 2006a). This consistency results from a basaltic or transitional basaltic–ultramafic composition, thermal emission models more appropriate for lava fountains, and magma superheating. Ultramafic magmas may be present, but probably not at temperatures in excess of 1800 K, and are not expressly required to explain observations to this date.

If magmas are either ultramafic or superheated basalt, in either case they would be of very low viscosity, which would explain the paucity of volcanic shields and the great areal extent and low relief of lava flow fields.

9.5 PPR observations

During the close fly-bys between 1999 and 2002, PPR measured thermal emission from volcanoes and passive (non-volcanically active) areas of Io. Almost all of Io was mapped at resolutions better than 250 km pixel^{-1}, with extensive regional and local coverage at higher resolutions (Rathbun *et al.*, 2004). Both daytime and nighttime maps were obtained. Passive daytime temperatures reached a peak of ≈130 K, whereas nighttime temperatures generally dropped down to ≈90–100 K (Rathbun *et al.*, 2004).

Many hot spots were readily identified (see Appendix 1), such as Pillan and Loki Patera. Despite its high level of volcanic activity, Pele was seen to emit less energy at 17 μm than many other volcanoes (Spencer *et al.*, 2000b). This was additional evidence that the emitting surfaces at Pele were hotter and younger than at most other locations on Io, and consistent with Pele's being an overturning lava lake (Davies *et al.*, 2001).

With many point sources, some large thermal anomalies stood out. One of these was the massive flow field at Lei-Kung Fluctus. This lava field observed by *Voyager* was unchanged 16 years later. The flows had a nighttime temperature of 135 K, as high as the passive surface daytime peak temperature. Lei-Kung Fluctus is too warm for SO_2 to condense on the surface, which explains its persistence. Another large thermal source appeared at Dazhbog (303°W, 55°N) between November 26, 1999 (Orbit I25), and August 6, 2001 (Orbit I31). PPR was most likely detecting the thermal emission from pyroclasts or lava flows emplaced by this eruption (Rathbun *et al.*, 2004). Other low-temperature thermal sources seen by PPR were not detectable by NIMS. One, at 130°W, 25°S, was not associated with a known hot spot or low-albedo feature.

In the absence of volcanic heat, surfaces are heated by incident solar radiation and temperature is controlled by re-radiation. Pole-to-pole PPR scans revealed a small nighttime temperature dropoff to ≈80 K toward the poles (Rathbun *et al.*, 2004). This temperature drop was less than expected from purely solar insolation, where incident radiation drops off as a function of $\cos^{1/4}$ (latitude). Why Io's poles are anomalously warm is still not known. It may be the result of additional endogenic thermal emission from lava flows preferentially emplaced at high latitudes (Veeder *et al.*, 2004) or some undefined property of the surface material in these regions.

Estimates from PPR data of Io's thermal emission were close to previous estimates (see Table 4.2). Io's passive background heat flow was less than 1 W m^{-2}

(Rathbun *et al.*, 2004), affirming that thermal output was dominated by hot spots (Veeder *et al.*, 1994).

9.6 *Cassini* and *Galileo* observe Io

At the end of 2000, the *Cassini* spacecraft passed close to Jupiter to receive a gravity assist on its long journey to Saturn. Although its closest approach to Io was not as close as either of the *Voyager* spacecraft, *Cassini*'s high data return rate allowed thermal emission from powerful volcanoes to be observed on short timescales. Observations from December 31, 2000, to January 1, 2001, were made in conjunction with *Galileo*, which was also observing Io during this time (Orbit G29). *Cassini*'s closest approach to Io took place on December 28, 2000, at a distance of about 1 million km. *Galileo* and *Cassini* dayside images revealed a huge new plume, about 400 km high, at Tvashtar Paterae, which left a 1200-km-diameter red deposit similar to that seen at Pele (Porco *et al.*, 2003; Milazzo *et al.*, 2005). The high data return capacity of *Cassini*, unaffected by technical glitches, allowed the return of nearly 500 observations of Io by the *Cassini* ISS narrow-angle camera (Porco *et al.*, 2003). Of particular interest were four sets of multiple-filter observations of Io in Jupiter's shadow, in which several hot spots were identified, the most prominent being Pele (see Radebaugh *et al.*, 2004). Faint emission was also detected from Loki Patera by ISS, SSI, and in the infrared by NIMS, indicating temperatures of at least 700–990 K, an indication of active silicate volcanism (Davies, 2003a).

The *Cassini* data provided a unique temporal dataset, allowing variability of thermal emission at visible wavelengths to be charted. Pele showed considerable variability at visible wavelengths on a timescale of minutes while maintaining steady thermal emission at longer ($\approx$5-μm) wavelengths (Radebaugh *et al.*, 2004). This temporal emission pattern was wholly consistent with what was thought to be taking place at Pele, highlighting the strengths of different instruments in observing different facets of volcanic activity at the same volcano. SSI and ISS were observing the hottest areas, which were very short-lived, whereas NIMS was detecting the bulk of thermal emission from cooled crust, which changed on a much longer timescale.

9.7 Adaptive optics and Hubble observations

The infrared observations obtained by *Galileo* NIMS of Io from Jupiter orbit were supported by observations from Earth and Earth orbit. During the *Galileo* mission, new observational techniques, most notably the use of adaptive optics (AO), produced multi-spectral observations from Earth-based telescopes at spatial resolutions of $\approx$160 km, comparable to many NIMS observations (e.g., Marchis *et al.*, 2001). Such is the pace of technological development. NIMS was essentially late-1970s technology, whereas AO was developed in the 1990s. These multi-wavelength AO observations of Io proved to be invaluable for augmenting *Galileo* data, monitoring known volcanoes, identifying new activity, detecting outbursts (the best example was a massive eruption at Surt [Marchis *et al.*, 2002]), and charting thermal emission from individual eruptions over several nights, something that was beyond the capabilities of *Galileo*.

Io was also observed by the Hubble Space Telescope from Earth orbit. The most significant discovery during the *Galileo* epoch was of gaseous sulphur in the Pele plume (Spencer *et al.*, 2000a), providing the best evidence to date of elemental sulphur on Io. Other observations during the *Galileo* epoch detected the Prometheus plume and determined that the sulphur dioxide atmosphere at latitudes less than $\pm 30°$ on the anti-jovian hemisphere appeared to be controlled by equilibrium with surface frosts, supported by the sublimation of these frosts (Jessup *et al.*, 2004).

9.8 Other discoveries

Other relevant discoveries of importance to Io's volcanism made by *Galileo* included the absence of a strong intrinsic magnetic field (Kivelson *et al.*, 2001), which indicated that Io does not have a convecting molten core. The analysis of *Galileo* close fly-by trajectories, having determined Io's moment of inertia, revealed that Io has a large Fe or Fe–FeS core comprising about 20% of Io's mass (Anderson *et al.*, 2001). Fly-by results also were consistent with a relatively thin (≤ 50 km) crust overlying a thicker (100–200 km), melt-rich aesthenosphere, and a silicate-composition mantle. This picture of Io's interior is described in Chapter 4.

9.9 Summary

To summarize, *Galileo* revealed that Io was essentially unchanged since 1979 on scales larger than ≈ 10 km (Geissler *et al.*, 2004a), which was surprising given the level of volcanic activity. A self-correcting mechanism appears to maintain this long-term photometric stability. The implication is that on a decadal scale, effusive surface activity is confined to areally small centers, whereas plume deposits remain relatively constant and revert to an Io "background" combination of composition, color, and albedo.

The primary magmas dominating Io's volcanism are silicate in composition. Very-high-temperature ultramafic volcanism cannot be ruled out in some locations, although temperatures may not have been as high as first thought. Sulphur volcanism plays a secondary role in effusive volcanism but is an important volatile in some explosive activity.

10

The lava lake at Pele

Because of its persistent red deposits, Pele is the most distinctive of Io's volcanoes (Plates 9a, b). Appropriately named after the Hawaiian goddess of volcanoes, Pele is the source of a giant plume more than 300 km high, as seen by *Voyager* (Strom *et al.*, 1981). The reddish deposits laid down by the plume are ≈1200 km across and are rich in sulphur and sulphur dioxide (Geissler *et al.*, 1999; Spencer *et al.*, 2000a). Closer to the vent, dark pyroclastic material, most likely of silicate composition, streaks the surface (Strom *et al.*, 1981; Geissler *et al.*, 1999). SSI and the Hubble Space Telescope (Plate 9f) have shown that the Pele plume can exceed 400 km in height (Spencer *et al.*, 1997b; McEwen *et al.*, 1998a).

Voyager observed dramatic changes in the shape of the plume deposits between encounters when the deposits changed from a "cloven hoof" shape (almost certainly caused by a vent obstruction of some kind) to a circular appearance (Figures 1.3d, e). The change in the shape of the plume deposits took place in the four months between *Voyager* encounters, so the resurfacing for some period of time was relatively rapid. Although the Pele plume was not seen by *Voyager 2*, it may have changed to a more tenuous, gas-rich form – a "stealth" plume (Johnson *et al.*, 1995). Nevertheless, the plume deposits persisted throughout the *Galileo* mission, covering deposits laid down by eruptions at nearby Pillan in a remarkably short space of time (Plate 9c). No other giant plume on Io has shown the persistence of Pele; other such plume deposits are the result of transitory explosive activity. Plumes are discussed in more detail in Chapter 16.

10.1 Setting

Pele is located at 257°W, 18°S, almost at the center of Io's trailing hemisphere. The main thermal source is located at the eastern end of an east–west-trending graben north of Danube Planum (Plate 9d) and abuts the southwestern edge of Pele Patera, a feature ≈30 km by 20 km in size (Radebaugh *et al.*, 2004). The position of the thermal source, comprised of a large hot spot and some much smaller hot spots detected by NIMS (Davies *et al.*, 2002; Lopes *et al.*, 2004), apparently is strongly tectonically controlled. The highest resolution *Voyager* images show an elongated low-albedo feature within the graben ≈24 km long and 8 km wide (Plate 9d). The graben has formed at the northern margin of a fractured plateau (Danube Planum) ≈200 km across. The plateau is not a volcanic structure but instead seems to be a fractured uplift, with the location of eruptive vents controlled by fractures (Strom *et al.*, 1981).

10.2 Observations of thermal emission

From *Voyager* IRIS data, it was evident that Pele was not like other hot spots detected on Io. The peak of thermal emission from Pele was at shorter wavelengths than at other locations, indicative of a larger proportion of relatively high (>400 K) surface temperatures. Ground-based observations showed that Pele was a long-lived hot spot on a decade-length timescale, and an important contributor to Io's volcanic heat flow (Veeder *et al.*, 1994). Detailed analysis of the IRIS data with models of silicate lava cooling indicated that the style of activity at Pele was consistent with the exposure of young, hot areas of lava (Carr, 1986; Howell, 1997, 2006). To generate the observed spectrum, such activity was limited to events of new, rapid flow emplacement (Carr, 1986; Howell, 1997); the rapid overturning of a lava lake that exposed molten lava (Howell, 1997); or lava fountains (Davies, 1996). Further constraint on eruption style was not possible from a single IRIS observation and had to await multi-wavelength data obtained over a suitable period of time to establish the eruption's thermal signature. Such data were obtained by *Galileo* NIMS and SSI. Figure 10.1 shows raw NIMS data for Pele and Pillan, and Figure 10.2 shows modeled NIMS spectra of Pele obtained from 1996 through 1999. These NIMS data showed that Pele's thermal output always peaked at a wavelength <5 μm, and the total thermal emission was consistently 200 GW to 230 GW (Figure 10.3). Almost all of this energy was emitted at wavelengths <10 μm (Davies *et al.*, 2001).

Two-temperature blackbody model fits to NIMS data revealed that most emitting surfaces were at relatively low temperatures in the range from 550 K to 770 K (the latter just hot enough to be detected by SSI) with an area ranging from ≈6 km² to ≈31 km². A smaller, hotter component at about 1400 K was also present, which was detected by SSI on numerous occasions (McEwen *et al.*, 1998a; Davies *et al.*, 2001). With an area varying from 0.4 to 1.6 km², the hot component makes up only a few percent of the total emitting area. Fitting the NIMS data with a lava cooling model (see Chapter 7) indicated a high rate of resurfacing of the same magnitude as that calculated from model fits to *Voyager* IRIS data (Howell, 1997, 2006). The high level of vigorous activity at Pele had persisted since *Voyager*.

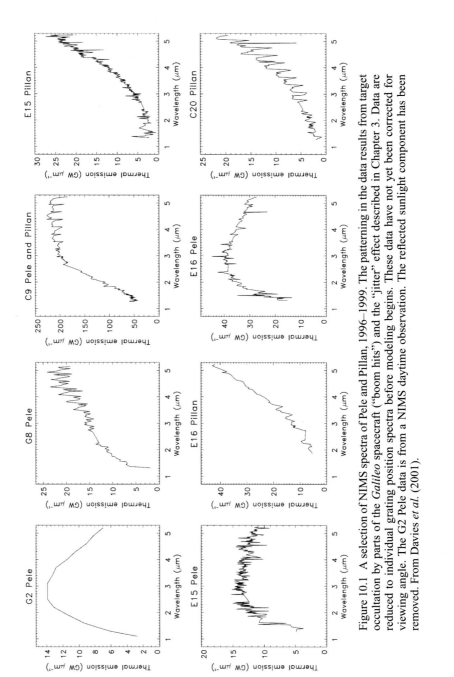

Figure 10.1 A selection of NIMS spectra of Pele and Pillan, 1996–1999. The patterning in the data results from target occultation by parts of the *Galileo* spacecraft ("boom hits") and the "jitter" effect described in Chapter 3. Data are reduced to individual grating position spectra before modeling begins. These data have not yet been corrected for viewing angle. The G2 Pele data is from a NIMS daytime observation. The reflected sunlight component has been removed. From Davies *et al.* (2001).

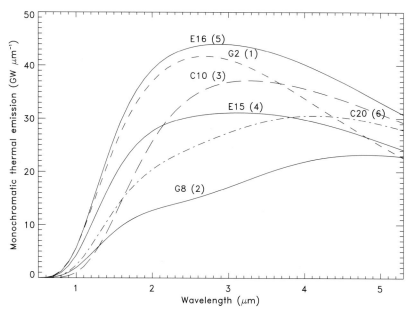

Figure 10.2 Thermal emission spectra from Pele, derived from two-temperature model fits to NIMS data. The numbers in parentheses indicate the order in which the data were acquired. Compared with other volcanoes, little change in spectrum shape or magnitude of thermal emission is detected. The conclusion is that Pele is an active lava lake. From Davies *et al.* (2001).

In higher-resolution *Galileo* data, several areas of high thermal emission were identified. Unfortunately, no high-resolution SSI or NIMS observations of the Pele hot spot and plume origin site were obtained in daylight. It was, however, possible to locate the nighttime and eclipse observation footprints with some degree of confidence based on available navigation information and through comparison of observations. High-resolution SSI images obtained on October 11, 1999 (Orbit I24) showed a line of hot spots, such as might be produced at the active margin of a lava lake or by skylights in the roof of a lava tube (McEwen *et al.*, 2000b). A high-resolution NIMS observation placed the most intense thermal source within the graben and also indicated that the inner edge of Pele Patera was thermally active (Davies *et al.*, 2001).

A NIMS observation obtained on October 16, 2001 (Orbit I32) at a range of ≈4000 km yielded a resolution of ≈1 km pixel⁻¹ and showed at least three hot spots in the graben, with the most intense spot in (as best as can be determined) the same location as in the I24 data (Davies *et al.*, 2002; Lopes *et al.*, 2004). During the same orbit, SSI obtained the best ever look at Pele, obtaining nightside images through five filters at an unprecedented 60 m pixel⁻¹ (Plate 9e). Detailed modeling of the data produced color temperatures of ≈1430 K to ≈1600 K, in the basaltic composition range (Radebaugh *et al.*, 2004; Keszthelyi *et al.*, 2006a).

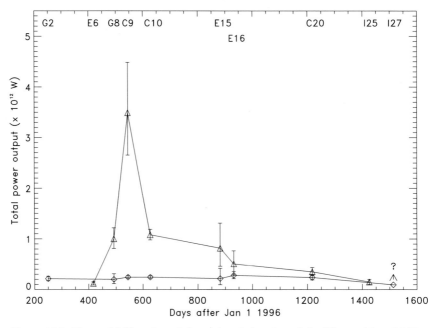

Figure 10.3 The variability of total thermal emission from Pele (diamonds) and Pillan (triangles). The datum at I27 is a minimum value. Pele's thermal emission is very steady. Pillan's rises to a peak in June, 1997 during a huge eruption and then declines over the next few years as the emplaced lava flows cool. From Davies *et al.* (2001).

At the same time *Galileo* was obtaining the I32 data, *Cassini* ISS was also observing Io at unprecedented temporal resolutions (Plate 9g). With *Galileo*, repeat observations of the same target were typically either a few minutes apart or weeks to months apart. With *Cassini*, observations were obtained several minutes apart, but for a 2-hour period that was repeated after 40 hours and again 40 hours after that. This observation schedule, integrated with other data, allowed the documentation of activity at Pele, at varying wavelengths, on timescales of minutes, hours, days, months, and years (Radebaugh *et al.*, 2004). The ISS data showed that visible-wavelength thermal emission varied considerably on a timescale of minutes.

Thermal emission at longer wavelengths remained stable. Multi-wavelength adaptive optics (AO) observations also allowed quantification of Pele's thermal emission during and after the *Galileo* mission and showed that the spectral shape in the infrared and intensity of emission were the same as seen by NIMS during the *Galileo* epoch (Marchis *et al.*, 2005).

10.3 A lava lake at Pele

Piecing together the above information leads to an inevitable conclusion: a very vigorous style of eruption is taking place at Pele, yet there is no evidence in NIMS spectra or SSI images (albeit at low spatial resolution) of the emplacement of lava flows. Nor is there evidence of cooling lava flows in PPR data at temperatures below the detection limit of NIMS (Spencer *et al.*, 2000b; Davies *et al.*, 2001). With no flows being emplaced, yet thermal activity indicative of newly exposed lava, the different possibilities proposed by Howell (1997) are reduced to one through the analysis of time-series data (Davies *et al.*, 2001). The conclusion is that Pele is an active lava lake, where at least part of the surface crust is being continually ripped apart, exposing lava at liquidus temperatures. The cause of the disruption is the degassing of sulphur and SO_2 from the magma. Variations in the amount of crust being disrupted cause the variability observed in the ISS data. This is a localized phenomenon; the rest of the surface remains relatively undisturbed or is quiescently replaced, yielding long-term stability of thermal emission at infrared wavelengths. The escaping volatiles form the Pele eruption plume and also entrain silicate particles. The total area of the emitting surfaces detected by NIMS was typically ≈ 17 km^2, equivalent in area to a circular lava lake with a diameter of ≈ 4.5 km. The actual area of the lake is almost certainly larger.

10.4 Importance of temporal and spectral coverage

The analysis of NIMS data highlights the importance of temporal coverage. Single observations at low spatial resolution (IRIS and most NIMS data) can often be used to somewhat constrain eruption style, but with a sequence of observations eruption style can be more tightly constrained (Davies and Keszthelyi, 2005). It is not possible to determine even from the high-resolution, nighttime SSI images of Pele what the emplacement mechanism is. The distribution of hot pixels could just as easily be from lava flows on the surface. In the absence of high-spatial-resolution daytime imaging, deriving eruption mode from remote-sensing data is ultimately constrained by the temporal evolution of thermal emission spectrum, temperature, and emitting area.

A deep understanding of volcanism at Pele is attained from observations with multiple instruments across a wide wavelength range. Each instrument is capable of detecting different facets of the eruption taking place. Visible imagers such as SSI and ISS are most sensitive to the hottest, youngest, relatively small areas of activity, with temperatures typically greater than 1000 K and a surface age measured in seconds. NIMS and many AO observations are sensitive to a middle range of temperatures, typically down to 300 K; the bulk of these emitting surfaces have ages of minutes to weeks or longer. IRIS and PPR are most sensitive to lower temperatures and older surfaces, and the absence of extensive, cooling flows at Pele in PPR data is the final piece of corroborating evidence for Pele's being a lava lake.

10.5 Lava lakes

An active lava lake is a volume of circulating magma exposed to the atmosphere. It is directly linked to and replenished by a deeper magma source. As such, a lava lake is a window into the interior of a planet and is therefore a valuable tool for volcanologists. Terrestrial lava lakes have been classified into two types: active and inactive (Swanson *et al.*, 1979). It is important to separate these two classes because they have different formation processes and thermal evolutions.

Active lava lakes can develop at the summit or on the flanks of a volcano and can be considered the top of a column of magma (Swanson *et al.*, 1979; Tilling, 1987). An active lava lake is therefore an open system, where surface activity is directly related to processes taking place deep underground, and, as such, yields insights into the magma supply mechanism and plumbing system of the volcano. Pele is such a system. Terrestrial examples include the lava lakes at Erta'Ale, in the Afar region of Ethiopia (Le Guern *et al.*, 1979; Tazieff, 1994) – incidentally in one of the hottest places on Earth – and Mt. Erebus, Antarctica (Giggenbach *et al.*, 1973), in the coldest.

Inactive lava lakes, on the other hand, are rootless and stagnant and do not form directly on top of the magma column (Swanson *et al.*, 1979). Instead, such lakes form by topographic ponding of lava (e.g., where lava flows into a crater). Examples on Earth include the Kilauea Iki, 'Alae, and Makaopuhi lakes on Kilauea (Peck *et al.*, 1966; Richter *et al.*, 1970; Peck *et al.*, 1977; Wright and Okamura, 1977; Peck, 1978; Swanson *et al.*, 1979). Ponds also form behind natural levees (Wilson *et al.*, 1993).

It is therefore important to determine whether a lava lake is active or inactive. Pele fits every requirement for classification as an active lava lake. With the exception of Loki Patera (Chapter 13), it is not possible to conclusively determine from available data whether other proposed lava lakes on Io (Radebaugh *et al.*, 2002; Lopes *et al.*, 2004) are active or whether they are stagnant, ponded flows. The thermal emission seen by NIMS and SSI at Tupan and Emakong Paterae, for example, does not show Pele-like thermal signatures, but instead the spectra and temperatures (see Davies, 2003b; Lopes *et al.*, 2004) indicate predominantly cool crust. This cool crust could be on either a quiescent lava lake or a lava flow (Davies, 2003a).

The only other hot spot on Io that displays similar occasional vigorous activity, but without an accompanying plume and without accompanying surface change, is Janus Patera. Located at 42°W, 3°S on the Jupiter-facing hemisphere of Io, Janus Patera was detected from the ground and by NIMS on several occasions (Spencer *et al.*, 1997c; Lopes-Gautier *et al.*, 1999; de Pater *et al.*, 2004; Marchis *et al.*, 2005). SSI detected persistent short-wavelength thermal emission (Radebaugh and McEwen, 2005). No plume was observed over Janus Patera in any *Galileo* observation, and there was no evidence of a surface change (Geissler *et al.*, 2004a).

Janus Patera was the probable location of a vigorous eruption detected in December, 2001 using the Keck Interferometer (de Pater *et al.*, 2004). Thermal emission came from a small area of about 4 km^2 with temperatures, from model fits, covering a range from 600 K to 1450 K (de Pater *et al.*, 2004). Several eruption scenarios fit this profile, including the possibilities that the observers were witnessing the early stages of an eruption, where lava flows were being emplaced with lava fountain activity or that a lava lake was vigorously overturning.

10.6 Implications for magma supply and interior structure

What makes the Pele lava lake unique among Io's volcanoes is the role played by volatiles. The lava erupting at Pele is so rich in gas that the release of these gases as magma rises and pressure decreases both disrupts the surface crust of the lava lake and forms the Pele plume. It is likely that some of Pele's volatiles are dissolved in the magma at depth. It is also probable that incorporation of volatiles (primarily sulphur and sulphur dioxide) takes place as the magma rises through volatile-rich layers of the crust. The interior of Io should be depleted of volatiles if volcanic activity has persisted over geologic time, but, on the other hand, the surface is being buried at such a rate that it is not unfeasible that recycling of volatiles and magmas is taking place at great depth.

The disruption of the crust on the lava lake leads to the preponderance of thermal emission at short infrared wavelengths observed by NIMS. Although the Pele plume is often hard to detect, the red plume deposits persist, denoting a more-or-less steady supply of sulphur, supplied at a rate faster than the transition time of S_2 and S_4 to S_8.

10.7 Plume composition and implications for volatile supply

The ease with which the Pele plume can be detected can vary on a timescale of days. With steady thermal output in the infrared implying a steady supply of lava, plume variability is caused by varying volatile content. Sometimes the plume is very hard to detect, a situation where the plume is comprised mostly of SO_2 with no particulates – a "stealth" plume (Johnson *et al.*, 1995). Because volcanic activity is areally restricted and any local surface volatiles are long gone, the variability in plume volatile content is derived from variability in the volatile content of the magma itself. The ratios of various gas species in the plume are consistent with equilibrium between the plume and mafic magmas at 1440 K (Spencer *et al.*, 2000a; Zolotov and Fegley, 2000).

Estimates have been made of the volatile content of the Pele plume. Spencer *et al.* (1997b) calculated a SO_2 mass in the plume of 1.1×10^8 kg. The supply rate to the plume is therefore this mass divided by the ballistic travel time for a particle in the plume (1320 s for a 300-km-high plume), yielding a SO_2 mass flux of 76 000 kg s^{-1}. With a steady supply of magma, $\approx$640 000–890 000 kg s^{-1}, the magma volatile content is therefore $\approx$10% by weight. Zhang *et al.* (2003) modeled the Pele plume and derived a supply rate of 1.1 × 10^4 kg s^{-1}, yielding a lower mass flux of 1.2 wt % to 1.7 wt %, still more than five times that calculated for saturated basalt originating from a depth of 60 km (Chapter 5). More recent estimates of Pele plume SO_2 fluxes are even higher, of order 10^5 kg s^{-1} (Kandis Lea Jessup, pers. comm., 2006). Assuming that the silicate eruption rate has not significantly changed (a stance supported by Pele's steady thermal emission), the latest SO_2 fluxes approach parity with silicate mass flux.

These mass fractions of sulphur and sulphur dioxide are too great to be completely dissolved in the magma at any likely temperature or pressure. Because there are no spreading lava flows to mobilize near-vent surface deposits of sulphurous materials to be incorporated into the plume, most of the volatiles have to be added to the system after magma genesis and before eruption.

One explanation is that volatiles are being added to the magma close to the surface, via a process called reservoir pumping. The magma column supplying the lava lake intersects a local "aquifer," only the liquid encountered is not water but SO_2, which is three times less viscous than water. The low viscosity allows easy movement through the relatively unconsolidated pyroclastic material making up the near-surface crust.

It may be that buried plume deposits feed this reservoir, so that material is recycled through the volcanic system, maintaining plume activity and explaining the persistence of activity and the uniqueness of Pele's plume.

10.8 Calculation of mass flux and flux densities

Using the method described in Chapter 7, it is possible to estimate lava lake mass flux from total thermal emission (Harris *et al.*, 1999a). This has been determined for Pele from NIMS data (Table 10.1, Figure 10.2). Pele mass fluxes are given in Table 10.2, where, for context, the mass fluxes from Pele during the *Galileo* and *Voyager* epochs are compared with fluxes at the terrestrial Kupaianaha, Hawai'i, lava lake (Davies *et al.*, 2001). The thermal output from Pele during E16, 280 GW, yields a mass flux of 6.44 × 10^5 kg s^{-1} to 8.89 × 10^5 kg s^{-1}, assuming a basalt composition (Table 7.1). These masses equate to a volumetric rate of 250 to 340 m^3 s^{-1}, considerably larger than mass fluxes at terrestrial lava lakes (up to 7 m^3 s^{-1} at Erta'Ale, in January, 1973 [Harris *et al.*, 1999a]). Chapter 8 described how mass density (mass per unit time per unit area) and thermal flux density (thermal emission per unit time per unit area) could also be used to constrain eruption style (Davies and Keszthelyi, 2005). Pele has a flux density of 17 kW m^{-2}, very similar to Kupaianaha

Table 10.1 *Two-temperature fits and power output from NIMS Pele data*

Orbit	Date	Hot component			Warm component		
		Temperature (K)	Area (km²)	Thermal emission (GW)	Temperature (K)	Area (km²)	Thermal emission (GW)
G2	Sept. 6, 1996	1394 ± 25	0.46 ± 0.05	99	769 ± 25	5.8 ± 0.6	116
G8	May 7, 1997	1362 ± 78	0.20 ± 0.07	39	549 ± 19	31.0 ± 10.0	161
C10	Sept. 18, 1997	1077 ± 13	1.6 ± 0.12	122	544 ± 7	24.4 ± 1.8	121
E15	May 31, 1998	1369 ± 141	0.42 ± 0.27	84	625 ± 43	15.4 ± 3.6	133
E16	July 20, 1998	1353 ± 46	0.63 ± 0.11	120	649 ± 16	15.9 ± 1.7	160
C20	May 2, 2000	1300 ± 71	0.41 ± 0.11	66	588 ± 8	25.4 ± 1.8	172

From Davies *et al.* (2001).

during periods of vigorous overturning (Flynn *et al.*, 1993). Although the areal scale of activity is vastly different, as evidenced by comparing the effusion rate from Pele (above) to that of Kupaianaha at 1.5×10^3 to 18.2×10^3 kg s^{-1}, mass flux densities are within a factor of two (0.04–0.05 kg s^{-1} m^{-2} at Pele and $\approx$0.02–0.03 kg s^{-1} m^{-2} at Kupaianaha). It is worth noting that *Voyager* IRIS found a similar volumetric rate, $\approx$350 m^3 s^{-1} (Carr, 1986), 17 years before *Galileo*. Pele has been active, seemingly in a more-or-less steady state, for decades. Even after *Galileo*, ground-based observations continue to show Pele erupting in a similar style (e.g., Marchis *et al.*, 2005).

10.9 Further comparison with lava lakes on Earth

The terrestrial analogues for this style of eruption are the long-lived, currently active lava lakes at Erta'Ale and Mt. Erebus. Other well-documented examples include Halema'uma'u, in the summit caldera complex of Kilauea, Hawai'i, which was an active lava lake from $\approx$1790 to 1924, and the Kupaianaha lava lake.

Erta'Ale is a shield volcano that has undergone at least seven eruptions in the past 125 years, erupting more or less continuously since at least 1967. There is evidence that this lake is permanent. It was observed in the 1960s (Barberi and Varet, 1970), again by field observers in the early 1970s, and by satellites from

Table 10.2 *Lava lake power output, mass eruption rates, power and
mass densities: Kupaianaha and Pele*

Lake	Date	Total thermal emission, F_{tot} (GW)	Mass flux[a] M (kg s^{-1})	Flux density F_{tot}/m^2 (kW m^{-2})	Mass density, M/m^2 (kg s^{-1} m^{-2})
Kupaianaha Stage 1[b]	Oct. 1987/ Jan. 1988	0.018	41–57	$\approx$22	0.018–0.025
Pele, E16, NIMS	July 20, 1998	280	644 000– 889 000	17	0.04–0.05
Pele, IRIS[c]	Mar. 1979	1160	2.7×10^6– 3.7×10^6	10	0.02–0.03
Pele, IRIS[d]	Mar. 1979	1333	3.1×10^6– 4.2×10^6	6	0.01–0.02
Kupaianaha Stage 2[b]	Oct. 1987/ Jan. 1988	0.014	32–44	5.3	0.014–0.019
Kupaianaha Stage 3[b]	Oct. 1987/ Jan. 1988	0.011	25–35	4.9	0.011–0.015

[a] Mass flux calculated using Harris *et al.* (1999a); see Chapter 7.
[b] Flynn *et al.* (1993) describe the level of activity in a lava lake as follows: Stage 1 is marked by lava fountaining and overturning. Stage 2 is marked by active rifting events. Stage 3 is a quiescent period when a thick crust covers the lake. Stage 1 and 2 observations almost always need a three-temperature model to fit the data. Stage 3 requires only a two-temperature model. Kupaianaha lake area = 2300 m^2.
[c] Pearl and Sinton (1982). A component of 113 km^2 at 654 K was used. The component of 20 106 km^2 at 175 K would not be detected by NIMS. NIMS has a theoretical low-temperature detection limit of 180 K so long as the source fills the NIMS pixel (Smythe *et al.*, 1995). The 654 K and 175 K components yield the best fit to the IRIS Pele data.
[d] Pearl and Sinton (1982). Another two-component fit to IRIS data yields 29.7 km^2 at 854 K and 181.5 km^2 at 454 K. Noise in the data at large wavenumbers renders this fit unreliable.

1965 to 1996. Recent satellite observations have often shown thermal activity at Erta'Ale (Wright *et al.*, 2004b; Davies *et al.*, 2006a). Erta'Ale has, at times, two lava lakes in pits in the floor of the summit caldera. The larger of the lakes has reached a size of $\approx$140 m across and was seen in a pit $\approx$80 m deep. Occasionally, the lakes overflow onto the floor of the caldera, but most of the time activity is confined to one pit. Oppenheimer and Francis (1997) believed that a higher magma density, caused by the cooling of the magma in the lake, inhibited eruptions. As a result, a large amount of heat was released from the lava lake with little newly erupted material. Oppenheimer and Francis further proposed that the magma for the most part accumulated in the underlying crust in the form of dikes and sills; the Afar region of Ethiopia is undergoing tectonic extension, so conditions are favorable for such intrusive activity at Erta'Ale. On Io, the rifts associated with Danube Planum and Pele may be subjected to similar extensional tectonic forces (Strom *et al.*, 1981; Davies *et al.*, 2001; Radebaugh *et al.*, 2001, 2004; Jaeger *et al.*, 2003).

Such activity, with little or no change in hot-spot surface area, must also take place on Io if volcanic activity is similarly confined. Resurfacing of Io by emplacement of lava is minimal under these circumstances, even though heat loss ($\approx$280 GW from Pele) is high. Resurfacing from Pele is nevertheless taking place, over a very large area, from deposition from volcanic plumes.

Another permanent, active, terrestrial lava lake resides in the summit crater of Mt. Erebus, on Ross Island, Antarctica (Plate 3). This lava lake was first observed during the early 1970s (Giggenbach *et al.*, 1973) and has been active ever since. A review of the remote sensing of this volcano from Earth orbit is found in Harris *et al.* (1999b). Volcanic activity is characterized by a persistent, convecting lake of somewhat unusual magmatic composition, an anorthoclase phonolite, which feeds a plume of acidic gases and aerosols (Kyle *et al.*, 1990, 1994). As a location of a permanent lava lake, Mt. Erebus is also a prime target for study of this relatively rare volcanological feature, both on the ground and using remote-sensing techniques (Rothery and Oppenheimer, 1994; Harris *et al.*, 1999a; Davies and Kyle, 2006; Davies *et al.*, 2006a; see Plate 3). Mass eruption rates have been estimated in the range 30–76 kg s^{-1} (Harris *et al.*, 1999a). There is evidence, in the form of multiple growth phases of phenocrysts (Dunbar *et al.*, 1994), to support the circulation model of magma proposed by Harris *et al.* (1999a), where magma rises via a conduit to the surface, cools, increases in density, and sinks back down to the magma chamber, where reheating takes place. The magma at Mt. Erebus is more viscous than basalt. As a result, gas finds it difficult to escape from the lava, and the lake undergoes explosive Strombolian eruptions on a regular basis.

Better terrestrial analogues to Pele are the basalt lava lakes that form during eruptions of Kilauea, Hawai'i. Some are lava ponds formed when lava flows into and fills a crater. The lava pond created by the Kilauea Iki eruption in 1959 was more than 100 m deep. One of the most closely studied examples of an active lake was the Kupaianaha lava lake (Plate 15b), which was active from July, 1986 to August, 1990. The lava lake, of tholeiitic basalt, formed as a result of the rupturing of the conduit beneath the Pu'u 'O'o cone, an active vent on the southeastern flank of Kilauea, and the subsequent eruption of lava through a new fissure. A lava lake 140 m by 300 m in size formed over the new vent, and frequent overflows built a broad, low shield 1 km in diameter and $\approx$56 m high (see summary in Heliker and Mattox, 2003). A smaller, stable lava lake $\approx$55 m in diameter formed by mid 1986, supplied via a lava tube (Flynn *et al.*, 1993).

The Kupaianaha lava lake underwent semi-periodic overturning, followed by the formation of a thick, stable, often-immobile crust. At times, the lake surface was covered by a slow-moving crust, and for shorter periods of time vigorous lava fountaining took place, which disrupted the crust on the lake (Flynn *et al.*, 1993).

The mass and thermal emission flux densities at Pele (see Table 10.2) and the shape of the integrated thermal emission spectrum (Figure 10.2) indicate that some part of the lake surface is being constantly disrupted: the style of activity at Pele is comparable with that seen at Kupaianaha (see Figure 8.3, Plate 15b) during vigorous overturning (Davies *et al.*, 2001). Comparing the volumes of lava moving through each lava lake shows that the Pele effusion rate dwarfs the peak volumetric flux at Kupaianaha ($\approx$6–7 m^3 s^{-1}). The scale of activity at Pele is, literally, unearthly. Yet, per unit area, the two lava lakes are very similar, indicating similar composition of lava and resurfacing mode. At Pele, the active, disrupted area is relatively small and most of the crust area is stable.

10.10 Summary

Volcanism at Pele can be summarized as follows: Pele is a site of constant, high thermal emission, and was seen in every appropriate *Galileo* and *Cassini* observation (e.g., Spencer *et al.*, 2000b; Davies *et al.*, 2001; Keszthelyi *et al.*, 2001a; Radebaugh *et al.*, 2004).

Thermal emission from Pele peaks at a wavelength shorter than that of most ionian volcanoes. Most of Pele's thermal emission is at wavelengths between 1 and 10 μm, implying relatively high surface temperatures. Equating a high temperature with a short exposure time implies that exposed lava is being rapidly replaced with new lava. Temperatures are indicative of a basaltic composition, although ultramafic temperatures cannot be ruled out.

There is no evidence of large expanses of cooling lava flows in NIMS or SSI data. The absence of cooling lava flows, especially the low intensities measured at long infrared wavelengths by PPR (Spencer *et al.*, 2000b; Davies *et al.*, 2001), strongly constrains eruption style to one where flows are not emplaced.

Pele is most likely an active, overturning lava lake. Pele is, in this respect, a unique feature on Io. No other hot spot on Io exhibits this particular style of activity coupled with a general absence of change over time.

Pele's being a lava lake explains the shape of the *Voyager* IRIS thermal emission spectrum (Howell, 1997). Reinforcing the lava-lake explanation is the similarity in thermal emission and mass eruption rate *per unit area* between Pele and an active, overturning basalt lava lake on Earth, at Kupaianaha, Hawai'i (Davies *et al.*, 2001).

Observed thermal emission at visible wavelengths (indicative of surfaces at temperatures >700 K) varies considerably on a timeframe of a few minutes. This variability would be expected from the constant exposure of different areas of lava near the liquidus temperature as the crust on the lava lake is disrupted. Thermal emission at infrared wavelengths covered by NIMS (0.7–5.2 μm) is much steadier, because it is dominated by cooling crust on the lake. At even longer infrared wavelengths, thermal emission drops off. PPR does not see evidence of large, cool flows but can apparently detect the surrounding cool pyroclastic blanket (Spencer *et al.*, 2000b).

The sulphur and sulphur dioxide in the Pele plume are from degassing of lava. Silicates are entrained into the resulting plume. The silicate particles detrain and fall out of the plume closer to the vent to form the dark inner plume deposits.

Pele contributes $\approx$200 GW to $\approx$230 GW to Io's thermal budget, or about 0.2% of Io's total output (10^{14} W). Pele contributes $\approx$2% of Io's 4.8-μm flux (see Veeder *et al.*, 1994).

11

Pillan and Tvashtar Paterae: lava fountains and flows

The eruptions at Pillan and Tvashtar Paterae[1] are very different, in terms of lava emplacement, to that seen at Pele. At Pillan, in particular, the eruption began in a spectacular explosive style and then transitioned to effusive activity. As at Pele, the explosive eruptions at these locations were driven by low-viscosity silicate magmas. At Pillan and Tvashtar Paterae, however, the explosive phases were short-lived, with thermal emission rapidly building up to a peak and then gradually subsiding as the eruption style changed. Finally, lava effusion came to an end and the emplaced flows cooled and solidified. The early stages of the eruptions were dominated by lava-fountain activity that fed large flows. Extensive pyroclastic deposits were laid down in both cases.

An interesting result from the study of the Pillan eruption was the indication of the presence of ultramafic magmas on Io. These magnesium-rich flows dominated volcanism on Earth aeons ago. If they were being erupted on Io, then the mechanisms of eruption and evolution of a process long extinct on Earth were being revealed. This possibility raised an intriguing question: could Io be an analogue for the early Earth (Matson *et al.*, 1998; McEwen *et al.*, 1998b)? The merits of the case for ultramafic magmas on Io were discussed in Chapter 9. Regardless of whether these lavas are found on Io, *Galileo* unexpectedly provided the opportunity to observe the emplacement of large silicate flows on a scale rarely seen on Earth in historical times, but which was once the dominant volcanic eruption process on Earth.

11.1 Lava fountains: outbursts explained?

The absence of an atmosphere on Io is the main driver behind explosive volcanic activity (Kieffer, 1982; Wilson and Head, 1983, 2001, 2003). Ionian magmas do

---

[1] The term "Catena," a chain of craters, was replaced with "Paterae" after review by the International Astronomical Union Planetary Nomenclature Committee in 2006. Tvashtar Catena became Tvashtar Paterae at that time. Mazda Catena became Mazda Paterae.

not require a large volatile content or high viscosities to produce explosive activity. Instead, the expansion of a small amount of gas is sufficient to generate high eruption velocities, ≈ 1 km s^{-1}, to form a plume 350 km high.

Lava fountains on Io should behave in similar ways to lava fountains on the Moon, a body of similar size, bulk density, gravity, and surface environment (the absence of a thick atmosphere). On the Moon, the driving volatile behind explosive activity was most likely CO_2; on Io, the driving volatile is SO_2, with some additional sulphur. Work on the dynamics of high-eruption-rate lunar eruptions predicted the formation of optically thick eruption clouds ≈ 4 km wide for fissure eruptions (Wilson and Head, 1981, 1983). The lava is ejected from relatively narrow fissures at high volumetric rates in relatively narrow jets. The eruption cloud then spreads due to expansion of the gas phase. As gas density decreases, the clasts become decoupled from the gas and follow essentially ballistic trajectories. The size of the fountain and distribution of the pyroclastic material depend on the material properties of the lava and the vent geometry, but primarily on the volatile content (Wilson and Head, 1981, 1983). On Io, with lower gravity and essentially no atmosphere, a low volatile content (< 0.01 wt %) is sufficient to produce ejection velocities that result in the formation of a plume that covers the observed area (Davies, 1988). Considering only ballistic trajectories, an eruption of magma on Io at a velocity of 60 m s^{-1} will produce a lava fountain 1 km high and deposits up to 2 km from the vent. In actuality, gas expansion will increase the area of deposition. On Earth, a similar eruption velocity would produce a lava fountain 180 m high; the deposition area is limited by a thick atmosphere, which inhibits expansion.

Large basaltic eruptions on Earth often commence with the opening of a linear fissure, the result of a dike, propagating laterally and vertically, intersecting the surface. Because the propagation of the dike is partly driven by exsolved gas fracturing country rock, the initial phases of activity are relatively gas rich, with gas expansion accelerating magma as it erupts. Additional impetus is given to erupting lava if the magma is positively buoyant. The early stages of fissure eruptions are often marked by the continuous eruption of lava along the entire length of the fissure, forming a "curtain of fire" (e.g., Wolfe *et al.*, 1987; see Plate 15a). The chilling of magma against dike walls restricts magma flow and the narrowest parts of the fissure become blocked. Activity becomes concentrated at a few locations or a single vent. After the more gas-rich magma is erupted, lava-fountain activity stops, and the eruption becomes effusive. A huge volume of lava can be erupted from a fissure in a relatively short time. A 25-km-long fissure with an effusion rate of 5 m^3 s^{-1} m^{-1} would erupt 1 km^3 of lava in a little more than 2 hours (see Chapter 7).

11.2 Pillan 1997: flood lavas and the emplacement of long flows

The 1997 Pillan eruption is one of the best-documented eruptions observed by *Galileo* (McEwen *et al.*, 1998b; Davies *et al.*, 2001; Keszthelyi *et al.*, 2001a; Williams *et al.*, 2001a; Davies *et al.*, 2006d). Analysis was greatly aided by having, at the height of the eruption, near-contemporaneous observations in eclipse by NIMS and multi-filter observations by SSI (Plate 10). NIMS obtained a sequence of observations between 1996 and 1999 that caught the early stages of activity, the buildup to peak thermal emission, and the aftermath, watching as the emplaced flows cooled over the years. Most of the "Pillan story" was derived from low-resolution data. The thermal sources detected by both NIMS and SSI were sub-pixel. High-resolution data obtained after the main eruption merely confirmed what had already been deduced.

11.2.1 Galileo *observes Pillan*

During the *Voyager* era, there was little to distinguish Pillan Patera, a small caldera with no low-albedo features that indicated silicate activity (McEwen, 1988). NIMS detected a small hot spot in the vicinity of Pillan in 1996 (Lopes-Gautier *et al.*, 1997), but it was not until May, 1997 (Orbit G8), as activity at Pillan began to ramp up, that NIMS detected a high-temperature eruption taking place. A two-temperature fit to those data (Table 11.1) indicated an upper temperature exceeding 1500 K, at the time the highest temperature derived from NIMS data (Davies *et al.*, 2001). The shape of the NIMS spectra (Figure 11.1) indicated a preponderance of thermal emission at short wavelengths, such as would be produced by lava fountains.

When NIMS and SSI observed Io on June 28, 1997 (during Orbit C9), a massive eruption was in progress. The most intense hot spot seen by SSI was located at 244°W, 8°S, very close to Pillan Patera (Plate 10a). Precise geolocation was difficult because, at a distance of 1.44 million km, SSI spatial resolution was 15 km pixel^{-1} and NIMS spatial resolution was 362 km pixel^{-1}. SSI obtained observations with both the clear and 1-µm filters. The intensity of thermal emission was so great that the detectors saturated (at a distance of 1.4 million km!) (Keszthelyi *et al.*, 2001a). A plume more than 140 km high was imaged by SSI (Plate 10c). In the lower-resolution NIMS data, Pele and Pillan were within the same pixel (Plate 10b). The steadiness of thermal output from Pele from orbit to orbit, however, enabled the subtraction of the Pele flux from the integrated spectrum, isolating Pillan's thermal emission (Davies *et al.*, 2001).

The existence of a dust-rich plume, the intense short-wavelength emission detected by SSI, and the evolution of the eruption thermal signature seen by NIMS (Figure 11.1) are consistent with large lava fountains feeding extensive lava flows.

Table 11.1 *Two-temperature fits to Pillan NIMS data*

Orbit	Date	Component 1		Component 2	
		Temperature (K)	Area (km²)	Temperature (K)	Area (km²)
G8[a]	May 7, 1997	1509 ± 78	0.20 ± 0.06	857 ± 18	31 ± 3
C9[a,b]	June 28, 1997	922 ± 12	27.5 ± 1.91	387 ± 12	1855 ± 379
C10	Sept. 18, 1997	1321 ± 15	1.19 ± 0.18	470 ± 2	335 ± 17
E15	May 31, 1998	1004 ± 153	0.44 ± 0.41	378 ± 16	684 ± 212
E16	July 20, 1998	986 ± 65	0.71 ± 0.33	392 ± 13	350 ± 105
C20	May 2, 1999	1138 ± 50	0.2 ± 0.06	399 ± 13	231 ± 15

From Davies *et al.* (2001).

[a] Pillan data isolated by subtracting E16 Pele flux from the combined Pele and Pillan spectrum, having corrected for emission angle. Pele and Pillan were both resolved in Orbits E15, E16, and C20.

[b] The two-temperature fit to the C9 data produces a synthesized clear/1-μm ratio far below that measured by SSI. In McEwen *et al.* (1998b), a three-temperature model was fitted to the NIMS spectrum of thermal emission from both Pele and Pillan, producing a high-temperature component at 1825 K and a synthesized clear/1-μm ratio closer to that seen by SSI.

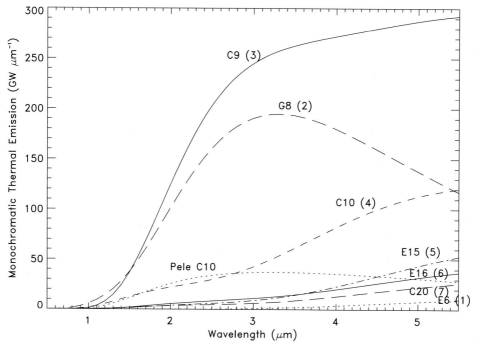

Figure 11.1 Thermal emission spectra from Pillan, derived from two-temperature model fits to NIMS data. The numbers in parentheses indicate the order in which the data were acquired. From Davies *et al.* (2001).

Observations during subsequent orbits revealed a huge "black eye" more than 400 km across on the surface of Io (Plate 9b). This dark feature consisted of deposits from a plume rising above the Pillan lavas. As described in Chapter 9, these deposits may be rich in orthopyroxene, a major constituent of basalt and ultramafic lavas (Geissler *et al.*, 1999). The September 17, 1997 (Orbit C10) SSI observations showed signs of interactions between the plumes from Pillan and Pele. The dark Pillan deposits were rapidly buried by the Pele plume deposits.

It is not known for certain from where the 1997 Pillan lavas erupted. They did not originate from within the patera, unlike activity at many other paterae on Io. Instead, the surface flow field abuts a mountain, Pillan Mons, that is riven by fractures (Plate 10d). Magma may have exploited an extension of these fractures to reach the surface (Keszthelyi *et al.*, 2001a).

The large volumes of lava emplaced at Pillan (described in the next sections) could have erupted quickly from a fissure tens of kilometers long (Williams *et al.*, 2001a; also see Chapter 7). By September, 1997, lava had covered an area of more than 3100 km^2. Sometime between September 17, 1997 (Orbit C10), and March 29, 1998 (Orbit E14), lava flows plunged into Pillan Patera and covered its floor, an area of 2500 km^2. Plate 16b shows this process taking place at 'Alo'i crater, Hawai'i.

NIMS watched as the flows emplaced in the summer of 1997 cooled over the next 4 years. The thermal emission spectrum changed as would be expected for the transition from lava fountains to insulated flows (Figure 11.1), and flows cooled in a manner consistent with the solidification of flows $\approx$10 m thick (Davies *et al.*, 2005). The large temperature contrast between the cooling flows and the cold Io background enabled NIMS and PPR to chart the decline of thermal emission.

The Pillan observation sequence is a particularly fine (and perhaps unique) set of observations that shows how this particular eruption style evolved thermally over several years. Even when NIMS could barely detect these flows, PPR could easily detect them (*e.g.*, in February, 2000; see Plate 8a; Spencer *et al.*, 2000b; Rathbun *et al.*, 2004).

Finally, this understanding of the eruption was accomplished almost entirely from low-resolution NIMS and SSI data. The most important factors were good temporal and spectral coverage rather than high spatial resolution.

11.2.2 Flow thickness

During passes close to Io (Table 3.2), SSI obtained some of the highest resolution images of the entire mission at Pillan, as high as 9 to 20 m pixel^{-1}. These images show a rough surface with pits and domes (Plate 10d). Other images also showed several flow margins and, for the first time, shadow measurements were made to

Table 11.2 Q_F, Q_E, and maximum dA/dt rates for Pillan 1997

Duration[a,b] (days)	Max Q_F (m³ s⁻¹)	Q_E (m³ s⁻¹)	Max dA/dt[c] (m² s⁻¹)
52	3.3×10^4	0.7×10^4	3.3×10^3
99	1.8×10^4	3.6×10^3	1.8×10^3
167	1.0×10^4	2.1×10^3	1.0×10^3

After Davies *et al.* (2006d).
[a] Assumes waxing period of 5 days.
[b] For an erupted volume of 31 km³ (excluding flows covering the floor of Pillan Patera).
[c] Assumes 10-m-thick flows (Williams *et al.*, 2001a).

determine the thickness of newly emplaced lava flows on a planet other than Earth. The Pillan flows were found to be 8 m to 11 m thick (Williams *et al.*, 2001a). The flows erupted by September, 1997 therefore had a volume of 31 km³.

11.2.3 Total volume erupted, and eruption and effusion rates

Covering 5600 km² with flows 10 m thick requires 56 km³ of lava – by recent terrestrial standards a volume almost beyond belief and most likely the largest outpouring of magma ever witnessed (Davies *et al.*, 2006d). For comparison, in 1980, Mt. St. Helens erupted ≈0.25 km³ of magma (dense rock equivalent). The volume quoted for Pillan is a *minimum* volume because the depth of flows covering the entire floor of Pillan Patera and the depth of the pyroclastic blanket covering an area of 125 000 km² are not known. For example, if the pyroclastic deposits have an average thickness of 1 cm, these deposits represent another 1.25 km³ of lava.

For the 31 km³ of lava emplaced as flows before September, 1997, the volumetric eruption rate Q_E necessary to produce the flows is between ≈2200 m³ s⁻¹ and ≈6900 m³ s⁻¹, assuming 10-m-thick flows (Davies *et al.*, 2006d). Remember, this is the *average* volumetric rate. As shown in Chapter 7, it is possible to estimate the likely maximum effusion rate by considering the Wadge (1981) effusion rate profile. Recall that Wadge plotted effusion rate against time for several eruptions and concluded that the decay rate of the waning phase of an eruption was exponential. The area under the curve is the total volume erupted. Applying the Wadge model for an erupted volume of 31 km³, peak effusion, eruption, and areal coverage rates are given in Table 11.2 for an eruption duration constrained between 52 and 167 days. Peak effusion (Q_F) values of 1.0×10^4 m³ s⁻¹ (167-day eruption) to 3.3×10^4 m³ s⁻¹ (52-day eruption) are reached (Davies *et al.*, 2006d).

Table 11.3 *Range of advance rates for Pillan 1997 flows imaged
in September, 1997*

Number of days	Advance rate per day to cover 60 km	Source
Min. 52	1.15 km/day	Williams *et al.* (2001a)
Min. 99	0.61 km/day	Davies *et al.* (2001)
Max. 167	0.36 km/day	Davies *et al.* (2001)

A separate issue from the total areal coverage rate is the rate at which the flows advanced across a flat plain toward Pillan Patera. If erupting from a fissure extending from the end of the fractured mountain (Pillan Mons) to the northwest of Pillan Patera, the flows at Pillan traveled, at most, about 60 km, with some individual flows attaining a width of $\approx$60 km (Keszthelyi *et al.*, 2001a; Williams *et al.*, 2001a). As shown in Table 11.3, the minimum rate of advance is not particularly fast, in the range $\approx$0.3 km day^{-1} to $\approx$1 km day^{-1}, depending on the eruption duration. The rate of advance for most of the eruption may therefore have been a crawl across the ionian surface. Still, the leading edges of the flow would have yielded heat to the mobilization of surface ices and would have cooled, thereby inhibiting faster flow progression. This would result in the formation of a "rubbly" flow surface texture, caused when faster-moving lava is impeded by a slow-moving flow front, causing the surface crust to crumple up (Keszthelyi *et al.*, 2004b).

11.2.4 Emplacement regime

High-resolution images of Pillan show several flow margins and a possible lava channel (Plate 10d). Unfortunately, downlink constraints meant that no context images were obtained at mid-range spatial resolutions. The available data are either at many kilometers per pixel or a few meters per pixel, so pinpointing the locations of the high-resolution images is difficult. Nevertheless, the flow geomorphology seen in high-resolution SSI data has been compared with terrestrial and planetary flow surfaces. Williams *et al.* (2001a) found that the Pillan materials did not easily conform with other terrestrial lava flows, although there were some similarities with rough-textured pahoehoe sheet flows from the 1783–1784 Laki, Iceland, eruption (Williams *et al.*, 2001a). The rough, domed, and pitted surface of the Pillan flows may have been caused by a fractured lava crust that formed during rapid emplacement and surging of low-viscosity flows (Williams *et al.*, 2001a). The disruption of the crust may have been due to a turbulent flow regime or explosive interaction between sub-surface ices and hot lava. Disruption of the crust certainly reinforces the interpretation of the spectral distribution of thermal emission seen by

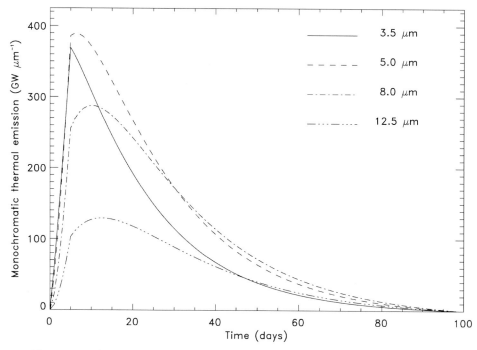

Figure 11.2 The evolution of thermal emission as function of wavelength for 99-day eruption, for 10-m-thick flows using the effusion profile shown in Figure 7.12. At Pillan in 1997,
at least 56 km³ of lava was emplaced as flows, 31 km³ before September 17, 1997.

NIMS during Orbit C9 (June, 1997). Alternatively, flows may have been emplaced
in a laminar flow regime.

As described in Chapter 7, the Wadge model charting change in erupted volume
also yields the change in areal coverage rate. With data on the area covered
each day, and on how surface temperature changes with time (Figure 7.8), the
evolution of integrated thermal emission and thermal emission as a function of
wavelength over the duration of the eruption, and subsequent cooling, can be
charted. Figures 8.2 and 11.2 show the evolution of total thermal emission, and
emission as a function of wavelength, for a 99-day eruption that yields the effusion
profile in Figure 7.12. The model thermal emission output for this eruption
(Figure 8.2) compares well with NIMS July, 1997 data (Figure 11.1), reproducing
not only the shape of the spectrum but also the magnitude of thermal emission
(Davies *et al.*, 2006d). Note also that flow emplacement is laminar, allowing an
insulated crust to form. Neither turbulent flows nor lavas at ultramafic liquidus
temperatures (>1800 K) are required to reproduce observed thermal emission and
intensities at Pillan in 1997 (Davies *et al.*, 2006d).

Generally, many long flows are inflated flows, but this is not the only way in which
long flows are emplaced (Keszthelyi *et al.*, 2006b). High-resolution imaging of some

martian lava flows shows remarkably fresh flood lavas with a surface morphology uncommon on Earth (Keszthelyi *et al.*, 2000, 2004b), but which is somewhat akin to the mixture of a'a and pahoehoe surface textures seen on the Laki, Iceland, flows (see Section 11.5.1). It may be that the Pillan flows, like the martian flood basalts, were inflated flows but with extensive mobile crusts (Keszthelyi *et al.*, 2000, 2004b).

11.3 Tvashtar Paterae

Another spectacular lava-fountain episode was observed by SSI in November, 1999 (during Orbit I25) at Tvashtar Paterae, a chain of nested calderas located at 160°W, 63°N (McEwen *et al.*, 2000b; Keszthelyi *et al.*, 2001a). Just before this episode, on November 26, 1999, ground-based astronomers detected a massive eruption in progress at Tvashtar. The magnitude of thermal emission indicated that a rare ionian outburst event was taking place (Howell *et al.*, 2001). The I25 SSI daytime observation resolved this event. The thermal emission caused image saturation during overexposure and "bleeding," where energy overflows into adjacent detectors. Examination of the bleeding process and modeling of the data indicated that lava was erupting from a fissure 25 km long and forming 1-km-high fountains, with lava flows at the eastern end of the fissure (Keszthelyi *et al.*, 2001a; Wilson and Head, 2001).

By February, 2000 (Orbit I27), the fissure eruption and lava fountains had ceased (in keeping with the short-lived nature of lava-fountain activity on Earth, where fountaining events typically last on the order of hours because the relatively gas-rich magma is soon erupted). Activity at Tvashtar had actually moved to a new location some tens of kilometers to the west. SSI observed that the floor of an adjacent patera was hot, indicating temperatures in excess of 700 K (Milazzo *et al.*, 2005).

During Orbit G29 at the end of December, 2000, a global Io observation by SSI showed a large red plume deposit around Tvashtar Paterae (Geissler *et al.*, 2004a). The deposit, >900 km in diameter, was very similar to the plume deposit at Pele, indicating the probable presence of sulphur as a plume volatile and similar vent conditions. At the same time, *Cassini* ISS observed the plume itself, which was >400 km high (Porco *et al.*, 2003). Over a year after the lava-fountain episode, material was being erupted that was rich enough in volatiles to produce a Pele-like plume (Milazzo *et al.*, 2005).

Observations from ground-based telescopes utilizing AO showed Tvashtar Paterae was still active in February, 2001 (Marchis *et al.*, 2002). When *Galileo* next had an opportunity to observe Tvashtar during Orbit I31 in August, 2001, the plume had ceased. By December, 2001, no thermal emission from Tvashtar Paterae was detectable using AO (Marchis *et al.*, 2005), indicating that after 2 years of varying activity Tvashtar was once again dormant. After a period of quiescence, perhaps during which the magma chamber was recharging (as suggested by Milazzo *et al.*,

2005), Tvashtar again erupted in the first half of 2006 and was observed by ground-based astronomers (Franck Marchis, pers. comm., 2006).

Deriving likely temperatures from daylight SSI data, let alone data where bleeding has taken place, is complex, but analysis of Tvashtar data obtained during Orbit 125 and subsequent orbits indicated that the highest reliable color temperatures were $\approx$1300 K (Milazzo *et al.*, 2005). Although higher temperatures could not be ruled out, they are not necessary to fit the data.

Analysis of the Tvashtar Paterae lava fountain episode observed by SSI in November, 1999 yielded volumetric eruption rates per length of fissure in the range of 0.7 to 7 m^3 s^{-1} m^{-1}. Multiplying this rate by the length of the fissure (25 km) yielded a volumetric effusion rate of $\approx 2 \times 10^4$ to $\approx 2 \times 10^5$ m^3 s^{-1} (Wilson and Head, 2001). The upper end of this range is very close to the largest rates inferred for such eruptions on Earth (Wilson and Head, 2001). Theoretical models of dike systems feeding fissure eruptions show that, for eruptions of positively buoyant magma, the elastic properties of rocks are such that the maximum rate derived for Tvashtar Paterae ($\approx$7 m^3 s^{-1} m^{-1}) is the largest value that can *ever* be realized in such eruptions. Interestingly, because this upper limit is independent of the magma-crust density difference and of acceleration due to gravity, this critical value should hold (for a given magma composition) on any planetary body (Wilson and Head, 2001).

11.4 Lava fountains on Io

Even before the *Galileo* mission, lava fountains had been proposed as the cause of ionian outbursts (Davies, 1996). This conclusion resulted from the examination of the January, 1990 outburst, located in the vicinity of Loki Patera (Veeder *et al.*, 1994; Blaney *et al.*, 1995), which appeared to capture the dying-down of lava-fountain activity even as clastogenic flows spread across the surface during the approximately 2 hours covered by observations. Model-based lava liquidus temperatures were in the range 1200 K to 1600 K (Davies, 1996), indicative of a basaltic composition. The January, 1990 lava eruption rate was estimated at 7.7 $\times$ 10^5 m^3 s^{-1}, assuming a flow thickness of $\approx$10 m. If the eruption lasted just 6 hours at this average rate, a volume of 11 km^3 was erupted. The area covered by these flows ($\approx$1670 km^2) is equivalent to about 8% of the dark floor area of Loki Patera, for example, and argues for the rapid emplacement of large, areally extensive flows. The erupted lava contained some 10^{20} J of specific and latent heat energy, most of which would be lost over a period of years as the flows solidified and cooled. This scenario, derived only from spectral data with no context imagery, is strikingly similar to what was observed at Pillan and Tvashtar Paterae.

Other multi-wavelength observations interpreted as being caused by lava fountains are described by Stansberry *et al.* (1997). In February, 2001, an outburst at Surt (the site of a Pele-like eruption between *Voyager* encounters) produced 100 times the peak infrared thermal emission of Pillan in 1997 (Marchis *et al.*, 2002). The Surt outburst may well have been the most energetic volcanic eruption ever witnessed *anywhere*. The peak integrated thermal output ($7.8 \pm 0.6 \times 10^{13}$ W) almost matched Io's total global thermal emission.

Lava-fountain events remove large amounts of heat from the interior of Io. The heat content of the Pillan lavas is 6×10^{19} J. The volume of material emplaced during $\approx$30 such events would be sufficient to cover Io to a depth of 1 cm, if the material were distributed evenly.

11.5 Terrestrial analogues: flood basalts and fissure eruptions

The eruptions at Pillan and Tvashtar Paterae provide a look at lava flow emplacement on a scale absent on Earth in historical times. The accepted largest recent lava flow event, and some of the largest that survive in the geologic record, are, respectively, the Laki, Iceland, eruption in the eighteenth century and the emplacement of continental flood basalts millions of years ago.

11.5.1 Laki (Skaftár Fires), 1783–1784

Larger lava flows than those erupted at Pillan have been emplaced on the terrestrial planets throughout geologic time and also quite recently on Io. Additionally, larger volumes of lava have been erupted from terrestrial volcanoes in the past 100 000 years. The eruption at Pillan in the summer of 1997, however, was the largest emplacement of lava flows ever witnessed on Earth or Io. The lava volume at Pillan (at least 56 km^3) exceeds the largest explosive volcanic eruption in human history: $\approx$30 km^3 (dense rock equivalent) erupted at Tambora in 1815 (Self *et al.*, 2004).

The peak effusion rate at Pillan is comparable to the greatest fissure eruption of historical times, at Laki, in southern Iceland, an eruption known to Icelanders as the Skaftár Fires.

The following description comes in great part from the classic paper by Thordarson and Self (1993). Between June, 1783 and February, 1784, 15 km^3 of tholeiitic basalt lava erupted from more than 140 vents along a 27-km-long complex of 10 *en echelon* fissures. An additional 0.4 km^3 of tephra was also erupted, with a prodigious amount of sulphur dioxide and other gases. The Pillan eruption bears a striking resemblance to the eruption at Laki, named after a mountain that was split by the fissure.

The first stage of the Laki eruption, lasting a few days, was characterized by lava fountains up to 1400 m high. Over the next 8 months, voluminous lava flows surged along the Skaftár and Hverfisfljot valleys during a series of rifting and eruption phases. More than 565 km^2 of land were covered, with lava flow fronts often covering 5 km per day. In the first few months of activity, more than 90% of the total lava volume was erupted.

The Laki eruption consisted of ten main eruptive episodes, inferred to have been the result of an irregular supply of magma through the dike system feeding the eruption. Each new episode began with a seismic swarm of increasing intensity as magma forced its way up through solid rock, leading to the formation of a new fissure. The opening of the fissure was followed by a short phase of phreato-magmatic activity caused by the high water table around the eruption site. Activity changed to violent explosive activity, Strombolian or sub-Plinian, followed by Hawaiian lava fountaining and effusive activity as the volatile content of the magma, the effusion rate, and the supply of available groundwater dwindled.

During the opening phases, effusive rates (Q_F) may have exceeded 10^4 m^3 s^{-1}, and mean eruption rates (Q_E) for the first two phases have been estimated at 8.5×10^3 and 8.7×10^3 m^3 s^{-1} from fissures 2.2 km and 2.7 km in length or – in terms of volumetric rate per unit length of fissure – 3.9 to 3.2 m^3 s^{-1} m^{-1} (Thorarinsson, 1969; Thordarson and Self, 1993).

At the same time, massive volumes of magmatic volatiles were released: an estimated 122 Tg of SO_2, 15 Tg of HF, and 7 Tg of HCl (Thordarson *et al.*, 1996; Wignall, 2001; Grattan, 2005). Those gases would subsequently have a devastating environmental, and possibly climatic, effect on Iceland and northern Europe. The volume of SO_2 erupted has an equivalent solid volume of ≈ 0.1 km^2. The total mass of released SO_2, HF, and HCl makes up 0.34% of the mass of the erupted lava. On Io, the volatiles released during a similar eruption (mostly SO_2) would mostly end up on the surface as frost. If the magma at Pillan had the same SO_2 content as the magma at Laki, then the SO_2 volume erupted at Pillan in 1997 is ≈ 0.1 km^3, equivalent to an SO_2 layer of thickness ≈ 1 mm across the Pillan "black eye."

11.5.2 Flood basalts: the Columbia River Flood Basalt Province

Larger lava flows have been emplaced during Earth's history on a scale far exceeding that at Pillan. The largest volcanic events in Earth's history formed flood basalt provinces, characterized by enormous volumes of homogeneous tholeiitic basalts erupted in a geologically short time span, but one that nevertheless was on the order of 1 million years (White and McKenzie, 1995). Individual flow units within the sequence may have been emplaced on timescales of months to years, not too different from the few months of activity at Pillan. Tectonically, flood basalts were not confined to plate boundaries but were instead associated with hot spots, probably generated by mantle plumes (Hooper, 2000). The initial-stage basalts had a distinctly mantle-derived isotope signature, but subsequent basalts had a geochemical makeup indicating derivation from the lithosphere. The large volumes of rapidly erupted basalt could only be formed if temperatures in the mantle were higher than normal. A widely accepted model of a mantle hot spot is a plume with a large head and much narrower tail rising from the core–mantle boundary or the lower–upper mantle boundary (see Chapter 4). With no plate tectonics on Io, mantle hot spots are a possible cause of Pillan-type volcanic activity.

The Columbia River Flood Basalts (CRBs), perhaps the closest terrestrial analogue to the Pillan flows in terms of volumetric eruption rate and mode of emplacement, formed exceptionally large plateaus of flat-lying flows. They formed about 16 million years ago as a probable result of a mantle plume that burned its way through the North American plate, resulting in the eruption of the CRBs; the basalts of the Snake River Plain; and the formation of Yellowstone, under which the hot spot resides today.

The total volume of CRBs erupted was 1.8×10^5 km^3, covering 1.6×10^5 km^2. Individual flows are huge. On the Columbia Plateau, for example, individual sheet flows with volumes of at least 700 km^3 can be traced for 600 km (Thordarson and Self, 1998). Eruption was through massive fissures, some tens of kilometers long. A few shield structures and occasional small cones, including spatter cones, formed in the later stages of the eruptions.

Of the layers of flows that make up the CRB sequence, the Roza lava flows are some of the best preserved and most closely studied. They cover a vast area (more than 40 000 km^2), extend for more than 150 km, and the thickness of the entire sequence ranges from 11.5 m to 67 m, with an average thickness of 36 m (Thordarson and Self, 1998). Four main lava flows make up the Roza. Individual flow lobes range in thickness from 0.5 m to 52 m, with an average thickness of ≈ 16.7 m. Fifty-seven percent of measured thicknesses fall into the range 5 m to 25 m (Thordarson and Self, 1998). The Pillan flows (8–11 m thick) also fall into this range.

The emplacement mechanism of CRBs, and the Roza in particular, has been the subject of much debate. Shaw and Swanson (1970) suggested eruption durations of a few days, with huge eruption rates and turbulently emplaced flows. More recently, the Roza flows were recognized as large inflated pahoehoe sheet flows, with estimates of duration of emplacement ranging from days to years (Thordarson and Self, 1998). Protected by an insulating crust, voluminous lavas could flow over a hundred kilometers without significant heat loss to impede movement.

11.6 Pillan comparisons with terrestrial eruptions

It is possible to construct the story of the 1997 Pillan eruption and at the same time highlight its similarities with the Laki eruption and other smaller fissure eruptions (such as the 1984 eruption of Mauna Loa). The main event at Pillan took place in a region where other volcanic activity had recently taken place (Laki was the site of an old volcano, and magma supply was linked to the simultaneous eruption of the nearby volcano Grimsvötn). In the spring of 1997, a dike, with gas-rich magma at its leading tip, forced its way to the surface at Pillan. Explosive activity began immediately with magma fragments being propelled great distances by the expansion of the gas-rich phase. On Earth, activity might have been Strombolian, throwing out great showers of clasts, or sub-Plinian, generating a convecting plume. On Io, the high eruption velocities, driven by gas expanding into a vacuum, would have thrown tiny silicate clasts tens or even hundreds of kilometers. This phase of the eruption may have lasted only a few hours.

The eruption fissure developed with a rapidly increasing effusion rate as the fissure widened and lengthened and the system opened down to the magma supply region. Optically thick lava fountains fed lava flows that spread away from the fissure across the apparently flat inter-vent ionian surface. During the early stages of eruption, magma would interact with volatiles in the upper crust, just as the Laki magmas interacted with groundwater, leading to more explosive activity. The lava fountaining may have followed a Laki-like eruption profile over the next few months, but the effusion rate would have built up to a peak probably within a few days to a few weeks. The *Galileo* dataset, although sparse, showed an increase in thermal emission and its more gradual decrease, which was consistent with the effusion profile for basalt eruptions observed at Laki and other terrestrial locations (Wadge, 1981; Thordarson and Self, 1998). By June, 1997, lava fountaining at Pillan was still taking place. A plume was present, indicating the eruption of some gas-rich magma, but it was probably driven primarily by the mobilization of surface volatiles by the advancing lava flows.

As the magma supply became depleted, the effusion rate died down. The eruption did not entirely stop for some time, indicated by the persistence of a hot component (> 980 K) detected by NIMS in observations obtained in 1998 to May, 1999 (Davies *et al.*, 2001). As last seen by *Cassini* in December, 1999, Pillan again appeared as an active, bright spot in nighttime data, although not as bright at visible wavelengths as Pele.

That Io has flow fields considerably larger than those emplaced at Pillan and Tvashtar is intriguing. These are the *fluctūs*, which cover many tens of thousands of square kilometers. Active emplacement of lavas on this scale has not yet been observed on Io, although Lei-Kung Fluctus is still warm (Spencer *et al.*, 2000b; Rathbun *et al.*, 2004). Fluctūs may also be close kin to terrestrial and martian flood basalts.

11.7 Summary: activity at Pillan in 1997 and at Tvashtar in 2000

These eruptions were characterized by basalt or ultramafic fissure eruptions with an initial phase of lava fountaining and the subsequent emplacement of extensive lava flows. The variation in effusion rate at Pillan matches theoretical models of basaltic eruptions (e.g., Wadge, 1981) and terrestrial field observations (e.g., Thordarson and Self, 1993). The flow emplacement style was probably large, single-lobe, pahoehoe sheet flows.

The Pillan 1997 eruption is the best candidate to date for ultramafic magmas on Io, although liquidus temperatures may not be as high as first thought (Chapter 9). The composition of the magma produced unusual surface textures, which share some characteristics of pahoehoe sheet flows.

The Tvashtar Paterae 2000 eruption was resolved in SSI imagery, albeit with a great deal of saturation. The modeled volumes erupted at Tvashtar Paterae are similar per unit fissure length to those postulated at the height of the Laki 1783–1784 eruption.

Large fissure eruptions with associated lava-fountain activity are the most likely explanation of ionian outbursts, during which the Io global 4.8-µm output doubles. The thermal emission from Pillan at the height of the 1997 eruption nearly tripled Io's 4.8-µm flux.

Possible terrestrial analogues of the Pillan lavas, the Columbia River Flood Basalts, were probably emplaced over decades, as long as high volumetric fluxes were maintained. The thermal source generating this activity may have been a mantle plume that quickly mobilized lithospheric material. The Laki 1783–1784 eruption took place at a plate boundary, a zone of extension; Iceland is part of a mid-ocean ridge, a spreading center. A plume rising from the ionian mantle may also have been the source of the Pillan lavas, but the relatively short duration of the eruption, where lavas appeared to exploit a fault to reach the surface, suggest a smaller, though still significant, source of magma than that which produced terrestrial flood basalts. The high level of heating within Io accelerates the pace of volcanic activity. Eruption may therefore take place before large volumes of magma have had time to form at the head of a mantle plume.

The reservoir and conduit systems beneath Pillan and Tvashtar Paterae are no doubt complex. The degree of complexity can be illustrated by considering the plumbing systems under terrestrial volcanoes. Magma was supplied to the Laki fissures via injection of magma from a reservoir under the nearby Grimsvötn volcano, which was also in eruption at that time. Kilauea and Mauna Loa were simultaneously in eruption in 1984. It is known that the shifting of stresses within one volcano, caused by movement and eruption of magma, can have an effect on another nearby. At Tvashtar Paterae, volcanic activity moved across the chain of paterae over the course of 2 years before activity died down, apparently entering a quiescent period while the magma system recharged (Milazzo *et al.*, 2005). To explain possible episodic activity at Tvashtar Paterae, a Hawaiian-scale hot spot could reside under the paterae, feeding a near-surface magma chamber that goes through a cycle of pressurization, failure, eruption, sealing, and recharging, leading to repressurization.

Finally, considering the restrictions imposed on *Galileo* data acquisition, it is remarkable that two lava-fountain episodes were observed in the act. Such activity on Io may be more common than previously thought. As on Earth, lava fountaining on Io will occur on different scales, primarily controlled by magma gas content.

12

Prometheus and Amirani: effusive activity and insulated flows

Effusive volcanism takes place where magmas are fluid and volatile-poor. The density difference between magma and crust and the pressures driving the ascent of magma are insufficient to create lava fountains. On Io, the most closely studied volcanoes where effusive volcanism is taking place, leading to the emplacement of extensive lava flow fields, are Prometheus and Amirani. At Prometheus in particular, study of the thermal emission and calculation of the effusion rate reveals the interior mechanisms and structure beneath the volcano by which magma is supplied to the surface. The closest terrestrial analogue to this style of activity is the current Pu'u 'O'o-Kupaianaha eruption of Kilauea, Hawai'i, where the typical mode of emplacement is that of inflating pahoehoe flows after magma transport through lava tubes over several kilometers.

12.1 Volcanic activity at Prometheus

Prometheus (Plates 6 and 7) is one of the most persistent of Io's volcanoes and the site of a volcanic plume that has been observed in every appropriate observation by *Voyager* and *Galileo* (e.g., Lopes-Gautier *et al.*, 1999; Davies *et al.*, 2006c). Study of the persistent eruption at Prometheus provides an insight into the mechanism of magma supply and heat transfer from the interior of Io to the surface and invites comparison with terrestrial analogues. Located on Io's anti-jovian hemisphere at 154°W, 1.5°S, Prometheus is best known for its long-lived plume, first seen by *Voyager*, and an extensive lava flow field first seen by *Galileo*. The Prometheus lava flows cover >6700 km^2, not including those covering Prometheus Patera, an area of some 800 km^2 (Davies, 2003b). The entire flow field was emplaced in the 16 years between *Voyager* and the first *Galileo* encounters in 1996 (McEwen *et al.*, 1998a). Throughout this time, the Prometheus plume has also persisted, with its characteristic annular plume deposit, rich in sulphur dioxide, migrating westward more than 80 km and keeping pace with the advancing flows (Plates 6b, c).

Plate 1 Io, the most volcanically active body in the Solar System, as seen by the *Galileo* Solid State Imaging experiment (SSI) on July 3, 1999 (Orbit C21). The image shows part of the leading and anti-jovian hemispheres from 220°W to 40°W. The center of the image is at 130°W. North is up. This mosaic uses the SSI near-infrared, green, and violet filters and has been processed to show Io as it would appear to the human eye. Most of Io's surface has pastel colors, punctuated by black, brown, green, orange, and red areas near active volcanic centers. Image resolution is 13 km pixel^{-1}. Courtesy of NASA (PIA02308).

Plate 2 Io as seen by *Voyager*. These six images, from a global mosaic produced by the U.S. Geological Survey, show (a) the Jupiter-facing hemisphere (90°W–0°W–270°W) of phase-locked Io (sub-jovian point is longitude 0°W); (b) leading hemisphere (180°W–90°W–0°W); (c) anti-jovian hemisphere (270°W–180°W–90°W); (d) trailing hemisphere (0°W–270°W–180°W); (e) North Pole view, longitude 0°W is down; and (f) the South Pole, longitude 0°W is up. The color of this mosaic is redder than Io would appear to the human eye.

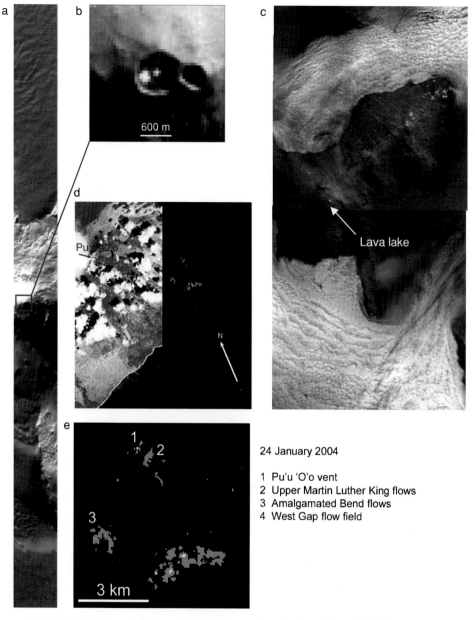

a

b

600 m

c

Lava lake

d

Pu'u 'O'o

N

e

1
2

3

3 km

24 January 2004

1 Pu'u 'O'o vent
2 Upper Martin Luther King flows
3 Amalgamated Bend flows
4 West Gap flow field

Plate 3 (a) Hyperion visible-wavelength observation of the lava lake at Mt. Erebus, Antarctica, on December 13, 2005. Each Hyperion swath is 7.7 km wide. (b) Enlargement of
the Hyperion image in the short-wavelength infrared (SWIR) at a spatial resolution of
30 m pixel^{-1}. (c) ASTER thermal infrared observation of Mt. Erebus on same date. The
ASTER image is 68 km across at a resolution of 90 m pixel^{-1}. (d) Hyperion data of pahoehoe
flows on Kilauea, Hawai'i, January 24, 2005. The panel on the right shows hot spots detected
by the Autonomous Sciencecraft Experiment (ASE) software on the *EO-1* spacecraft.
(e) Detail of active lava vents and flows in the Hyperion data.

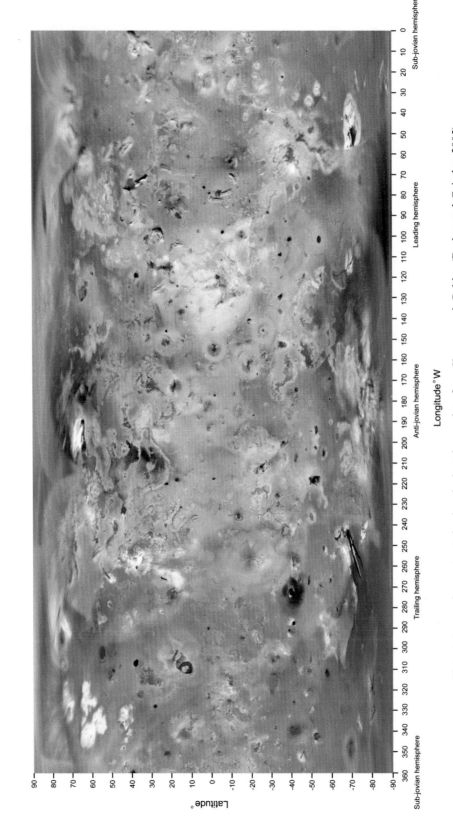

Plate 4 Io mosaic constructed using the best image data from *Voyager* and *Galileo* (Becker and Geissler, 2005). Courtesy of the U.S. Geological Survey.

270 W

360 W   180 W

b

Southern Hemisphere

180 W   360 W

Northern Hemisphere

Plate 5 Io, views centered on (a) the north pole and (b) south pole, reprojected from the mosaic in Plate 4.

90 W

a

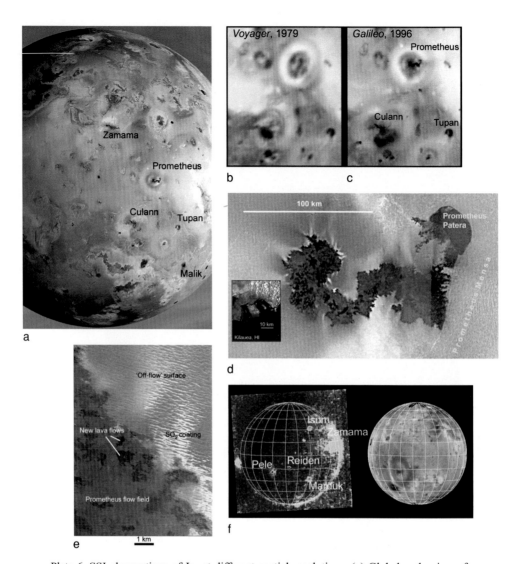

Plate 6 SSI observations of Io, at different spatial resolutions. (a) Global-scale view of Io against the clouds of Jupiter, obtained on March 29, 1998 (Orbit E14). Resolution is 3 km pixel^{-1}. Some major hot spots are labeled. Courtesy of NASA (PIA01604). (b–c) Changes at Prometheus between *Voyager* and *Galileo*, using global-scale images. The circular plume deposits are >200 km across. Courtesy of NASA (PIA00495). The images in 6a–c are false color, enhanced to accentuate albedo and color variation. (d) True-color high-resolution image (170 m pixel^{-1}) of the Prometheus flows emplaced between 1979 and 1996 (image obtained February 22, 2000). Courtesy of NASA (PIA02565). Kilauea, Hawai'i, 1983–2000 flows are shown to the same scale in this *Landsat* image (inset). From Davies *et al.* (2006c). (e) High-resolution (12 m pixel^{-1}) SSI observation of Prometheus obtained on February 22, 2000 (Orbit I27). Courtesy of NASA (PIA02557). New lava flows have low albedo. SO_2 gas jets from the edges of the flow field. (f) Eclipse observation of Io obtained in June, 1996 (Orbit G1), showing intense thermal emission (red) from active hot spots. A daylight image is shown on the right for context. Courtesy of NASA (PIA00739).

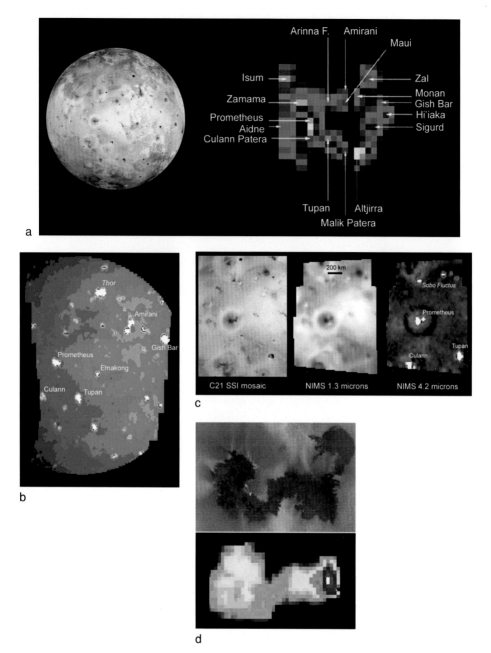

Plate 7 NIMS observations of Io, and Prometheus in particular, at different spatial resolutions. As resolution improves, more hot spots are detected. (a) Global view at 4.8 μm (right) at low spatial resolution (350 km pixel^{-1}), obtained June 28, 1996 (Orbit G1). From Davies *et al.* (2000a). (b) 4.7-μm daytime map with spatial resolution ≈30 km pixel^{-1}, obtained October 16, 2001 (Orbit I32). From Lopes *et al.* (2004). (c) Regional NIMS daytime maps at 1.3 and 4.2 μm, obtained October 11, 2001 (Orbit I24). Spatial resolution is ≈23 km pixel^{-1}. An SSI image is shown for context. Courtesy of NASA (PIA02515). (d) High-resolution daytime NIMS mosaic of the Prometheus flow field, showing the most intense region (red-white), presumably the vent where magma reaches the surface, at the eastern end of the flow field. An SSI context image is also shown. From Lopes *et al.* (2001).

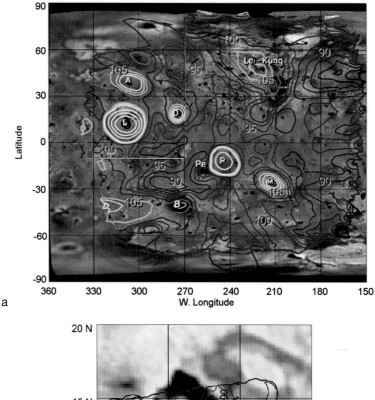

a

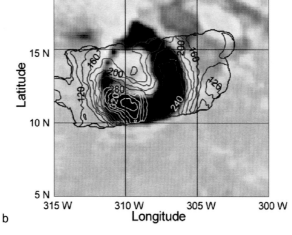

b

Plate 8 PPR maps of Io. PPR is most sensitive to low-temperature areas and can measure thermal emission not only from the non-volcanic Io surface but also lava surfaces at temperatures below the detection levels of NIMS and SSI. (a) Nighttime temperature map of Io from data obtained in 1999 and 2000. Contours are shown in K. Key: P = Pillan Patera. Pe = Pele. M = Marduk. B = Babbar Patera. L = Loki Patera. A = Amaterasu Patera. D = Daedalus Patera. Courtesy of NASA (PIA02548). (b) High-resolution nighttime temperature map of Loki Patera, Io's most powerful volcano, showing that the highest temperatures were in the southwestern portion of the patera when the image was obtained in October, 1999 (Orbit I24). Courtesy of NASA (PIA02524).

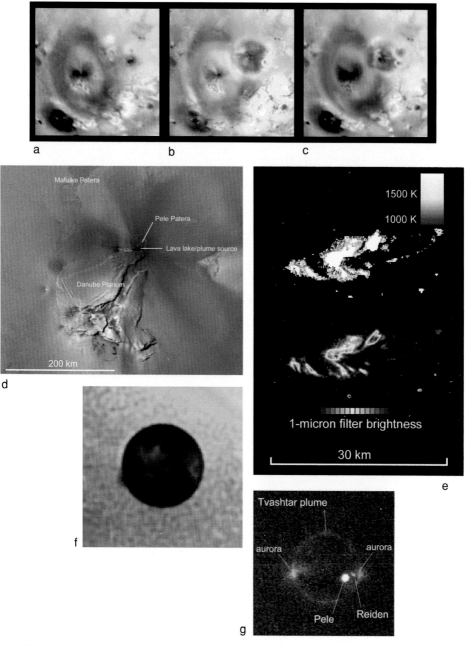

Plate 9 Observations of Pele. (a) Enhanced-color images of the red Pele plume deposit, >1200 km across, in April, 1997 (Orbit G7); (b) September, 1997 (Orbit C10), showing the black deposits from the Summer 1997 eruption of Pillan; and (c) July, 1998 (Orbit C21). Courtesy of NASA (PIA02501). (d) *Voyager* image NASA PIA02232 of the Pele vent area (reprojected by Paul Schenk, LPI). The Pele plume source and main hot spot are located in a graben, abutting (and possibly encroaching into) Pele Patera. (e) Pele temperature map derived from a high-resolution SSI observation obtained in October, 2001 (Orbit I32). From Radebaugh *et al*. (2004). (f) Hubble Space Telescope detection of the Pele plume on June 19, 1997. Courtesy of J. Spencer/Lowell Observatory/HSTI, NASA (PIA01256). (g) *Cassini* ISS observation of Io on January 1, 2001, which detected auroral glow and hot-spot thermal emission. Courtesy of NASA (PIA02882).

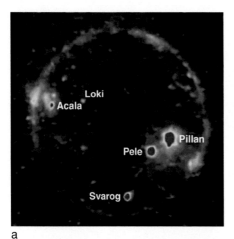

a

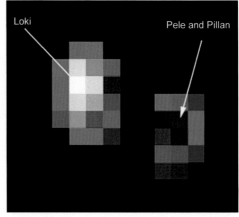

b

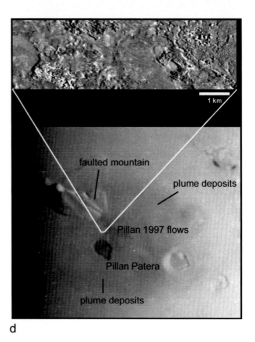

c

d

Plate 10 Observations of Pillan. (a) SSI clear-filter observation, obtained in eclipse, of the explosive eruption at Pillan in June, 1997 (Orbit C9). Courtesy of NASA (PIA01635). (b) NIMS observation of the eruption within a few minutes of the SSI observation in (a). The lower NIMS resolution meant that Pele and Pillan were within the same pixel. From Davies *et al.* (2001). (c) SSI observation of Io, obtained during Orbit C9. The dust-rich Pillan plume is on the limb. The shadow of the Prometheus plume points toward the terminator. Courtesy of NASA (PIA01081). (d) Bottom: 10-m-thick lava flows from the 1997 eruption covered 3100 km^2 and also flooded Pillan Patera (an additional area of 2500 km^2), as seen in this SSI observation obtained in March, 1998 (Orbit E14). Top: high-resolution (9 m pixel^{-1}) SSI images obtained on October 11, 1999 (Orbit I24), showed lava channels, domes and pits, and a strange lava flow surface texture. Courtesy of NASA (PIA02507).

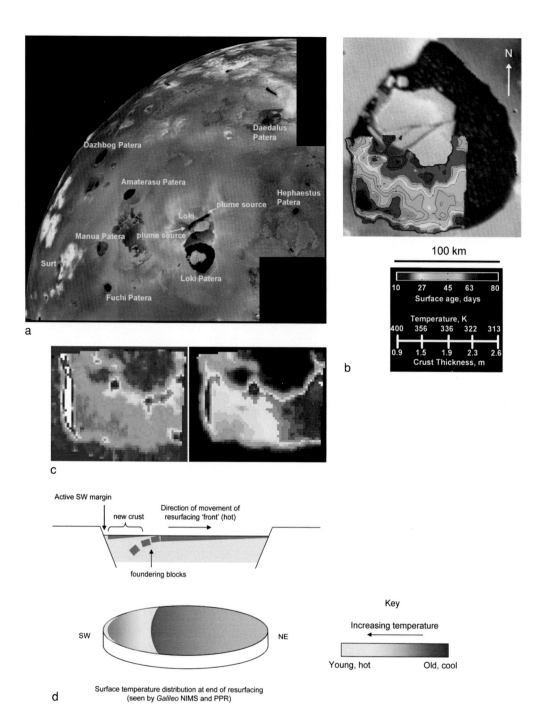

a

b

100 km

10   27   45   63   80
Surface age, days

Temperature, K
400   356   336   322   313

0.9   1.5   1.9   2.3   2.6
Crust Thickness, m

c

Active SW margin

new crust   Direction of movement of
resurfacing 'front' (hot)

foundering blocks

Key

Increasing temperature

SW   NE

Young, hot   Old, cool

d   Surface temperature distribution at end of resurfacing
(seen by *Galileo* NIMS and PPR)

Plate 11 Loki Patera, Io's most powerful volcano, was best observed by *Voyager 1* in 1979. (a) Loki Patera region. Base image courtesy of NASA. (b) Loki Patera temperature, surface age, and crust thickness map from NIMS observation 32INTHLOKI01 obtained in October, 2001 (Orbit I32). From Davies (2003a). (c) NIMS thermal emission maps at 2.44 μm (right) and 4.8 μm (left) from the same observation, where red and white denote the greatest thermal emission. Courtesy of NASA (PIA02595). (d) The preferred resurfacing mechanism for Loki Patera, the foundering of the crust on a "magma sea." From Matson *et al.* (2006b).

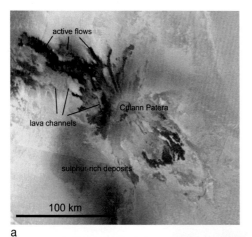

a

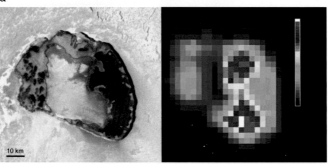

b

c

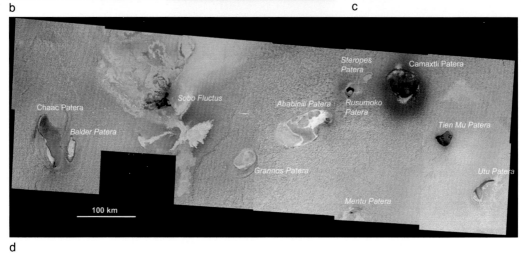

d

Plate 12 (a) SSI false-color image of Culann Patera in November, 1999 (Orbit I25). Base image courtesy of NASA (PIA03885). (b) (left) This magnificent SSI image of Tupan Patera was obtained at a resolution of 132 m pixel^{-1} in October, 2001 (Orbit I32). (right) NIMS data showed that the dark areas in the SSI image were also the warmest. Courtesy of NASA (PIA03601). (c) Amirani, the longest active lava flow field in the Solar System. Courtesy of NASA (PIA02567). (d) The Chaac-Camaxtli region of Io, which contains many different patera geomorphologies. Base image courtesy of NASA (PIA02566).

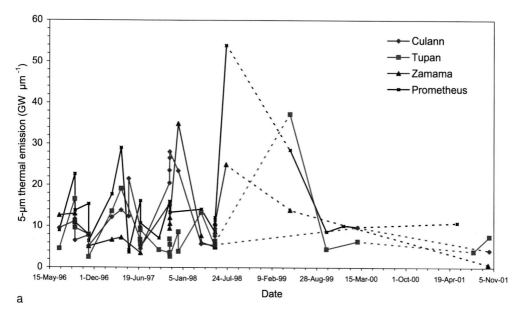

a

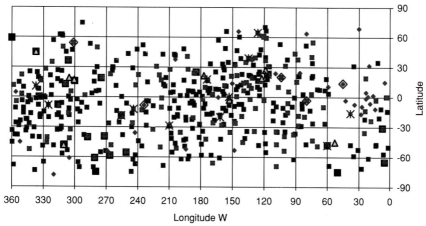

- ■ Hot spot (Voyager)
- ▪ Patera
- ◆ Hot spot (Galileo, ground-based, HST)
- △ Voyager plume
- ✳ Galileo active plume
- ◇ Galileo plume deposits

b

Plate 13 (a) Variability of hot-spot thermal emission from Prometheus (Davies *et al.*, 2006c), Culann Patera, Tupan Patera, and Zamama (Ennis and Davies, 2005). The dotted lines are gaps of one year or more between observations. (b) Locations of hot spots (see Appendix 1), paterae (from a list supplied by J. Radebaugh), and plume sources on Io.

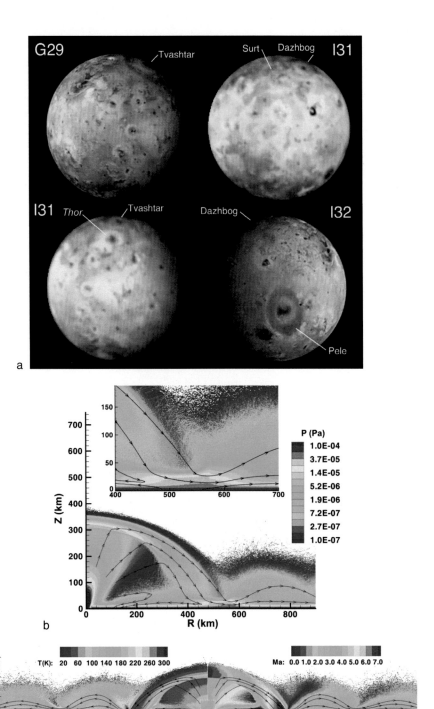

Plate 14 (a) Red and white deposits from several large plumes as imaged late in the *Galileo* mission, in a mosaic from Turtle *et al.* (2004). (In all images, north is up.) (b) Pressure contours of a dayside Pele-class plume. The whole flow region is shown. Details of the "bounce" region are shown in the inset. (c) Temperature (left) and Mach number (right) for a dayside Prometheus-class plume. (b) and (c) from Zhang *et al.* (2003).

a

b

c

d

Plate 15 Terrestrial volcanic eruption styles for basaltic lava of relatively low viscosity and low silica content. (a) Lava fountains, 20 to 50 m high, from a fissure eruption, feeding clastogenic flows, Krafla, Iceland, September, 1984. Courtesy of E. Tryggvason. (b) The Kupaianaha lava lake, Kilauea, Hawai'i. Courtesy of the U.S. Geological Survey, E. W. Wolfe. (c) Open channel flow, Mauna Loa, Hawai'i, 1984. Courtesy of the U.S. National Park Service, Scott Lopez. (d) Pahoehoe flows on Kilauea, Hawai'i, August 24, 1994. The flow is about 2 m wide. A cooling crust rapidly forms on lava flows, but the molten interior can be seen through cracks in the crust. Courtesy of Peter Mouginis-Mark.

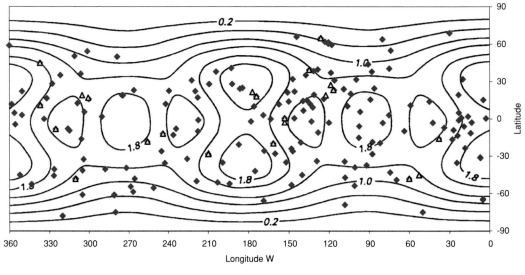

◆ Hot spot   △ Plume

a

b

Plate 16 (a) Distribution of Io's hot spots and active plumes plotted on the expected heat flow resulting from heating in an aesthenosphere and upper mantle, as modeled by Ross *et al.* (1990). Based on an image from Lopes-Gautier *et al.* (1999). (b) Lava cascades into the 'Alo'i crater during an eruption of Kilauea in 1969. The lava fountains in the distance are about 30 m high. Courtesy of the U.S. Geological Survey, D. Swanson.

This migration is necessary to maintain the Prometheus plume, which apparently results from the boiling off of SO_2 ices by the advancing lava flows (Kieffer *et al.*, 2000; Milazzo *et al.*, 2001). The SO_2 in the Prometheus plume is not endogenic to the magma. Regional-scale SSI images (Plate 6a; McEwen *et al.*, 2000b) show a wisp of red surface deposits extending >50 km eastward from the vent, most likely resulting from a trace amount of sulphur degassing from the lava, and consistent with sulphur-rich gas from the vent being swept outward by gas flow from the main SO_2 plume located to the west.

12.1.1 Galileo *observes Prometheus*

Peak temperatures derived from NIMS G1 (June, 1996) data (>1400 K) are indicative of active silicate volcanism, and the shape of the integrated NIMS thermal emission spectrum is indicative of a laminar flow regime, consistent with the emplacement of insulated flows (Davies *et al.*, 1997, 2000b). NIMS thermal spectra obtained both at night and from deconvolved dayside data (L. Soderblom, pers. comm., 2002) show that the spectral shape from 1 to 5 μm – the "thermal signature" of the Prometheus eruption – does not change greatly even with large variations in thermal emission. This consistency indicates that it is not eruption style that is changing with time but the areal extent of volcanic activity (Davies *et al.*, 2006c).

Comparison of high-resolution SSI images obtained in October, 1999 (Orbit I24) and February, 2000 (Orbit I27) revealed the emplacement of new flow lobes on top of older ones, mostly in the westernmost regions of the flow field (McEwen *et al.*, 2000b; Keszthelyi *et al.*, 2001a). The new flows covered about 60 km^2 in just over 4 months, implying an average coverage rate of 5 m^2 s^{-1} (Keszthelyi *et al.*, 2001a). It should be noted that this is a minimum coverage rate. Only part of the Prometheus flow field was included in both 1999 and 2000 SSI observations, and newer pahoehoe flows overlap older ones. Estimates of instantaneous areal coverage rates can vary considerably from time-averaged rates, just as effusion rate can differ from eruption rate (see also Davies *et al.*, 2005).

12.1.2 *Thin flows at Prometheus*

Evidence for relatively thin flows at Prometheus, compared with the 10-m-thick flows at Pillan, is as follows: NIMS data analyses yielded instantaneous areal coverage rates of 35 m^2 s^{-1} during June, 1996 (G1), with volumetric eruption rates for mafic magma of ≈21 m^3 s^{-1}. The implied flow thickness is therefore of order ≈1 m (Davies, 2003b). This number should not be interpreted as the thickness of the entire flow field but rather the thickness of the individual flow units currently being emplaced on top of older flows. Additional evidence of flow thickness comes from the speed at which new flows cool to a temperature at which volatiles can condense on them, increasing albedo. The high-resolution SSI images used to determine the area covered by new lava also showed that a similar area ($\approx$60 km^2) had faded in the same time period (4 months) (Keszthelyi *et al.*, 2001a). Sulphur would begin condensing when flow surface temperature drops below $\approx$390 K, and SO$_2$ would condense at temperatures below 130 K. As shown in Chapter 7, thin (<1 m) flows cool and solidify relatively quickly, completely solidifying in less than four days on Io. After solidification, surface cooling proceeds faster than with thicker, partially molten flows, where heat loss is buffered by the release of latent heat. The temperature of a 1-m-thick flow decreases below the effective low-spatial-resolution NIMS detection level ($\approx$220 K) after about 70 days (Davies *et al.*, 2005), consistent with the rate at which flows change albedo at Prometheus, as seen by SSI.

Using a qualitative albedo–age relationship, at least three populations of flows can be identified at Prometheus (Plate 6e), evidence of increased periods of activity separated by times of relative quiescence (Davies *et al.*, 2005).

12.1.3 Style of activity at Prometheus

The thin lava flows and laminar style of lava emplacement, consistent with a quiescent emplacement mechanism and indicative of pahoehoe-type lava flows, fit other observations of Prometheus. High-spatial-resolution NIMS data show above-background thermal emission from all of the Prometheus flow surfaces, with the most intense thermal source, probably the site where magma is reaching the surface, at the eastern end of the flow field close to the edge of Prometheus Patera (Lopes *et al.*, 2001; see Plate 7d). High-resolution SSI I27 (February, 2000) data show two incandescent pixels in this location. The surface temperature distribution, as best as can be determined from dayside data, indicates an overall cooling trend toward the west (Lopes *et al.*, 2001), before rising slightly where the flows are spreading out. A low-temperature area in the middle of the flow field appears to be relatively free of low-albedo (i.e., young and hot) flows. Surface temperatures in this "new flow-free" area must be lower than $\approx$400 K for volatile condensation to occur. Lava transport through this area must be in tubes or sheets with thick roofs. The wide distribution of breakouts at the western end of the flow field suggests an intricate network of tubes. The western end of the flow field appears to be filling a shallow depression (Paul Schenk, pers. comm., 2005). The morphology of the individual breakouts, with lava spreading in many directions simultaneously, indicates very shallow slopes. On such shallow slopes, terrestrial flood basalts have formed broad sheet flows rather than discrete lava tubes (Keszthelyi and Self, 1998).

Thus, the surface of the Prometheus flow field is covered in overlapping flows, forming a compound flow field, with most flow emplacement from 1999–2000 almost entirely confined within the pre-existing flow field. The thermal spectra of Prometheus are consistent with quiescent lava emplacement in the laminar regime, typified by compound inflating pahoehoe flow fields.

12.1.4 Magma supply at Prometheus

The variability of 5-μm thermal emission from Prometheus is shown in Plate 13a. Assuming that change in intensity is a function of effusion rate, episodes of increased activity have been identified (Davies *et al.*, 2006c). From a peak in the Prometheus 5-μm flux between November 6, 1996 and May 7, 1997 (182 days), the methodology described in Section 7.3 was used to determine a total erupted volume of ≈ 0.8 km^3, a peak effusion rate (Q_F) of ≈ 67 m^3 s^{-1}, and an eruption rate (Q_E) of ≈ 50 m^3 s^{-1} (Davies *et al.*, 2006c).

Eruption rate constrains the geometry of the conduit supplying magma to the surface, be it a circular conduit or a fissure (Section 7.5). Applying Equations 7.35 through 7.42 (see Figure 7.15), the above value of Q_E is met by a conduit just over 3 m in diameter with a magma velocity at the surface of ≈ 6 m s^{-1}. From a fissure with an aspect ratio of 1:100, the width is ≈ 0.75 m and the length is 75 m, with a surface velocity of ≈ 0.9 m^2 s^{-1}.

The persistence of activity at Prometheus and the episodic nature of magma supply are similar to those seen at Kilauea, Hawai'i, where magma is supplied from a shallow magma chamber. Blake (1981) derived the relationship between the volume of the magma chamber and the volume of magma that had to be injected into it to cause the chamber to overpressurize to the point where the chamber fails. Magma is raised to the surface by the excess pressure. An erupted volume of ≈ 0.8 km^3 (Davies *et al.*, 2006c), assuming the volcano is constructed of interbedded basaltic silicates and pyroclasts (a weak basalt volcano), implies a magma chamber with a volume of ≈ 1100 km^3 (Blake, 1981). If the areal extent of the magma chamber is similar to the areal extent of Prometheus Patera, then the chamber is about 1 km deep. To ensure normal stability of its roof, the depth of the roof of the reservoir would have to be at least ≈ 2–3 km (Rowan and Clayton, 1993), making the depth of its center ≈ 3–4 km. Alternatively, if the reservoir has a more ellipsoidal shape, its radius would be ≈ 2.5–5 km and its center would need to be located at a depth of ≈ 8–10 km to place the roof at least 2 km to 3 km below the surface. A possible surface and subsurface structure for Prometheus is shown in Figure 12.1.

12.1.5 Prometheus bound?

The magma eruption rate (Q_E) needed to produce the flow field at Prometheus, if 30 m thick as suggested by Kieffer *et al.* (2000), is 400 m^3 s^{-1} (Davies *et al.*, 2006c).

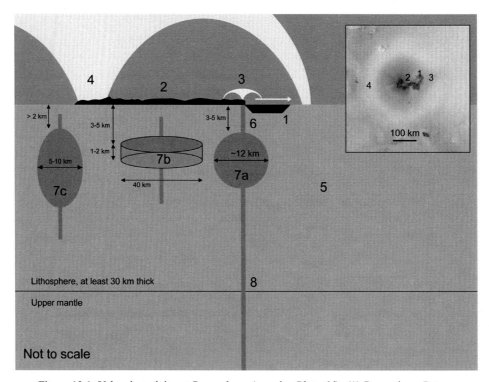

Figure 12.1 Volcanic activity at Prometheus (see also Plate 6d). (1) Prometheus Patera, which is about 40 km in diameter; (2) the Prometheus flow field, emplaced between 1979 and 1996; (3) a small plume of endogenic volatiles, rich in short-chain sulphur, entrained with outflowing SO_2, and deposited toward the east (3 in inset image) (McEwen *et al.*, 2000b); (4) the SO_2-rich Prometheus plume, formed by interaction between hot lava and the volatile-rich substrate (Kieffer *et al.*, 2000; Milazzo *et al.*, 2001), forming the distinct annular plume deposit (4 in inset image); (5) the upper crust of Io, consisting of an unconstrained mix of silicates, pyroclasts, and other volatile-rich plume deposits; and (6) the conduit that supplies magma to the surface at an effusion rate $Q_F \approx 50$ m^3 s^{-1}. 7a–c are possible magma storage chamber geometries, all with relatively shallow roof depths (see Section 7.6.2). (7a) is a spherical chamber, (7b) is broad and shallow. (7c) is ellipsoidal, with a deeper center. (8) The magma chamber is fed by a conduit from the mantle. From Davies *et al.* (2006c).

Rates derived from observations over the duration of the *Galileo* mission were much lower than 400 m^3 s^{-1}. *Galileo* may have witnessed the tail end of a long-duration eruption, where effusion rate has decreased considerably from the early stages of activity (Davies *et al.*, 2006c).

This postulated decreasing effusion rate raises an intriguing point regarding plume activity at Prometheus. With lavas pooling in a regional depression and the volume of erupting lava at levels below what must have been a peak of activity well before the *Galileo* epoch, the supply of heat necessary to maintain the plume is under attack on two fronts. The Prometheus plume will cease when either the local volatile supply is exhausted (Kieffer *et al.*, 2000) or the heat being supplied by an ever-diminishing magma supply rate can no longer mobilize sufficient volatiles (Davies *et al.*, 2006c). Unless a much greater influx of magma to the system takes place, re-invigorating the mobilization of volatiles or breaking out into previously untapped volatile-rich regions, the Prometheus plume is destined for extinction. Long-term monitoring of SO_2 and thermal emission from the Prometheus region will reveal whether the plume does die out.

12.2 Comparison with Pu'u 'O'o-Kupaianaha, Hawai'i

The current Pu'u 'O'o-Kupaianaha eruption (hereinafter referred to as the Pu'u 'O'o eruption) of Kilauea, Hawai'i, is probably a suitable terrestrial analogue for Prometheus. There are certainly many similarities between these volcanoes, even though activity at Prometheus takes place on a greater areal and volumetric scale than at Pu'u 'O'o (Davies *et al.*, 2006c).

12.2.1 Magma and eruption style

Broadly, lava flows in Hawai'i form with pahoehoe or a'a surface textures, although there are sub-classes of texture (e.g., "glassy" pahoehoe and "ropy" pahoehoe). Flow movement along tubes is common.

The temperatures of lavas at Prometheus are poorly constrained but appear to be consistent with basalt. On eruption, viscosities appear to be lower at Prometheus than for Hawaiian basalt, possibly because of composition or superheating effects (Keszthelyi *et al.*, 2005a). No volcanic shield, the existence of which might be used to constrain viscosity, has been observed at Prometheus, but the eruption may not have persisted long enough for such an edifice to form under the current emplacement regime (Davies *et al.*, 2006c). During their advance downslope to the west, the flows have probably cut deeply into the sub-flow volatile-rich surface.

As described earlier, based on NIMS and SSI observations, the current style of emplacement at Prometheus is indicative of a compound pahoehoe flow field in terms of area and temperature distribution, areal coverage rates, implied flow regime, and the mass and flux densities (Davies *et al.*, 2006c).

12.2.2 Plumes

Prometheus has two main plumes: a small sulphur-rich plume emanating from the magma conduit and the main SO_2-rich plume centered on the westernmost flows. At Pu'u 'O'o, a more complicated but similar situation exists, although this is probably coincidental. There are often three main sources of volcanic plumes at Kilauea. The first is the Pu'u 'O'o vent, where (currently) magma first reaches the surface. Degassing of the magma at the vent yields a plume of mostly H_2O ($\approx$90 tons per day), CO_2 ($\approx$300 tons per day), and SO_2 ($\approx$100 tons per day), with lesser amounts of H_2, CO, H_2S, and HCl (Casadevall *et al.*, 1987; Symonds *et al.*, 1994).

The second plume, akin to the process forming the main Prometheus plume, results from the flow of degassed lava, after a journey of some 10 km through a lava tube system, into a volatile-rich environment: the Pacific Ocean. Large plumes of water vapor often form, which contain entrained silicate particles.

The third, much smaller plume is located at Halema'uma'u, the summit caldera of Kilauea, where CO_2-rich gas escapes.

12.2.3 Episodic activity and magma supply

Volcanic activity is episodic at both Prometheus and Kilauea. Pu'u 'O'o has undergone several "episodes" since the current eruption centered on Pu'u 'O'o began in 1983 (see summary in Heliker and Mattox [2003]). Each new episode denotes vigorous new eruptive activity either from a different vent or after a slowdown or pause in activity. For example, the current episode at the time of this writing, Episode 55, began on February 27, 1997, and is characterized by effusion from Pu'u 'O'o flank vents and mainly tube-fed flows.

Magma supply at Pu'u 'O'o is currently more or less continual from a shallow magma chamber with a vertical extent of $\approx$3 km, centered at a depth of $\approx$3 km. The gravity ratio between Earth and Io is a factor of 5.5, and a proportional ratio in the vertical extent of magma reservoirs would be expected (Wilson and Head, 1994). This proportional ratio could imply that on Io possible vertical extents of reservoirs are $\approx$16 km, with centers at $\approx$10 km depth and the tops of such systems at a depth of $\approx$2 km. These values are similar to those found in Section 12.1.4 for an ellipsoidal reservoir at Prometheus (Davies *et al.*, 2006c).

Estimates of the volume of the Pu'u 'O'o magma chamber range from 0.08 to 40 km^3 (Cervelli and Miklius, 2003). Individual eruptive episodes produce volumes typically of <0.01 km^3 (HVO, 2005). Episode 55 is somewhat of an anomaly, being so long-lived. Using the work of Blake (1981), the typical erupted volume implies a magma chamber volume of $\approx$10 km^3, neatly in the range derived by Cervelli and Miklius (2003). Recent work interpreting changes in the tilt of Kilauea and Mauna Loa over several eruption cycles revealed that the shallow magma system at Kilauea consists of two distinct magma reservoirs (Cervelli and Miklius, 2003): a small reservoir located 0.5 km east of Halema'uma'u, with its top 500 m to 700 m below the surface, and the south-caldera reservoir with its top about 4 km below the surface. The south-caldera reservoir is deflating slowly, at a rate not exceeding

2.6×10^6 m^3 yr^{-1} – about 2% of the total volume of lava erupted at Pu'u 'O'o every year (Cervelli and Miklius, 2003).

The eruption rate Q_E at Pu'u 'O'o between 1983 and 2004 was $\approx$4 m^3 s^{-1}. From February, 1997 through December, 2004 (part of Episode 55), 1.23 km^3 of lava (of dense rock equivalent 3000 kg m^{-3}) were erupted, yielding $Q_E = 5$ m^3 s^{-1}. Peak eruption rates have been as high as 490 m^3 s^{-1}, which occurred during Episode 38 in October, 1985. During that episode, more than 1.5×10^7 m^3 of lava were erupted in 8.5 hours. Peak Q_F was almost certainly larger.

Q_F values for Prometheus range from $\approx$15 to >125 m^3 s^{-1}. During the *Galileo* epoch, Q_E was $\approx$40 m^3 s^{-1} (Davies *et al.*, 2006c).

12.2.4 Areal coverage rates

A total volume of 2.6 km^3 of magma was erupted at Pu'u 'O'o from 1983 to 2004, covering a land area of 117 km^2. A large proportion of the erupted volume flowed or ultimately collapsed into the sea (HVO, 2005). Taking a period of time when flow emplacement was confined only to land (Mattox *et al.*, 1993), Keszthelyi *et al.* (2001a) estimated an areal coverage rate at Pu'u 'O'o of 0.05 km^2 day^{-1}. In a setting like Prometheus, therefore, where there is no massive slope break at the coast and no ocean, one might expect the Pu'u 'O'o flow field to have covered 400 km^2 in 21 years, giving a flow stack thickness of about 6.5 m. This area and volume are dwarfed by Prometheus (Plate 6d). As described earlier, flows from Prometheus covered >6700 km^2 in more or less the same time. Prometheus Q_E values are likely two orders of magnitude larger than at Pu'u 'O'o (Davies *et al.*, 2006c).

12.3 Amirani flow field

Another flow field, one considerably larger than that found at Prometheus, is found at the persistent hot spot Amirani, which is actually several hot spots along a huge lava flow field (Plate 12c). Located at $\approx$116°W, 23°N, the Amirani flow field is the longest active lava flow in the Solar System, at more than 300 km long with an average width of $\approx$60 km (Keszthelyi *et al.*, 2001a). SSI I25 (November, 1999) and I27 (February, 2000) data revealed 620 km^2 of new flows (in 23 separate breakouts), emplaced in 134 days at an average areal coverage rate of $\approx$60 m^2 s^{-1} (Keszthelyi *et al.*, 2001a). NIMS G1 (June, 1996) data yielded surface temperatures in excess of 1200 K, areal coverage rates of $\approx$80 m^2 s^{-1}, effusion rates of $\approx$60 m^3 s^{-1}, and implied flow thicknesses of $\approx$1 m (Davies *et al.*, 2000a; Davies, 2003b). The shape of the thermal emission spectrum is indicative of insulated flows. As at Prometheus, NIMS 4.5- to 5-μm data show elevated thermal emission along the entire main flow but not the flows extending to the west (Lopes *et al.*, 2001). A thin red channel connects a reddish, low-albedo patera to the southernmost extent of the flow field. Diffuse red deposits indicate that lava is degassing from an active vent. Amirani exhibited intermittent Prometheus-type plume activity, with the source of the SO_2 plume centered on the southern quarter of the flows (Keszthelyi *et al.*, 2001a). It is difficult to determine just how many truly separate active centers make up Amirani. Observed as a single hot spot in low-spatial-resolution data, Amirani NIMS data revealed large variations in thermal emission (Davies, 2002).

12.4 Discussion and summary

The Prometheus eruption has several similarities with volcanic activity currently taking place on the flanks of Kilauea: namely, a long-lived but episodic, effusive eruption with the emplacement of laminar, insulated flows, punctuated by periods of increased activity (Heliker and Mattox, 2003). Activity at Prometheus is on a much greater areal scale than at Kilauea in general and Kupaianaha-Pu'u 'O'o specifically, a hallmark of ionian activity compared to that of Earth (Carr, 1986; Davies, 2001; Davies *et al.*, 2001; Keszthelyi *et al.*, 2001a; Davies *et al.*, 2006c). As at Pu'u 'O'o, magma at Prometheus originates from a deep source (at Kilauea, this is at a depth >60 km [Thornber, 2003]). The magma is transported to near-surface storage chambers, where some degassing takes place, before rising and erupting at the surface. Eruption results in flow emplacement and thermal interaction with easily vaporized *in situ* solids and liquids.

Given that the style of emplacement observed at Prometheus is also seen in other locations on Io (although usually without the formation of a large SO_2 plume), it may be that relatively shallow magma chambers are common (Davies *et al.*, 2006c).

13

Loki Patera: Io's powerhouse

Located at $\approx310°$W and 12°N, Loki Patera (Plate 11a) is Io's most powerful and most intriguing volcano and one of the most prominent features on Io. Loki Patera appears as a low-albedo (and, by inference, relatively hot), sub-circular feature more than 200 km in diameter. A feature that looks like an "island" or "raft" takes up $\approx25\%$ of the area of the patera. This "island" is fractured, and pieces appear to have broken off in a manner akin to the calving of a terrestrial ice shelf. The "island" feature did not change appearance in the years between *Voyager* and *Galileo* so it is almost certainly immobile. It is likely to be either a resurgent dome (similar to that found in the terrestrial Long Valley Caldera) or possibly even a foundered mountain block (T. V. Johnson, pers. comm., 2005). The appearance of the Loki Patera region changed between *Voyagers 1* and *2* (Smith *et al.*, 1979c) but, by the time of *Galileo*, looked very similar to its appearance as seen by *Voyager 1*, another indication of Io's tendency to exhibit long-term surface color and albedo stability (McEwen *et al.*, 1998a). Additionally, the eruption mechanism at Loki Patera – the manner in which considerable volumes of lava yield their heat – has to maintain this appearance.

Loki Patera is therefore something of a paradox, showing little visible change while undergoing a high level of volcanic activity. Thermal output from Loki Patera can vary between $\approx10\%$ and 25% of Io's *total* thermal emission (Veeder *et al.*, 1994) and between 2% and 50% of Io's 4.8-μm output, yet this emission comes from less than 0.1% of Io's surface area. This high thermal emission means Loki is easily observed and monitored from Earth (Veeder *et al.*, 1994). Although the initial *Voyager* IRIS and ground-based IRTF observations had insufficient spatial resolution to conclusively associate the dark material with the Loki hot spot's thermal emission, subsequent studies of the relationship of albedo and volcanic heat sources (McEwen *et al.*, 1985) and spatially resolved *Galileo* data from NIMS and PPR demonstrated that the visually dark regions are indeed the source of thermal emission (Lopes-Gautier *et al.*, 2000; Spencer *et al.*, 2000b; Davies, 2003a). About

Table 13.1 *Two-temperature fits to* Galileo *low-resolution NIMS Loki Patera data*

Observation	Date	Temperature of "hot" area T_H (K)	"Hot" area A_H (km^2)	Temperature of "warm" area T_W (K)	"Warm" area A_W (km^2)	Total thermal output (10^{12} W)
G7INVOLCAN05	Apr. 5, 1997	990	3	460	2 320	6.1
C9INCHEMIS06	July 25, 1997	962	5	373	11 700	13.0
16INHRSPEC01	July 21, 1998	612	27	345	3 830	3.3
22INHRSPRC01	Aug. 14, 1999	576	21	339	2 420	1.9
29INIWATCH03	Dec. 28, 1999	878	2	417	2 330	4.1
					Average	5.7

From Matson *et al.* (2006b).

200 km to the north of Loki Patera is the source of two plumes that were active during the *Voyager* epoch. No plume at or in the vicinity of Loki Patera was detected by *Galileo*.

Volcanism at Loki Patera is dominated by silicate magma, but it took a surprising length of time to determine this conclusively (see review in Matson *et al.* [2006b]). SSI observed faint thermal emission from Loki Patera in eclipse observations (McEwen *et al.*, 1998a), indicating areas at temperatures at or exceeding 700 K, and NIMS detected hot areas in excess of 900 K (Table 13.1). Any lingering worry that the low-spatial-resolution NIMS data were including thermal emission from other hot spots within the field of view, or that the SSI data were in fact detections of auroral glow, were dispelled when SSI and NIMS observed Loki Patera in darkness during C9 (July, 1997). NIMS detected temperatures in excess of 950 K, and SSI saw a faint glow from Loki Patera and not from any other hot spot in the NIMS field of view. The observed high temperatures were in Loki Patera, and the volcanic activity was silicate in nature.

13.1 *Voyager* to *Galileo*

Several observations obtained by *Voyager* and *Galileo* instruments revealed a changing pattern of albedo and temperature distribution across the patera. The highest resolution *Voyager 1* image of Loki Patera (still the highest resolution image obtained of the patera) showed that the southwestern margin of the caldera was relatively dark (Figure 9.3); *Voyager 2* imagery at lower spatial resolution showed that the dark area had expanded more than 100 km eastward. The implication was that

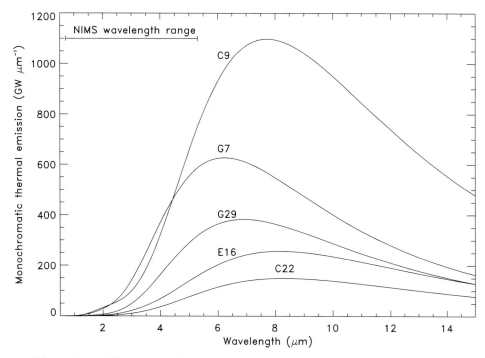

Figure 13.1 Loki Patera's evolving thermal emission spectrum as modeled from NIMS data during the *Galileo* mission. Spectra are labeled with orbit designation. The NIMS wavelength range is also shown. After Matson *et al.* (2006b).

the floor of the patera was being covered by newly erupted lava, moving eastward from the western margin. Assuming this resurfacing took place in the ≈120 days between *Voyager* encounters, the rate of advance was about 1 km per day.

During the *Galileo* epoch, high-spatial-resolution PPR images obtained during a brightening event at Loki Patera in 1999–2000 showed that thermal emission followed a similar pattern to that observed at visible wavelengths by *Voyager*. A PPR observation obtained in October, 1999 (Orbit I24) located the area of highest thermal emission in the southwestern portion of the patera (Plate 8b). An observation obtained in February, 2000 (Orbit I27) revealed that the warm area was now located to the east of the Loki island, indicating movement around the "horseshoe" in a counterclockwise direction (Spencer *et al.*, 2000b).

13.2 Style of activity

In low-spatial-resolution NIMS data where Loki Patera was sub-pixel (Figure 13.1), the thermal emission spectra appeared similar to those from Prometheus, but on a much larger scale. This spectral signature indicated a surface dominated by cooling crust, with a relatively small hot area. Two-temperature fits to the NIMS data are shown in Table 13.1. The ratio of hot area to warm area – the implied surface "crack fraction" – is usually much less than 1% at Loki Patera, similar to crack fractions derived at Prometheus and Amirani.

Despite large variations in the intensity of thermal emission with time (as seen by NIMS), the relative thermal emission at different wavelengths varied very little. Even when Loki Patera was being resurfaced during a brightening episode, relatively little incandescent material was being exposed; the eruption style is for the most part quiescent. The lava has to have a low enough volatile content to allow degassing at a rate that does not disrupt the lava surface, which would expose the hot interior and increase short-wavelength thermal emission. There generally are no massive Pillan-like outbursts – no lava fountains or open channel flows (Davies, 2003a).

A medium-resolution ($\approx$20 km pixel^{-1}) NIMS observation of part of Loki Patera obtained in October, 1999 (Orbit I24) showed that, compared to Io background temperatures, the patera floor was warm, as were the cracks – or fissures – across the island, but the island itself was cold (Lopes-Gautier *et al.*, 2000). The I24 observation also indicated that part of the margin of the patera exhibited higher than average thermal emission, hinting that the margins were active, perhaps the result of the crust on a lake or pond breaking up or perhaps where magma rising up faults was arriving at the surface.

As seen in a high-resolution (1.1 km pixel^{-1}) SSI near-terminator observation of Loki Patera obtained on October 16, 2001 (Orbit I32), the surface of the low-albedo patera floor is relatively smooth and exhibits specular reflectance (Turtle *et al.*, 2004). This specular reflectance has been interpreted as being due to forward scattering off glassy lava surfaces. The same SSI observation also provided the best constraints on the height of the patera rim as being no greater than 100 m (Turtle *et al.*, 2004).

13.3 Temporal behavior

Decades of observing Io revealed that Loki Patera underwent brightenings every 1 to 3 years, with increases in the 3.5-μm and 3.8-μm flux up to a factor of 10 over fluxes during quiescent times. Each brightening typically lasted for a few months (Veeder *et al.*, 1994; Spencer and Schneider, 1996; Howell *et al.*, 2001). At least eight brightening events were detected between 1987 and 2001 (Rathbun *et al.*, 2002). Loki Patera brightenings are not as large as Io's thermal outbursts and persist longer, suggesting a different mode of eruption than that of outbursts is responsible.

A statistical analysis of 3.5-μm and 3.8-μm data collected by ground-based and Earth-orbiting instruments and NIMS from 1987 through 2001 (Figure 13.2) revealed a fascinating aspect of Loki Patera's volcanism. Short-wavelength infrared

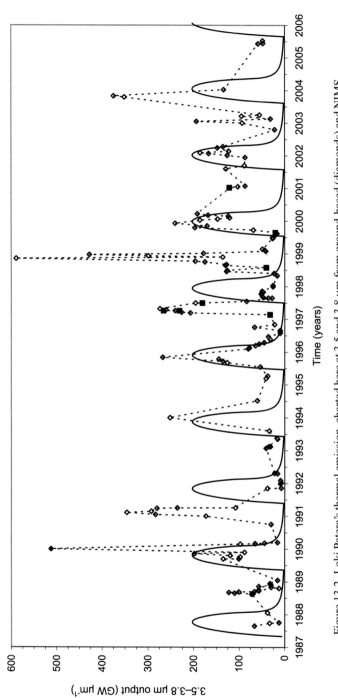

Figure 13.2 Loki Patera's thermal emission, charted here at 3.5 and 3.8 μm from ground-based (diamonds) and NIMS (black squares) data, showed periodic activity on a cycle of ≈540 ± 15 days for many years (Rathbun *et al.*, 2002). The spike in data in January, 1990 was an outburst in the Loki vicinity (Veeder *et al.*, 1994). The black undulating curve is the 3.5-μm output from the lava-lake-crust foundering model developed by Matson *et al.* (2006b). Base image from Rathbun and Spencer (2006).

thermal emission and, by inference, the volcanic activity producing it, were statistically periodic. Eruptions at Loki Patera were cyclic, with a 540 ± 15-day period to a high significance level (Rathbun *et al.*, 2002). This discovery of periodicity also demonstrated the value of decade-length ground-based observation programs, without which this revealing analysis would not have been possible.

13.4 Resurfacing of Loki Patera

As noted by Matson *et al.* (2006b), scientists have more or less exhausted the possibilities of styles of volcanic activity at Loki Patera using relatively simple models and morphological and volcanological comparison with terrestrial analogues (Lunine and Stevenson, 1985; Carr, 1986; Davies, 1996; Howell, 1997; Lopes-Gautier *et al.*, 2000; Davies, 2001; Rathbun *et al.*, 2002; Davies, 2003a; Gregg and Lopes, 2004; Matson *et al.*, 2006b).

Two models can be considered for resurfacing: the emplacement of new flows and lava-lake overturn (Davies, 1996; Rathbun *et al.*, 2002; Davies, 2003a; Matson *et al.*, 2006b).

13.4.1 Resurfacing with lava flows

The lava flow model envisioned magma ascending along a vent along the southwestern margin of Loki Patera (see Plate 11) and spreading eastward to flood the entire patera floor, a distance of more than 200 km. Flows would have to be at least 10 m thick to travel this distance, implying that between 1979 and 2001 a stack of flows at least 140 m thick was emplaced. (An aside: at this rate the patera, currently ≈ 100 m deep, will be in danger of overflowing sometime around 2015.) This resurfacing mechanism requires a huge volume to be erupted during each episode, >200 km^3, or a total of 2800 km^3 between 1979 and 2001 (Matson *et al.*, 2006b). Emplacement of such a vast volume as lava flows, with associated hot vents, cracks, flow fronts, and breakouts, should increase Loki's thermal emission at short wavelengths during eruption, something not observed by either *Galileo* or *Cassini*.

The lava flow model has some inherent difficulties, but these are more intuitive than highlighted by the 3.5-μm and 3.8-μm data. The resurfacing mechanism at Loki Patera has to be inherently simple to produce the observed temperature distribution and temporal behavior. It is difficult to envision a mechanism by which such large volumes of magma are *periodically* generated and erupted, although it must be said that Io is certainly a part of a forced dynamic system. The supply and storage of magma via a system of dikes, reservoirs, and magma chambers is complex, which is why episodic – not periodic – behavior is observed on Earth. Additionally, flows would have to propagate a great distance over a surface that is, to all appearances, flat (Matson *et al.*, 2006b).

13.4.2 Resurfacing by lava-lake overturn

The lava-lake overturn model proposed by Rathbun *et al.* (2002) had the advantage of simplicity over resurfacing by lava flows. Loki Patera was proposed to be a silicate lava lake, albeit five orders of magnitude larger than most terrestrial lava lakes (making the Loki lava lake the area of Massachusetts). Terrestrial active lava lakes have been observed to overturn in one of two ways. The first method is a process that at first glance looks like a small-scale version of the formation, movement, and destruction of oceanic tectonic plates. The second is by foundering of the surface crust. In both cases, the new lava exposed during resurfacing cools and solidifies rapidly to form new, rigid crust, underneath which is molten lava. Crust formation is initially rapid, forming a relatively low-density, pore-filled crust, but slows with time (Figure 7.4). Void space decreases with depth, and crust bulk density increases with time (Matson *et al.*, 2006b).

In the "mobile crust" model, the surface crustal plates move across the lake, driven by convection in the lava beneath. These plates can be destroyed either against the walls at the margins of the lake or by being sucked down in "subduction zones," where convecting magma rapidly descends. Lava exposed in the wake of the plate solidifies to form a new plate, and the process repeats itself. The surface crust survives typically on the order of minutes to hours before being recycled (Plate 15b).

The second mechanism is that of crustal foundering (Rathbun *et al.*, 2002; see Plate 11d). Here, the crust becomes too dense to be supported by the underlying liquid and sinks. Void-free solid lava is about 5% denser than liquid lava, but volatiles in the crust create pores that reduce bulk density. The bulk density of the crust, however, increases with time as more material is added to the crust because the porosity decreases with increasing depth. Eventually, a critical thickness is reached and the crust founders, to be replaced with new lava. This process has been observed at the Makaopuhi lava lake, Kilauea, Hawai'i, which overturned repeatedly in this manner in 1965 (Wright *et al.*, 1968). It may be significant that Makaopuhi was an inactive lava lake, where lava had flowed into a crater to form a lava pond 800 m wide and 84 m deep (Wright *et al.*, 1968). There was no injection of material into the lake from below, nor was there any deep heat source that could have created convection in the lava, leading to lateral surface crust movement.

As proposed by Rathbun *et al.* (2002), foundering at Loki Patera would start in the active southwestern margin of the patera and, to explain the *Voyager* and PPR observations, would propagate eastward across the patera (or counterclockwise around the horseshoe) at a rate between 1 and 2 km day^{-1}. The new crust that formed behind the "front" would solidify and cool. It would take about 200 days to completely resurface the patera. At that point, the surface crust would be oldest, coolest, and thickest in the southwest and youngest, hottest, and thinnest at the other extent of the horseshoe. The lake then remains quiescent for nearly a year. At the end of this time, when the southwestern portion of the crust has reached a thickness of about 7 m (see Section 7.2 and Figure 7.4), the bulk density of the crust reaches, and then exceeds, the density of the supporting liquid magma. The crust sinks, and the resurfacing cycle begins again. Loki Patera appeared to be a predictable volcano, something not encountered before.

13.4.3 The trickster

No sooner had the Rathbun *et al.* paper been published in May, 2002 than Loki Patera changed its temporal behavior. The authors of the paper ruefully joked about the perils of trying to predict the behavior of a volcano named after the Norse god of trickery. Instead of the periodic behavior seen for many years, activity became more chaotic before becoming steady (Rathbun and Spencer, 2006; see Figure 13.2). Any physical model of resurfacing and thermal emission would have to explain this new behavior as well as the old.

13.4.4 NIMS takes a close look

In October, 2001, one of the most impressive NIMS observations of the entire *Galileo* mission was obtained (Plate 11). The observation, named 32INTHLOKI01, covered most of the southern half of Loki Patera, with radiance data obtained at 12 wavelengths across the NIMS wavelength range. This observation was designed expressly to study the thermal structure of the surface of Loki Patera, and the Rathbun *et al.* foundering model received a boost when the data were analyzed (Davies, 2003a).

From a two-temperature fit to each spectrum, the warm-component temperature yielded surface age for each pixel and also crust thickness (Section 7.2, Plate 11b). The surface was oldest in the southwest and progressively younger to the east and north. Crust ages ranged from 10 to 80 days, implying that the resurfacing had begun sometime in late July or early August, 2001. When ages were plotted against distance (see Davies, 2003a, Figure 2), the rate of resurfacing across a wide front was $\approx$1 km day^{-1}, in the range predicted by Rathbun *et al.* The general speed and direction of resurfacing (counterclockwise around the patera) were confirmed by Howell and Lopes (2007), although lower crust temperatures were found. These temperatures implied an earlier date of the onset of resurfacing than that proposed by Davies (2003a). In either case, the position of any active resurfacing front was outside the coverage of the 32INTHLOKI01 observation.

From the Davies (2003a) analysis, implied crust thicknesses ranged from 0.9 m to 2.6 m. Additionally, some relatively intense pixels were identified around the margins of the patera. One interpretation was that these were locations where the crust on the lava lake was breaking up against the walls of the patera (Lopes *et al.*, 2004). With this temperature and area distribution to work with, the relative merits of lava flows and the two lava-lake models were investigated by Matson *et al.* (2006b), who concluded that the observed temperature distribution was reproducible only with the lava-lake "foundering crust" model.

13.5 Modeling the resurfacing process

The uncomplicated, predictable nature of the foundering resurfacing mechanism meant that it could be tested with a mathematical model (Matson *et al.*, 2006b). With a known surface area being replaced daily (during the active resurfacing phase) and with the relationship already established for how the lake surface will cool (Chapter 7), the evolving temperature distribution and both local and integrated thermal emission spectra were modeled. Figure 13.3 shows the evolution of the thermal emission at different wavelengths over the active phase and subsequent passive cooling phase, over five cycles (Matson *et al.*, 2006b). The evolution of total thermal emission is shown in Figure 8.4. It should be noted that a hot crack fraction is not included. The analysis is focused on the crust.

It is illuminating to consider the evolution of total thermal emission from Loki Patera over each cycle. Peak model thermal emission is reached at ≈ 150 days into the resurfacing cycle. The minimum model thermal emission, 5.7 TW, is 6% of Io's total (10^{14} W). The maximum thermal emission is 15.6 TW, 16% of Io's total. The average thermal emission is 9.6 TW, or $\approx 10\%$ of Io's total. The range is consistent with the observations of thermal emission from Loki Patera (typically 10%–15% of Io's thermal emission) (Veeder *et al.*, 1994). The total thermal emission over 540 days is 4.5×10^{20} W (Matson *et al.*, 2006b).

The derived fluxes also match the very low level of thermal emission at visible wavelengths observed by SSI and ISS, are within a factor of two of those seen by NIMS, and match the 2:5-μm ratios and fluxes seen by NIMS in low-spatial-resolution data (Davies and Keszthelyi, 2005).

A similar resurfacing modeling technique is used by Rathbun and Spencer (2006) to show that the observed 3.5-μm flux from Loki was quantitatively matched by the foundering model. Rathbun and Spencer also considered the change in behavior of Loki thermal emission in 2001 and demonstrated that the new thermal flux could be explained by making small variations in the speed of the resurfacing front (Rathbun

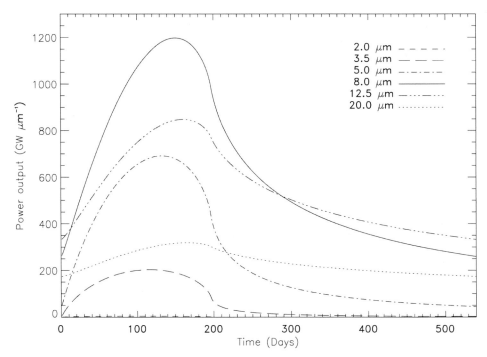

Figure 13.3 Thermal emission at different wavelengths from an elliptical, basalt lava lake 196 km across (the long axis) and 140 km across at its widest point, with a total surface area of 21 500 km^2 (equivalent in area to Loki Patera's dark floor), over a single resurfacing cycle. A resurfacing wave propagates across the lake at 1 km per day. For each surface element (1 km wide), the age of the surface is used to determine the surface temperature, using the basalt cooling curve for a semi-infinite case (Figure 7.8). The integrated thermal emission is determined by adding the spectra for all surface components. It takes 200 days to resurface the lake, after which the lake surface passively cools for a further 340 days. Total thermal emission variability for this event is shown in Figure 8.4.

and Spencer, 2005). Foundering ultimately depends on the contrast between the bulk density of the crust and the liquid magma beneath and, by making small variations in the porosity of the crust, the observed thermal behavior at 3.5 μm could be matched by the resurfacing model.

It appears that during 2000 or 2001, the volatile content of the Loki magma (already very small) changed very slightly. A decrease in volatile content meant that the forming crust increased in density faster than during the "periodic" phase, so that crust foundering began ahead of schedule and propagated more rapidly across the patera with varying speed, from <1 to ≈6 km day^{-1}, a function of the variation (never more than 1%) in crust bulk density (Rathbun and Spencer, 2005).

Rathbun and Spencer also ran the resurfacing model using values for a sulphur lava lake and determined that, although the peak thermal emission from Loki Patera could be reproduced, a lake of liquid sulphur could not sustain the observed thermal flux for the necessary periods. Volcanism at Loki is dominated by silicates.

13.6 Magma volume at Loki Patera

It is possible to determine how much heat energy is transferred through the Loki volcanic system by a straightforward energy balance calculation. The following methodology is taken from Matson *et al.* (2006b). The volume of crust, V (km^3), that forms and is subsequently destroyed during one 540-day cycle is given by

$$V = zA, \qquad (13.1)$$

where A is the low-albedo patera floor area (2.15×10^{10} m^2) and z is the thickness of crust after 540 days (≈ 7 m; see Chapter 7). V is therefore ≈ 150 km^3. If the average bulk density ρ of the crust after 540 days is 2600 kg m^{-3}, the crust mass m (kg) is

$$m = Az\rho, \qquad (13.2)$$

yielding $m = 3.8 \times 10^{14}$ kg.

The heat content of the crust (i.e., the latent heat [F_{latent}] liberated through solid-ification and the integrated sensible heat loss [$F_{sensible}$] through the crust, from the final surface temperature of ≈ 250 K [crust temperature after 540 days] to the liq-uidus temperature of 1475 K at the base) should be equivalent to the total observed heat loss (F_{total}).

For $m = 3.8 \times 10^{14}$ kg, latent heat loss F_{latent} (using a latent heat of fusion of 4×10^5 J kg^{-1}) is 1.52×10^{20} J. $F_{sensible}$ is determined from the temperature profile (determined using Equation 7.11) in the crust at the moment of foundering and is calculated for the entire area of the lake (A) using

$$F_{sensible} = A \sum_{x=0}^{x=z} \Delta T_x \, c_p V_{T,x} \, \rho, \qquad (13.3)$$

where ΔT_x is the temperature difference between liquidus temperature and crust temperature at depth x; c_p is the specific heat capacity (1500 J kg^{-1} K^{-1}); $V_{T,x}$ is the volume of the crust at temperature T at depth x; and ρ is the crust bulk density. Integrating over the depth of the crust, the sensible heat loss is 1.4×10^{10} J m^{-2}, or 2.9×10^{20} J over the entire surface of the lake (Matson *et al.*, 2006b).

This calculation shows that ≈ 2000 km^3 of magma has solidified and cooled – and ultimately foundered – at Loki Patera in the past 20 years, an average of ≈ 100 km^3 year^{-1}, which represents at least 10% of Io's total thermal emission over that time (Matson *et al.*, 2006b).

13.7 Summary: a class of its own

Loki Patera is another unique feature on Io, unparalleled in its enormous power output and areal scale. Lava lakes exist on Earth, but nothing contemporary matches Loki Patera. The "magma sea" model developed by Matson *et al.* (2006b) matches all observational data for Loki Patera and provides a framework for additional modeling. Matson *et al.* also suggested that the presence of a large liquid body, perhaps extending from the surface down to the aesthenosphere, would be the target of preferential tidal dissipation. This tidal dissipation would be a mechanism by which the magma in the sea could be continually heated while venting a vast amount of energy to space.

The foundering of surface crust as a resurfacing mechanism is a particularly attractive explanation for the excellent multi-spectral, multi-decade Loki Patera observation sequence, including the data obtained by *Voyager, Galileo*, and *Cassini*. The foundering process is quiescent, so no large areas at incandescent temperatures are generated; is predictable; generates temperature and area distributions that compare favorably with observations; does not rely on complex magma supply processes; is self-sustaining; and (perhaps the clinching argument) is a process that is inevitable anyway, if Loki Patera is indeed a lava lake (Matson *et al.*, 2006b).

Given its immense size, Loki Patera has been designated a "magma sea" rather than a "lava lake" (Matson *et al.*, 2006b). There is an absence of convection on the scale of the lake: like Makaopuhi, Loki Patera's surface crust passively sinks, without lateral movement. The crust on the sea does not move but forms and founders in place.

14

Other volcanoes and eruptions

The volcanoes described in the previous chapters are either unique on Io or are class types of silicate eruptions. Many other volcanic centers display diverse styles of eruption, and colors and geomorphologies indicative of other lava compositions. The tour of Io's volcanoes continues with a closer look at some of these features.

14.1 Tupan Patera

Tupan Patera (141°W, 19°S) (Plate 12b) is one of Io's most colorful features. The patera is 75 km × 50 km in area and about 900 m deep (Turtle *et al.*, 2004). Bright red material, probably short-chain sulphur allotropes, colors most of the patera floor and diffuse deposits are seen on the surface southeast of the patera (Keszthelyi *et al.*, 2001a; Turtle *et al.*, 2004). Black silicates cover the floor of the eastern half of the patera and appear in patches in the western half. In NIMS data these dark areas are the warmest areas, whereas the central "island" is cold (Lopes *et al.*, 2004). A relatively uniform black line traces the edge of the patera floor in the western half of the patera and may be a tide line like that seen at Emakong (Turtle *et al.*, 2004; see Section 14.5). The appearance of bright material in patches on the eastern patera floor is consistent with the melting of sulphur from the patera walls and patera margins; the sulphur then flows and pools on cooling silicates on the floor of the patera. Turtle *et al.* note that several areas on the eastern side of the patera are green, supporting the hypothesis that there is a chemical reaction between red sulphur deposits and warm silicates (McEwen *et al.*, 2004; Williams *et al.*, 2004).

NIMS G1 (June, 1996) spectra revealed a small area (<0.01 km^2) at >1100 K, indicative of active silicate volcanism, and a larger, cooler area of ≈30 km^2 at ≈440 K (Davies, 2003b). In high-resolution NIMS data, color temperatures along the margin of the patera were as high as 750 K (Lopes *et al.*, 2004). Tupan Patera is a persistent hot spot (Lopes-Gautier *et al.*, 1999), and the location of these hot pixels led to the suggestion that Tupan may be an active lava lake (Lopes *et al.*, 2004). No vigorous overturning episodes were observed in the visible or infrared by *Galileo* that could have confirmed this suggestion, but Tupan Patera may instead overturn quiescently, like Loki Patera (see Chapter 13). As shown in Plate 13a, during the *Galileo* mission Tupan Patera exhibited variable brightness at ≈5 μm (Lopes *et al.*, 2004; Ennis and Davies, 2005). The 5-μm flux varied considerably, from as low as 2.6 GW μm^{-1} (November, 1996, Orbit C3, and November, 1997, Orbit C11) to a maximum of 37.3 GW μm^{-1} in May, 1999 (Orbit C20).

Estimates of effusion rate needed to maintain the observed thermal emission vary accordingly, from a volumetric flux of ≈8 m^3 s^{-1} in November, 1997 to 114 m^3 s^{-1} in May, 1999. The average volumetric flux over the *Galileo* mission was ≈25 m^3 s^{-1}.

14.2 Culann Patera and environs

Just to the north of the mountain Tohil Mons, Culann Patera (161.5°W, 19.9°S) is another very colorful volcano (Plate 12a). Volcanic activity has resulted in the emplacement of deposits of many colors, indicative of both silicate and sulphur volcanism (Keszthelyi *et al.*, 2001a; Williams *et al.*, 2005). A complex relationship exists between different-colored units. Culann Patera itself has a highly irregular scalloped margin and a green floor: the green color may be the result of interactions between sulphur and silicate lava, as seen at Tupan Patera (Keszthelyi *et al.*, 2001a). Lavas have spilled out of the patera. A red, curving line extending to the northwest from the southwestern tip of the patera may mark the location of a lava tube that feeds dark silicate flows to the northwest. Culann is classed as a persistent thermal source (Lopes-Gautier *et al.*, 1999). In June, 1996 (Orbit G1) temperatures in excess of 1200 K were derived from NIMS data (Davies, 2003b), most likely from the dark, presumably silicate, areas. Again, the thermal signature of the eruption indicated insulated lava-flow emplacement or quiescent lava-lake overturning, with an estimated effusion rate of ≈20 m^3 s^{-1} (Davies, 2003b). Culann's 5-μm variability is shown in Plate 13a.

14.3 Zamama

Zamama, located at 173°W, 21°N, was one of the first hot spots detected by *Galileo* during Orbit G1 in June, 1996 (Davies *et al.*, 1997; Lopes-Gautier *et al.*, 1997; McEwen *et al.*, 1997). SSI and NIMS night and eclipse data (Figures 9.6 and 9.7) showed that temperatures of at least 1100 K were present. Those temperatures were one of the first confirmations of silicate volcanism on Io from *Galileo* data (Davies *et al.*, 1997). Daytime SSI images revealed small volcanoes (Figure 14.1), a new flow field, and associated red (sulphur-rich) and white deposits that had been emplaced since the *Voyager* encounters (Keszthelyi *et al.*, 2001a).

The Zamama region has been mapped in detail (Williams *et al.*, 2005). High-resolution SSI observations show that at least 15 flows emanate outward from the

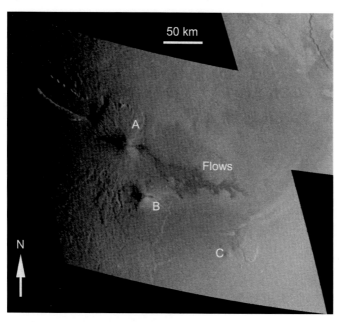

Figure 14.1 This SSI image of Zamama was obtained on October 16, 2001. Three shield volcanoes can be identified (*A, B, C*). Silicate lava flows emanate from the summit of *Zamama A*, mostly to the east, where they have formed a large flow field. Images of this region obtained in 1998 revealed an active plume, generated by hot silicate lava advancing over an ice-rich surface. Deposits to the west of *Zamama A* were red, evidence of sulphur escaping from the magma as it reaches the surface (see Chapter 12).

summit region of a low shield volcano, *Zamama Tholus A* (see Figure 15.4) (Schenk *et al.*, 2004; Williams *et al.*, 2005). The primary flows form an east–west-trending lava flow field more than 140 km long (Keszthelyi *et al.*, 2001a). Diffuse red (sulphur-rich) surface deposits emanate from this volcano, probably the result of sulphur degassing from erupting magma. The white deposits emanate from the flow field, where the lava flows have mobilized surface ice deposits, as at Prometheus. These shield volcanoes are examined in more detail in Chapter 15.

Zamama was thermally active in at least 26 NIMS observations (Ennis and Davies, 2005), exhibiting a highly variable 5-μm thermal emission (Plate 13a). Thermal emission from Zamama peaked in December, 1997, when the 5-μm flux reached levels about a factor of 10 greater than 6 months before (Ennis and Davies, 2005). This 5-μm flux increase coincided with a 75-km-high plume that was rich in SO_2 (Geissler *et al.*, 2004a; Williams *et al.*, 2005). Knowing the likely emplacement mechanism from the thermal signature (in this case, insulated flows), modeling of NIMS data indicated peak effusion rates at that time in excess of 100 m^3 s^{-1} (Ennis and Davies, 2005). The absence of new flows spreading across the surface in SSI images indicated that flows were mostly being emplaced on top of older flows, much in the style of Prometheus and Amirani.

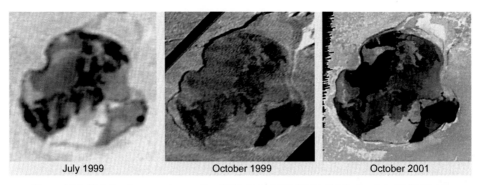

| July 1999 | October 1999 | October 2001 |

Figure 14.2 Gish Bar Patera as imaged by SSI in July, 1999 (Orbit C21); October, 1999 (Orbit I24); and October, 2001 (Orbit I32). The patera is 106 km by 115 km in size. North is up. Much of the variation from image to image is caused by different viewing and illumination geometries as well as different filters (Turtle *et al.*, 2004). The new dark flows in the western part of the patera and in the southeastern corner in the image on the right appear warm in NIMS data obtained in October, 2001 (Orbit I32) (Lopes *et al.*, 2004). Courtesy of NASA (PIA03884).

14.4 Gish Bar Patera

Gish Bar Patera (located at 89.1°W, 15.6°N) is an irregularly shaped patera about 106 km by 115 km in area. It is a persistent hot spot as seen by NIMS (Lopes-Gautier *et al.*, 1999); modeling of NIMS data indicated the presence of a relatively small hot spot (by Io standards) with an area of $\approx$65 $\pm$ 20 km^2, an areal coverage rate of $\approx$12 m^2 s^{-1}, and an effusion rate of $\approx$7–15 m^3 s^{-1} (Davies *et al.*, 2000a; Davies, 2003b). Flow thickness was estimated to be $\approx$1 m (Davies, 2003b).

Gish Bar Patera is noted as a possible location of an outburst in August, 1999 (see Table 2.1) that was observed by ground-based astronomers and NIMS (Howell *et al.*, 2001; Lopes *et al.*, 2004). SSI obtained high-resolution images of Gish Bar Patera several times (Figure 14.2). These images show that lava flows had covered $\approx$430 km^2 by October, 2001 (Orbit I32) (Turtle *et al.*, 2004). Assuming that the eruption began shortly after the August, 2001 encounter (Orbit I31), when NIMS saw little thermal emission from Gish Bar Patera (Lopes *et al.*, 2004), the implied average coverage rate was $\approx$230 m^2 s^{-1} (Turtle *et al.*, 2004), comparable to the average coverage rate estimate of $\approx$330 m^2 s^{-1} at Pillan in 1997 (Davies *et al.*, 2001). Peak rates were almost certainly higher.

14.5 Emakong Patera: sulphur volcanism?

Emakong Patera (119.1°W, 3.2°S) is a heart-shaped feature roughly 75 km by 65 km in area. It is intriguing because of its high potential as a site of extensive sulphur volcanism (Williams *et al.*, 2001b). During Io fly-bys, high-resolution images from

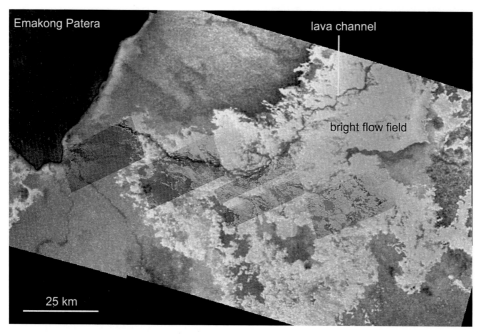

Figure 14.3 The bright flows on the flanks of Emakong Patera are prime candidates for extensive effusive sulphur volcanism. Silicate lava flows appear to have mobilized large deposits of sulphur. Dark, sinuous silicate flows cut channels into the sulphur. The base image was obtained on November 25, 1999 (Orbit I25) and the high-resolution images (30 m pixel^{-1}) on October 15, 2001 (Orbit I32). Courtesy of NASA (PIA02598).

SSI (140 and 33 m pixel^{-1}) and NIMS (3–5.5 km pixel^{-1}) were obtained of the patera and a flow channel beginning near the rim. The SSI October, 2001 (Orbit I32) observation (Figure 14.3) shows a dark, sinuous lava channel with widths ranging from ≈620 m to ≈1400 m, and a complex mixing of dark and light lavas, with islands of bright material (Turtle *et al.*, 2004). The new images confirmed that the flow channel originated from the overflow of lava from Emakong Patera (Turtle *et al.*, 2004). The dark channel feeds extensive whitish-yellow flows and gray-black flows, the largest at least 370 km long and ≈190 km wide at its broadest point.

NIMS data showed the patera floor was warm during February, 2000, at a temperature of ≈300 K, which is consistent with recent volcanic activity (Lopes *et al.*, 2001). By the I32 fly-by (October, 2001), the temperature of most of the patera floor had dropped below 230 K (Lopes *et al.*, 2004). Those NIMS data also showed a scattering of warmer pixels (>270 K) along the southern and eastern margins of the patera floor. Lopes *et al.* (2004) proposed that Emakong Patera was a lava lake and that the warm pixels were where the crust on the lake was breaking up against the patera walls. Supporting the lava-lake hypothesis, a dark ring appears to extend around the entire patera. This ring may be a tide mark, where the level of lava in the patera dropped (Turtle *et al.*, 2004). The cause of the drain-back, leaving the dark rim, may be withdrawal of lava or eruption through a vent in the flank of the patera. The amount of cooling observed, from $\approx$340 K to $\approx$190 K over 2 years, is more in keeping with cooling silicates than cooling sulphur (Chapter 7). Emakong Patera is most likely covered in cooling silicates, either flows or a crusted-over and mostly solidified lava pond.

The flows on the flanks of Emakong have complex morphologies that in part resemble the upper crusts of terrestrial silicate flows. However, their colors are suggestive of sulphur flows that have quenched at different temperatures (e.g., Sagan, 1979). Emakong may be a complex mixture of silicate and sulphur flows (Turtle *et al.*, 2004), with silicate flows melting sulphur and SO_2 deposits to generate sulphurous flows, in a manner similar to sulphur eruptions at Siretoko-Iosan (Watanabe, 1940) and Mauna Loa (Skinner, 1970; Greeley *et al.*, 1984) (see Chapter 5).

Other bright flows seen on Io (at Culann, for example) make up about 2% of Io's surface area. At Sobo Fluctus, located at 150°W, 14°N, extensive bright white flows that may be highly sulphurous in nature have spread across the surface for >150 km (Plate 12d). Darker flows emanating from an active hot spot lie on those bright flows (Williams *et al.*, 2002).

14.6 Balder and Ababinili Paterae: SO_2 flows?

The possibility also exists for effusive activity on Io dominated by liquid SO_2. At a few locations imaged at high resolution are deposits very rich in SO_2. Balder Patera, located at 155.8°W, 11.3°N (Plate 12d), has a very bright, white floor with a high abundance of SO_2 coverage (estimated at 70–90% by Douté *et al.*, 2001) that may be frost or even ice. The deposit is confined within the patera. Smythe *et al.* (2000) hypothesized that the deposit could have been emplaced rapidly as a liquid SO_2 flow. An alternative hypothesis is that SO_2 gas or liquid slowly seeped from the edges of the patera or through the patera floor and then froze within the confines of the patera. Deposits almost as rich in SO_2 are found in nearby Ababinili Patera, located at 141.8°W, 12.6°N (Plate 12d). An excellent description of the distribution of SO_2 within the Sobo Fluctus and Balder Patera region can be found in Douté *et al.* (2004).

14.7 The plumes of Surt and Thor

Based on total power output, a brief eruption at Surt (337.1°W, 44.9°N) is the largest eruption witnessed on Io to date. This eruption was detected by ground-based observers with the Keck Interferometer telescope on Mauna Kea, Hawai'i, using AO, and is described by Marchis *et al.* (2002). A small brightening was seen on February 20, 2001, modeled as a very small area (<8000 m^2) at very high temperature (>1400 K). This almost certainly marked the first stage of surface activity. Two days later, a huge eruption was under way. The total thermal emission at this time was 7.8×10^{13} W, almost doubling Io's total thermal emission (Marchis *et al.*, 2002). To give some idea of the scale of the thermal emission (which is discussed in Chapter 17), the 5-μm thermal emission at the peak of the eruption was 5000 GW μm^{-1}, ten times the intensity needed to be classified as an outburst eruption. A two-temperature fit to the data yielded an area of 95 km^2 at 1470 K and 1780 km^2 at 880 K. A multi-temperature model fit (Section 7.2.5; see Figure 7.11) yielded a range of areas totaling 880 km^2 with temperatures ranging from 1080 K (≈ 8.9 km^2) to 1450 K (≈ 0.5 km^2), meaning that the entire area was incandescent (Marchis *et al.*, 2002). Data were interpreted as being the result of a gigantic lava-fountain eruption. Indeed, when this area was imaged by SSI, a 1000-km-diameter red ring typical of explosive eruptions had been laid down (Plate 14a).

Surt was the site of an eruption between *Voyager* encounters in 1979 that laid down a similar Pele-like plume deposit (Figure 1.4) and is the only site where a repeat eruption of this type (sporadic, explosive) was seen from *Voyager* to *Galileo*. The 1979 eruption left a dark spot more than 50 km in diameter that had disappeared by February, 1997, in keeping with the tendency for surface changes to fade with time (Chapter 9). Despite the violence of the February, 2001 eruption, little change was seen in the Surt vent region post-eruption. Most likely, activity took place within the confines of a caldera that limited the spread of lava flows.

Thor (133.8°W, 40.7°N) is noted for an eruption that created the tallest plume seen on Io, more than 500 km high, which was observed during August, 2001 (Orbit I31). Lava flows were also emplaced during that eruption (Turtle *et al.*, 2004; Williams *et al.*, 2005). Unlike Pele, Tvashtar, and Surt, no red ring was laid down around Thor (Williams *et al.*, 2005). Instead, a white ring rich in SO$_2$ frost was laid down (Plate 14a), detected in NIMS data (Douté *et al.*, 2004).

Plume types, plume generation, and resulting deposits are discussed in Chapter 16.

Section 5

Volcanism on Io: the global view

15

Geomorphology: paterae, shields, flows, and mountains

Io boasts some of the most impressive topography in the Solar System, with mountains higher than Mt. Everest and volcano-tectonic depressions deeper than the Grand Canyon on Earth. The shapes of these volcanic depressions, associated shields, and volcanic cones and the morphology of lava flows on Io's surface yield important clues to the nature of the magma and the interior processes that generate the observed geomorphology.

15.1 Paterae on Io and calderas on Earth

The most common volcanic feature on Io, a patera, is currently defined by the International Astronomical Union, guardian of planetary nomenclature and feature names, as an "irregular saucer-like crater." This may be a misnomer because few saucers are perfectly flat, as the floors of many paterae appear to be. Io's paterae have steep walls and arcuate margins and, geomorphologically, are unique to Io. However, in some respects they do resemble some calderas formed on Earth, Mars, and Venus (e.g., Radebaugh *et al.*, 2001). In some cases, notably at Maasaw Patera (Figure 1.7c), the craters are nested and look strikingly similar to nested calderas at the summit of Mauna Loa, Hawai'i, on Earth, and Olympus Mons on Mars (Figure 15.1a, b).

Four hundred and twenty-eight paterae have been mapped within a region covering about 70% of Io's surface (Radebaugh *et al.*, 2001). Locations are shown in Plate 13b. The total number of paterae may exceed 500 (Carr *et al.*, 1998), with some low-albedo features in low-spatial-resolution *Galileo* and *Voyager* observations likely to be paterae.

Approximately 13% of paterae are found adjacent to mountains. Schenk *et al.* (2001) studied the *Voyager* and *Galileo* datasets and found that, on a global scale, regions with more volcanic centers tended to have fewer mountains, and vice versa.

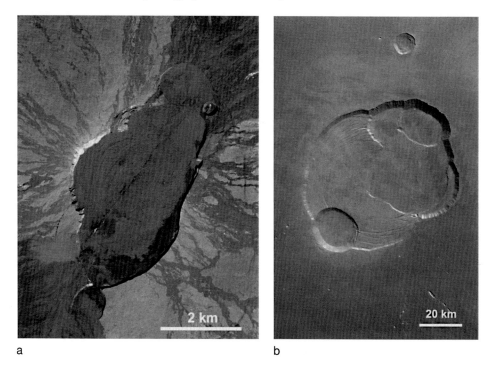

a   b

Figure 15.1 Caldera complexes on basalt shields on Earth and Mars. (a) Mokuaweoweo, the summit caldera of Mauna Loa, Hawai'i. Courtesy of NASA. (b) The summit caldera of Olympus Mons, Mars. Both elongated structures show evidence of several stages of collapse. Courtesy of ESA/DLR/G. Neukum (FU Univ).

However, Jaeger *et al.* (2003) noted that the number of mountains in direct contact with paterae (41%) is significantly higher than expected in a random distribution. SSI imagery showed sufficient cases to suggest that magma ascending along faults associated with mountain building is a relatively common occurrence (Jaeger *et al.*, 2003), as was seen with the Pillan 1997 eruption (Plate 10d).

Forty-two percent of paterae have irregular margins, and only 8% are found on low volcanic shields (Radebaugh *et al.*, 2001). Many smaller paterae, a large number of which exhibit continuous volcanic activity, are located at latitudes between 25°N and 25°S (Radebaugh *et al.*, 2001). Toward the poles, paterae are fewer in number (per unit area), but larger. Although the mechanism of formation is not well understood, this increase in size may result from a combination of possibly thicker polar crust (Tackley *et al.*, 2001; see Chapter 17), deeper magma origin, and larger eruption volumes. The distribution of paterae shows peaks at 330°W and 150°W longitude, a distribution possibly related to the direction of Jupiter's greatest tidal effects (Radebaugh *et al.*, 2001; Tackley *et al.*, 2001). A large number of hot spots (>70%) are found in or adjacent to paterae, indicating the importance of paterae to volcanism and heat transport. Paterae play a major role (perhaps *the* major role) in the transfer of heat from Io's interior to the surface. Loki Patera alone is responsible for typically 10% to 20% of Io's total heat loss.

With so much volcanic activity confined to paterae, it is inevitable that some paterae contain silicate or sulphur lava lakes (Lunine and Stevenson, 1985; Radebaugh *et al.*, 2002; Lopes *et al.*, 2004). Just how many paterae contain this volcanological phenomenon, rare on Earth, is not known; in most cases of patera volcanism, it is difficult to determine from available thermal emission data whether a patera is being resurfaced by lava flows or by crust replacement on a lava lake or a lava pond (Davies, 2003a). The long-lived Pele lava lake must be connected to an equally long-lived, voluminous magma chamber or some deeper, equally productive magma supply, which cannot be said with any certainty for other lava-lake candidates save Loki Patera (Matson *et al.*, 2006b).

Some paterae have filled with silicate lava and overflowed, leading to the emplacement of dark, low-albedo flows across the surface (e.g., at Prometheus). Others, such as Pillan Patera in 1997, have been flooded with lavas that have spilled into the patera after flowing across the surface from a distant vent.

15.1.1 Patera sizes

The mean diameter of paterae on Io is ≈ 42 km, and the largest is Loki Patera, at 200 km in diameter. This diameter is large in comparison with other Solar System volcanic depressions (Radebaugh *et al.*, 2001). The largest terrestrial calderas are ash-flow calderas, sub-circular collapse craters, the largest of which exceed 20 km in diameter. The largest terrestrial caldera is Yellowstone, which has a diameter of >70 km (Pike and Clow, 1981). These terrestrial ash-flow calderas, the closest in size to ionian paterae, formed as a result of sudden, rapid eruption, mostly along ring fractures, of several km^3 to >1000 km^3 of magma. In some cases, magma was siphoned away to erupt elsewhere, leaving a void that led to collapse. A well-documented example is the collapse of Katmai, Alaska, in 1912, when magma erupted at Novarupta, ≈ 10 km away (Hildreth and Fierstein, 2000).

Shadow measurements in near-terminator images obtained by both *Voyager* and *Galileo* show that paterae can be very deep; Chaac Patera is 2.7 km deep with very steep (70°) walls. In comparison, the summit caldera of Kilauea is about 6 km in diameter and 165 m deep, and the summit caldera of Mauna Loa, the largest terrestrial basaltic shield volcano, is roughly 8×6 km and 183 m deep (Newhall and Dzurisin, 1988). Unlike terrestrial calderas, most paterae appear to be simply inserted into flat plains and lack the broad shields characteristic of basalt volcanoes on Earth and Mars. This flat appearance may be caused by much lower viscosity lavas (e.g., McEwen *et al.*, 1998a); sudden, rapid, and short-lived eruptions that do not persist long enough for massive shields to build up; or a very different formation mechanism.

The Mokuaweoweo caldera complex on Mauna Loa (Figure 15.1a) and the summit calderas of Kilauea and other basaltic volcanoes were formed by incremental collapse into a partially drained magma chamber (e.g., Wood, 1984). The removal of the magma that supports the pre-caldera surface can be either by eruption of magma or draining of magma laterally or downward. Mokuaweoweo bears a striking morphological resemblance to the Olympus Mons caldera complex on Mars, except that the latter is >70 km in diameter (Figure 15.1b). On Io, nested paterae at Maasaw, Tvashtar, and Gish Bar resemble these incremental-collapse terrestrial and martian calderas (Carr *et al.*, 1979; Radebaugh *et al.*, 2001).

15.1.2 Lava composition and patera formation

Silica-rich ash-flow calderas have been suggested as the closest terrestrial analogues to ionian paterae because of their sizes and low, broad, shield-like morphologies (Carr, 1986; Radebaugh *et al.*, 2001). These calderas are formed during the eruption of copious volumes of silica-rich magma with a high volatile content. No large volcanic edifices are formed, and the largest calderas are comparable in size to many paterae. However, there is no evidence for high-silica-content eruptions on Io, at least from observed flow morphologies and volcano shapes. It is possible that high-silica-content, volatile-rich magmas erupt explosively into a vacuum, fragmenting so completely that the resulting pyroclasts are deposited thinly over a very large area and never form flows or steep-sided volcanic edifices (Wilson and Head, 1983, 2001). Those deposits would then be rapidly buried by other plume fallout. Large explosive eruptions certainly take place on Io, but none of the deposits studied at usable resolution have a significant felsic component (McEwen *et al.*, 1998b; Keszthelyi *et al.*, 2001a; Geissler *et al.*, 2004a). Instead, deposits from explosive events at Pillan and Tvashtar appear to be mafic to ultramafic (magnesium-rich) in composition.

Other mechanisms could form a patera. Thermal interaction between dense silicate intrusions that stall at the base of thick layers of frozen volatiles could remove so much material, erupted in plumes, that the surface collapses. Alternatively, having removed a large volume of volatile material, paterae could form by drain-back of magma into dikes (Davies and Wilson, 1988). These mechanisms rely on the presence of thick layers of sulphurous material. Such layers were initially thought to lack the strength to support the observed steep walls of paterae (Clow and Carr, 1980). Given a very high thermal gradient (once thought to be up to 200 K per kilometer), sulphur would be molten at the base of a 2-km-high wall, undermining slopes and causing collapse. However, the realization that resurfacing and burial are so rapid on Io that the lithosphere is essentially cold means that such topography-supporting, sulphur-rich deposits are not unreasonable. Thermal exchange between silicate lava flows and SO_2-rich surfaces will remove some SO_2, but such is the efficiency of heat removal at the silicate–SO_2 interface that a relatively thin layer of SO_2 is removed, with most heat loss from silicate flows being from the upper surface by radiation (Johnson *et al.*, 1995; Davies *et al.*, 2005). In this manner, deposits of sulphur and SO_2 can be buried under silicate lava flows (Johnson *et al.*, 1995).

The excavation of upper crust volatiles by silicate intrusions, leading to collapse, was again considered at the end of the *Galileo* mission (Keszthelyi *et al.*, 2004a). The process is shown in Figure 15.2. Silicate magma rises through the predominantly silicate lithosphere and stalls when it encounters a volatile-rich, lower-density layer. First, SO_2 and then sulphur are mobilized. The gas may escape as plumes or may flow away from the intrusion beneath the surface if the volatile-rich layer is porous. Mobilized sulphur and SO_2 may erupt effusively onto the surface as ionian equivalents of Icelandic *jökulhlaups* (caused by a sub-glacial eruption melting the ice above: melt water, heavily laden with silicate debris, eventually breaks out in a powerful, debris-laden flood). If silicate intrusive activity continues, the intrusion may eventually become unroofed and exposed. Alternatively, the removal of volatiles and withdrawal of magma also could cause collapse.

As noted by Keszthelyi *et al.* (2004a), this evolutionary process may have been revealed in a line of paterae in the Chaac-Camaxtli region of Io (Plate 12d). The first stage is represented by Sobo Fluctus. The bright flows at Sobo Fluctus (see Chapter 14) appear to be rich in SO_2, and at the center of the flows there is a small black feature and red deposits, indicative of a small amount of silicate activity and accompanying release of sulphur. The second stage is represented by Grannos and Steropes Paterae. These paterae have relatively shallow slopes and have formed in an area possibly rich in volatiles, where mobilized SO_2 has flowed away along a postulated porous layer, causing surface collapse. Similarly, Ababinili Patera, with bright irregular floors and ridges, is proposed to be the site of high rates of sub-surface volatile flow, which have deformed the patera shape. Ababinili's white and yellow coloration is possibly indicative of sulphur volcanism. The final stage is represented by Ruaumoko Patera, with a dark floor of unroofed silicates and mottling of the surface around the patera caused by ejection of silicate materials through degassing or volatile interaction (Keszthelyi *et al.*, 2004a).

After volcanic activity ceases, a patera will eventually be erased from the surface of Io. If the patera has not already filled with lava, then it will be gradually filled by deposits from nearby plumes and from the escape of volatiles within the patera. Such sapping may undermine the patera walls, hastening its erosion.

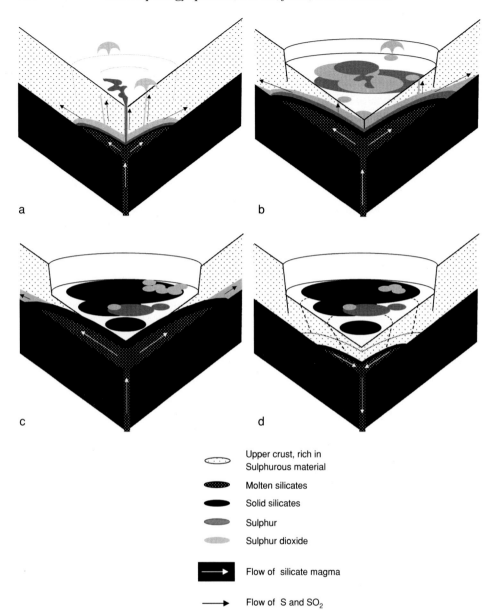

Figure 15.2 Possible mechanisms of patera formation and evolution. Diagrams a–c are based on a model by Keszthelyi *et al.* (2004a). (a) Silicate magma rises through the silicate crust until it encounters layers of less dense, easily mobilized volatiles. (b) Heat transfer drives off volatiles, especially if the layers are permeable. (c) If activity persists, the silicates may become exposed at the surface. Alternatively, magma withdrawal (d) after volatile removal could also cause the surface to collapse (Davies and Wilson, 1988).

Sub-glacial volcanic activity on Earth can form table mountains, with steep sides and flat tops, which are exposed when the overlying ice layer is removed. No such structure has been seen on Io, probably because of Io's lower viscosity lavas and limited lifetimes of eruptions, or because processes of burial outstrip processes of exposure.

15.2 Shield volcanoes

Voyager and *Galileo* imagery have long indicated that, although Io lacks any Olympus Mons-like towering volcanic edifices (Schaber, 1982), several shield-like volcanoes are present, most associated with central paterae (Moore *et al.*, 1986; Schenk *et al.*, 2004). These features exhibit little relief (<3 km) and have very low slopes (0.2°–0.6°). Several shield volcanoes appear to be associated with broad rises of 1 km to 3 km in height; only five shield volcanoes have been identified as having steep flank slopes between 4° and 10°. These steep slopes are restricted to within 20 km to 30 km of the summit, but it appears that on these volcanoes lava flows occur only from vents on the lower flanks where slopes are less than 1° (Schenk *et al.*, 2004).

From these shield volcanoes emanate long, radial flows, of which the best-known examples are Maasaw Patera and Ra Patera (Figures 1.7 and 1.8). The 45 shield-like volcanoes identified by Schenk *et al.* make up 8% of the ≈550 volcanic centers of all types identified on Io (Schenk *et al.*, 2001). Although this survey is incomplete because of uneven coverage, it identifies two morphological classes of shield volcanoes.

15.2.1 Low shield volcanoes

Many shield volcanoes are characterized by the absence of a steep summit edifice or positive topographic structure. Ra Patera, ≈450 km wide, serves as a type example. It has a large, central, volcano-tectonic depression (patera) and a surrounding low shield with mean slopes of ≈0.1° over scales of 10 km or more (Schenk *et al.*, 2004). Other examples of low shield volcanoes include Agni Patera, Daedalus Patera, Aten Patera, Culann Patera, and Maasaw Patera, the latter noted for its nested summit caldera.

A few features exhibit more substantial relief. For example, Ruwa Patera is a shield with dual nested summit paterae, each ≈30 km in diameter, whose edifice is 300 km wide and 2.5 km to 3 km high, with slopes of 0.7° to 1.1°. This slope is somewhat less than generally accepted terrestrial shield volcano slopes, such as Mauna Loa's average slope of 4°. Mbali Patera, however, is centered on a broad

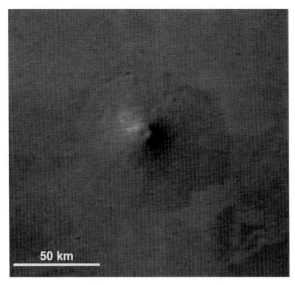

Figure 15.3 A small shield volcano at 246°W, 30°S, identified in *Voyager* data by Moore
et al. (1986). Courtesy of NASA.

100-km-wide, 1.5-km- to 2-km-high topographic dome, with steepest slopes around
3°, closer to Mauna Loa's slope.

15.2.2 Steep-sided shield volcanoes

Very few ionian shield volcanoes do exhibit significant relief and have roughly
symmetrical conical structures with slopes greater than 1°, more in common with
canonical shield volcano morphology (Schenk *et al.*, 2004). *Voyager* imaged only
one such feature (Figure 15.3), an as-yet-unnamed conical shield volcano ≈60 km
across at 246°W, 30°S, with a corrected relief of 1.9 km to 2.0 km, a summit
patera ≈5 km across, and an average slope of ≈4° (Moore *et al.*, 1986). However,
slopes increase toward the summit, where slopes are ≈6°–7°. Five other steep-sided
volcanoes were identified in *Galileo* data. Three are at Zamama (Figure 14.1),
provisionally designated *Zamama A*[1], *B*, and *C*. At *Zamama A*, maximum slopes
are 8°–9°. From the summit region of *Zamama A* emanate flows 20 km to 50 km
long and typically <5 km wide. The profiles for the *Voyager* shield and *Zamama A*
are shown in Figure 15.4a. Figure 15.4b compares shield sizes and shapes on Earth,
Mars, and Io. *Zamama B* is smaller than *Zamama A*, but both exhibit a conical shape
similar to the *Voyager* shield. *Zamama C* is even smaller; it is a small conical mound
15 km across and only ≈250 m high, with slopes of 3°. Nevertheless, it appears to
be a shield volcano (Schenk *et al.*, 2004). The other shield-like features in this class

---

[1] Names in italics are provisional, pending approval by the International Astronomical Union.

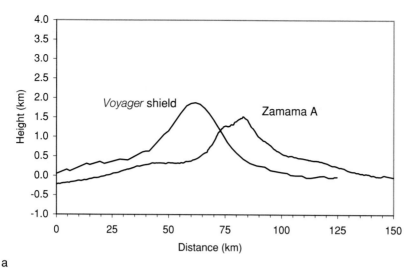

a

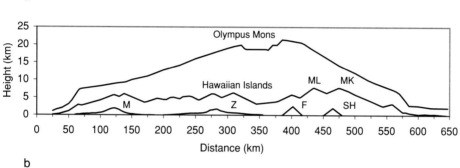

b

Figure 15.4 (a) Topographic profiles of the volcano *Zamama A* (Z), a low shield with increasing slopes toward the summit, and the shield (M) identified by Moore *et al.* (1986) from *Voyager* data (Figure 15.3). The vertical scale is greatly exaggerated. From Schenk *et al.* (2004). (b) Topographic profiles of the ionian volcanoes in (a) (M and Z) compared with shields on Earth and Mars. The largest shield in the Solar System is the martian volcano Olympus Mons. The Hawaiian Islands are shown from the sea floor (the islands break the surface at a level of ≈5 km). Mauna Kea (MK) and Mauna Loa (ML) are identified. Also shown for scale are Mt. St. Helens (SH) and Mt. Fujiyama (F). The vertical scale is greatly exaggerated.

are Tsui Goab Tholus (162°W, 0°N,), a feature about 800 m high with slopes from ≈2° to 4°–6° at the summit, and a nearby small shield-like volcano ≈125 km across, rising ≈300 m above the surrounding plains.

Of course, these slopes are steep only by ionian standards. By terrestrial standards, Mauna Loa and Kilauea have shallow slopes, typifying basaltic shields. Toward the end of the eruptive life of a Hawaiian shield, such as at Mauna Kea, more silicic lava is erupted. This results in modification of the summit caldera and the formation of a silicic cap on the volcano, leading to steeper slopes toward the summit. Terrestrial simple cone volcanoes, fortified with more silicic lava, may

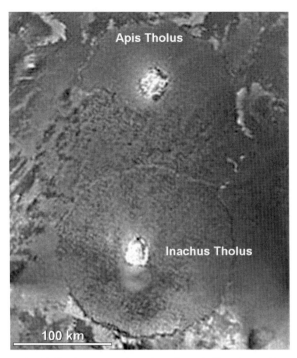

Figure 15.5 The shield-like features Apis and Inachus Tholi, as imaged by *Voyager*. These enigmatic features were to be imaged at high resolution by *Galileo*, but the opportunity was missed when the spacecraft temporarily shut down because of radiation damage. Courtesy of NASA (FDS 1639034).

have slopes approaching 40°, far greater than any slopes proposed for volcanic shields on Io. Ionian volcanoes are among the flattest in the Solar System – yet another indicator of low-viscosity lavas on Io.

15.2.3 Unusual shield volcanoes

Inachus Tholus (349°W, 16°S; see Figure 15.5) is an ≈150-km-diameter, circular, relatively flat mesa with a patera at the center that is ≈25 km wide and ≈2 km deep (Schenk *et al.*, 2004). Although no flows are visible, the circular, elevated morphology and summit depression are reminiscent of a shield volcano or an eroded shield remnant (Schenk *et al.*, 2004). Apis Tholus (Figure 15.5) is a similar but less well-defined feature found just to the north of Inachus Tholus (in planetary nomenclature, a *tholus* is a small dome-like mountain or hill). Both of these tholi were imaged by *Voyager*. High-resolution observations were planned for *Galileo* but never obtained because of the spacecraft entering "safe" mode (i.e., shutting down) as a result of radiation hits (Turtle *et al.*, 2004). Inachus and Apis Tholi remain unusual and enigmatic.

15.2.4 Volcanic shield evolution

An evolutionary sequence for ionian shield volcanoes was proposed by Schenk *et al.* (2004), with a direct bearing on the discussion of patera-forming mechanisms. Initially, effusive volcanism constructs a steep-sided shield. The central region then collapses to form a patera, not necessarily the result of an explosive event. Schenk *et al.* note that steep-sided shield volcanism has not been observed inside paterae on Io, suggesting that resurgent volcanism after patera formation does not include a relief-building phase. Such relief-building is a common occurrence within silicic calderas on Earth (see Francis and Oppenheimer, 2004). The absence of these steep-sided structures may be caused by subsequent changes in magma composition, volume, and rheology, or some other factor such as a change in supply mechanism or a failure in supply. Quasi-concentric mesas are found around some paterae, similar to those seen around some ionian shield volcanoes. These structures may be remnants of pre-existing shield volcanoes (Schenk *et al.*, 2004). If this interpretation is correct, then the sites of shield volcanoes today could be the future locations of major patera-forming eruptions (Schenk *et al.*, 2004).

15.3 Lava flow morphology

Effusive volcanic activity is the result of the eruption of relatively volatile-poor magma, where the gas content is insufficient to disrupt the magma into fragments. Instead, a lava flow forms, spreading across paterae floors and the inter-patera plains of Io, forming flow fields and carving channels into the ice-rich surface. Effusive activity on Io has emplaced huge volumes of lava of various compositions, covering areas on scales ranging from small effusions of lava at the very limits of detection of SSI and NIMS to tens of thousands of square kilometers. Flow emplacement at Pillan, Prometheus, Amirani, and other locations was discussed in Chapters 11, 12, and 14. Only at Pillan was a direct thickness measurement possible (Williams *et al.*, 2001a). At other locations, image resolutions were insufficient to carry out similar analyses, although limits were imposed on maximum flow thicknesses, typically $\approx$50 m or larger. This is a constraint, certainly, but probably not a particularly useful one, given the likelihood of low-viscosity lava forming thin flows, even on shallow slopes. Neither *Voyager* nor *Galileo* obtained stereo images at sufficient resolution to determine lava flow thickness (Keszthelyi *et al.*, 2001a; Schenk *et al.*, 2004).

Flow thicknesses at locations other than Pillan was inferred from estimates of areal coverage rate, volumetric flux, and rate of cooling and were found to be typically on the order of 1 m (Davies *et al.*, 2000a; Davies, 2003b; Davies *et al.*, 2005).

High-resolution SSI images, at resolutions of 6 m pixel^{-1} at best, show flow surfaces that are difficult to interpret, especially as some of these images were garbled (radiation damage again) and had to be carefully, but not completely, de-scrambled (see Turtle *et al.*, 2004).

Constraining slope is an important step toward modeling flow emplacement using Bingham-type relationships (Chapter 7), but efforts are greatly hampered by not knowing active flow geometry, especially channel width and flow advance rate, with the ideal case yielding these quantities *at the same time*. The best that can be done is to calculate the edges of the computational envelope into which solutions fall, using average areal coverage rates and the few quantitative flow parameters known with any accuracy. Lava densities are not known, but a value of 2600 to 2800 kg m^{-3} is a conservative choice, and probably suitable for both ultramafic and mafic lava. In using Bingham rheology, the least constrained parameter is that of flow velocity. For example, average areal coverage rates for Pillan are $\approx$300 m^2 s^{-1}, but flow advance rates are not known. Additionally, areal coverage rates probably varied by orders of magnitude greater and smaller than this average (see Chapter 11).

Schenk *et al.* (2004) examined the Pillan 1997 case and calculated an upper range for viscosity of 10^3–10^5 Pa s (comparable to terrestrial basalt) for lava flows 1 m to 10 m thick, on slopes of 0.1°. Yield strengths were estimated at 10^1–10^2 Pa, lower than for terrestrial basalts. The flow velocity used in these calculations was $\approx$0.01 to 2 m s^{-1}, for a 52-day to 132-day eruption. Viscosity estimates would come down by orders of magnitude to as low as 1 Pa s if flow velocity were higher (Schenk *et al.*, 2004).

In summary, the steep upper slopes of some ionian shields indicate lavas of viscosity and yield strength similar to terrestrial basalts. The analysis of slopes, long runout distances, and estimated rheological properties support arguments for low-silica-content magmas on Io (Schenk *et al.*, 2004).

Final stages of eruptions may involve a transition to higher-silica-content lavas, but activity on Io may start with lower-silica lava than what traditionally is found erupting at terrestrial basaltic shields.

15.4 Lava channels

Other interesting volcanic geomorphologies on Io include lava channels carved deep into the ice-rich surface. In most images at low spatial resolution, fine dark lines are difficult to interpret, being possibly either fault lines or thin channels, although some are clearly visible – as on the slopes of Emakong Patera, for example (Chapter 14).

Schenk and Williams (2004) discovered a prominent lava channel >190 km long and from $\approx$0.5 km to 6 km wide, thought to be associated with the active Tawhaki Patera hot spot (76°W, 3°N). The channel was sinuous, with interior islands. Photoclinometric analysis of the *Galileo* images indicated that this channel, named Tawhaki Vallis, was $\approx$40–65 m deep. Although a structural contribution (e.g., a buried channel) cannot be ruled out, the depth, morphology, and sinuosity of the channel were consistent with erosion by lava. Erosion by flowing silicate over silicate substrate or flowing sulphur over sulphur substrate would likely require eruption durations of at least days to months in order to form a ≈50-m-deep channel, whereas a thick, channeled, flowing silicate over a sulphur substrate or flowing sulphur over a frozen SO_2 substrate may require only hours to days (Schenk and Williams, 2004).

15.5 Mountains and formation mechanisms

Analysis of *Galileo* data shows that the mountains of Io tower up to 18 km above the surrounding plains (Turtle *et al.*, 2001; Schenk and Wilson, 2003). The 149 mountain features identified by Carr *et al.* (1998) and Schenk *et al.* (2001) were categorized by general morphology into mesas (elevated plains with a relatively smooth and flat top surface), plateaus (elevated plains with a rugged surface, without a steep or prominent peak), peaks (isolated promontories), ridges (elevated structures dominated by one or more prominent linear or arcuate rises), massifs (elevated structures with rugged and complex surface morphologies rising to one or more peaks), mixed types (complex structures featuring two or more of the preceding morphologies, classified by the dominant morphology), unclassified (because of low-resolution imagery or inability to discern morphologies), and volcanic mountains, described previously (Schenk *et al.*, 2001).

The formation of mountains is intimately linked to lithosphere structure and internal processes. Io's mountains appear to be tilted blocks of crust, and the impressive heights require a lithosphere at least tens of kilometers thick, not only to support the mountains but also to provide the material that has been uplifted (see, for example, Carr *et al.*, 1979; Masursky *et al.*, 1979; Carr *et al.*, 1998; Schenk and Bulmer, 1998; McKinnon *et al.*, 2001; Schenk *et al.*, 2001; Turtle *et al.*, 2001; Jaeger *et al.*, 2003).

How mountains form is not clearly understood, and several models have been proposed (see summary in Jaeger *et al.* [2003]). Schenk and Bulmer (1998) proposed that rapid resurfacing of Io at rates up to 1 cm year^{-1} (e.g., Johnson *et al.*, 1979) leads to a buildup of compressive stress in the lithosphere as concentric shells of material are continually buried, with the stress leading to thrust-faulting on a scale that could uplift mountains. This model suggests pervasive faulting of the lithosphere as the stresses that build up exceed the strength of the lithosphere below a depth of a few kilometers. These faults provide pathways for magma to reach the surface.

Turtle *et al.* (2001) found that thrust-faulting in a fractured lithosphere would produce parallel mountain ranges rather than isolated peaks, although lithospheric inhomogeneities would alter these features.

McKinnon *et al.* (2001) suggested that a decrease in the rate of volcanic resurfacing would lead to a buildup of heat at the base of the lithosphere on a local, regional, or global scale. The horizontal, compressive, thermal-induced stress would be taken up along lithospheric faults, and displacement along these faults would cause surface normal faulting.

Refining the analysis performed by Schenk *et al.* (2001), Jaeger *et al.* (2003) studied the relationship between mountains and paterae and found that of 92 apparently tectonic mountains, 38 abutted paterae (e.g., Gish Bar and Shamshu Paterae) – a proportion higher than to be expected for a random distribution of both. Further, they proposed that orogenic faults through the lithosphere provided conduits to the surface for magma. Mountain formation could be facilitated by aesthenospheric diapirs impinging on the base of the lithosphere, concentrating compressive stresses in the lithosphere above, and leading to orogenic tectonism.

15.6 Conclusions

In conclusion, the general form of Io's geomorphology indicates a preponderance of high-temperature and relatively low-viscosity silicate lava interacting with an upper crust that is, for at least a few kilometers in depth, relatively or abundantly rich in easily mobilized volatiles.

16

Volcanic plumes

Volcanic plumes are the most impressive manifestations of volcanism. It was fitting, therefore, that the first detection of active volcanism on Io was of a volcanic plume (Figure 1.2, Morabito *et al.*, 1979). In the wake of discoveries made by *Voyager*, the *Galileo* mission greatly advanced understanding of Io's plumes and revealed the importance of their role in the resurfacing of Io.

Large volcanic plumes, which can reach heights of hundreds of kilometers, and their resulting surface deposits are the most visible indicators of ongoing or recent volcanic activity on Io. A variety of mechanisms form these plumes. The largest plumes are the result of explosive volcanic activity with the greatest excess pressures produced by an abundance of volatiles in the magma. Smaller plumes are formed by the interaction of recently erupted lava and surface deposits of sulphur and SO_2. Even smaller plumes are formed by the relatively quiescent escape of volatiles from erupted lava and fumaroles.

16.1 Explosive activity on Io and Earth

Explosive volcanic activity is driven by the release of volcanic gases dissolved in magma at high pressure, the interaction of magma with external volatiles (e.g., with groundwater on Earth or a deposit of sulphur or sulphur dioxide on Io), or a mixture of both processes. (Note: *Volcanic Plumes* by Sparks *et al.* [1997] is highly recommended reading.) The effect of even a small amount of dissolved volatiles is exaggerated on Io, compared to Earth, because there is little discernible atmosphere to inhibit lava eruption and plume expansion. As noted by Kieffer (1982), if the Yellowstone geyser Old Faithful were transported to Io, the change in surface gravity alone would increase its height from 50 m to 300 m. The effect of expansion on the erupting column in the ionian vacuum, however, would increase this height to 38 km. Extensive deposits (sometimes more than 1000 km across) laid down at

Figure 16.1 The explosive eruption of Mt. St. Helens, Washington, on May 18, 1980. A large convective Plinian plume has formed above the vent. Courtesy of U.S. Geological Survey, J. Postman.

Pele, Surt, Tvashtar, and other locations (see Figures 1.3d and 1.4, and Plates 9 and 14) are the result of the eruption of gas-rich magma.

On Earth, large-scale volcanic plumes, caused by explosive volcanism, have produced awesome spectacles and considerable hazards (Figure 16.1). The largest plumes, formed by silicic volcanoes in exceptionally violent eruptions, can reach high altitudes, even to the troposphere ($\approx$20 km) on occasion. A plume from the June, 1991 Mount Pinatubo, Philippines, eruption climbed through an altitude of 13 km in <30 s, faster than a fighter jet (Thompson, 2000).

Huge volumes of lava can erupt in a short period of time once the conduit between magma chamber and surface has opened. Examples of eruption rates from large explosive terrestrial eruptions are shown in Table 16.1. Explosive activity can, for

Table 16.1 *Mean discharge rates (Q_E), column heights, and eruption durations of some large historic eruptions*

Eruption	Q_E (m³ s⁻¹)	Height (km)	Duration (hours)
Agung, 1963	650	10	5
Bezymianny, 1956	230 000	36–45	0.5
El Chicon, 1982	27 300, 18 800, 15 000	22–30	4, 7, 5
Hekla, 1947	17 000–33 000	27.6	2.0
Mt. Pinatubo, June 15, 1991	400 000	34.0	1.3
Mt. St. Helens, May 18, 1980	5 200	16.0	0.4
Nevado del Ruiz, 1985	13 000	24–29	0.3
Quizapu, 1932	60 000	27–30	18
Soufrière, 1902	11 000–15 000	15.5–17	2.5–3.5
Vesuvius, AD 79[a]	60 000	32	11

From Sparks *et al.* (1997).
[a] For the more ash-rich gray plume, as opposed to the white plume that also formed during the eruption.

a time, greatly exceed the largest observed effusive eruption rates (see Chapter 11). Explosive activity on such a scale is rare on Io, where low-viscosity lavas do not trap volatiles to the same extent as more silica-rich, viscous terrestrial lavas, and effusion rates (Q_F) greater than 10^5 m³ s⁻¹ are confined to the largest outburst eruptions (Blaney *et al.*, 1995; Davies, 1996).

The ascent and emplacement of basaltic magmas on Earth and the Moon (and, by extension, on Io) and the pivotal role played by exsolving gas in explosive volcanism were described in a classic research paper by Wilson and Head (1981). Magma ascending from a chamber follows one of three possible eruption scenarios, depending on volatile content (Figure 16.2). Gas-free magmas erupt effusively. At low levels of gas content, exsolution is insufficient to disrupt the magma column. At higher concentrations, gas will completely disrupt the column, fragmenting the magma and resulting in a spray of gas and pyroclasts at the surface. Gas exsolution takes place at shallow depths, <2 km (Wilson and Head, 1981).

Magma degassing that results in explosive activity is a complex process (e.g., Sparks, 1978; Wilson and Head, 1981). From the study of products from terrestrial explosive eruptions, fragments typically resemble a rigid froth, with abundant bubbles resulting from gas coming out of solution as the magma depressurizes (e.g., Sparks *et al.*, 1997). The degassing process is controlled by the magma gas content, the physical properties of the magma, and the rate at which the magma is rising. Explosive eruptions are driven in great part by the tremendous expansion of the exsolved gas phase. For example, a cubic meter of rhyolite at 1173 K (900°C) with 5 wt% of water dissolved at depth will occupy a volume of

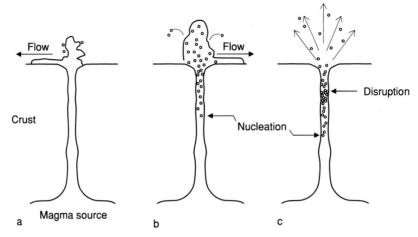

Figure 16.2 Gas exsolution from magma and effect on the magma column. Three cases
are shown: (a) with no gas content, magma eruption is effusive; (b) with a low gas content,
exsolution occurs but is insufficient to disrupt the magma column; and (c) where the gas
volume is so large the magma disrupts into a spray of gas and pyroclasts. From Wilson and
Head (1981).

670 m^3 as a mixture of water vapor and magma at the surface (Sparks *et al.*,
1997).

On Io, as on the Moon, the absence of a significant atmosphere capable of
inhibiting exsolution means that, close to the surface, the pressure acting on the
ascending column abruptly decreases by many orders of magnitude (from ≈1 bar
at a depth of ≈3 m to 10^{-9} bar at the surface), producing a violent and rapid
acceleration. Eruption products emerge at the surface at high speeds because the
effective viscosity of the clasts and gas mixture is several orders of magnitude less
than that of the liquid magma alone (Wilson and Head, 1981).

16.2 Io observations

At least 20 active plumes were detected by *Voyager* and *Galileo* cameras. Plume
observations during the *Voyager* epoch are described by Strom *et al.* (1981). Plume
observations during the *Galileo* epoch and the resulting deposits are described,
discussed, and summarized by Geissler *et al.* (2004a). Plumes consist of dust and gas
components, in varying relative concentrations. Figure 16.3 shows the distribution
of plume deposits as derived by Geissler *et al.* (2004a) from SSI images.

Micron-sized particles in dust-rich plumes (Collins, 1981) scatter light at shorter
wavelengths, giving the plumes a bluish color. Plumes rich in small particles of
dust or condensed particles (snow) are most readily detected against dark sky on
the sunlit limb, particularly when illuminated from behind at high solar incidence

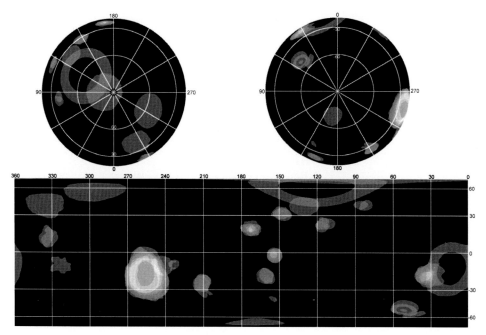

Figure 16.3 The distribution of plume deposits on Io as seen by *Galileo* SSI (Geissler *et al.*, 2004a). The brightness of the deposits shown in the figure is a function of the frequency of deposition, not the brightness of the resulting deposits on Io's surface.

angles (Geissler *et al.*, 2004a). Gas-rich plumes are hard to see at visible wavelengths and are more easily detected at ultraviolet wavelengths.

SSI did not have ultraviolet sensitivity and detected most plumes in daylight images. Additionally, some plumes were detected in eclipse and nighttime observations through auroral emission (Geissler *et al.*, 2004a). Ultraviolet images from *Voyager* and *Cassini* revealed that many plume envelopes were larger than shown in *Voyager* and SSI visible-wavelength images: dust particles detrain from the plume, while the remaining gas streams further outward (Collins, 1981; Strom *et al.*, 1981; Geissler *et al.*, 2004b). As with hot spots, additional plumes were most likely missed due to incomplete spatial coverage and less-than-optimum viewing geometries. *Voyager 2* imaged plume deposits at Aten and Surt (Figure 1.4) from short-lived eruptions that took place between spacecraft encounters. *Galileo* also imaged recently emplaced Pele-like plume deposits without ever seeing the plumes that formed them.

16.3 Plume types

From *Voyager* data, two main plume types were proposed: Pele-type and Prometheus-type plumes (McEwen and Soderblom, 1983). The two plumes seen at

Table 16.2 *Pele- and Prometheus-type plume comparison*

	Pele-type	Prometheus-type
Height (km)	300+	$\approx$100
Diameter of deposit (km)	1000–1400	200
Formation mechanism	Degassing of magma[a]	Lava–ice interaction[b]
Plume makeup	Mostly gas[c]	Mostly dust, particulates, SO$_2$ snow[c]
Main constituents	SO$_2$, S$_2$, silicate	SO$_2$, trace S$_2$
S$_2$/SO$_2$ ratio	0.08–0.3[d]	<0.005[e]
Deposit colors	Red, orange, black	White
Central core[f]	Diffuse	Optically thick
Eruption velocity[f] (km s^{-1})	>1	$\approx$0.5
Bright shock[f]	Seen at Pele	Not seen
Best observed in[f]	Ultraviolet	Visible
Locations	Trend to high latitudes	Trend to equatorial band
Primary silicate eruption style	Explosive	Effusive
Longevity	Generally short-lived (days)[g]	Can persist for years
Local resurfacing rate (cm yr^{-1})[h]	0.06	2.8[i]

[a] McEwen and Soderblom (1983); see also Davies *et al.* (2001).
[b] Kieffer *et al.* (2000); Milazzo *et al.* (2001).
[c] Geissler (2003).
[d] Spencer *et al.* (2000a).
[e] Jessup *et al.* (2004).
[f] Strom *et al.* (1981).
[g] Except Pele, which, although only intermittently detected, has persisted for at least two decades.
[h] Zhang *et al.* (2003), assuming frost density of 1.5 g cm^{-3}.
[i] Assuming SO$_2$ production rate of 10^5 kg s^{-1}, deposited in a ring 75 km wide, of outer radius 200 km. If the ring is 50 km wide, the local resurfacing rate is $\approx$4 cm year^{-1}.

Loki by *Voyager* appeared to be a hybrid class, and a fourth class, pure gas-phase "stealth" plumes, was proposed shortly before *Galileo* arrived at Jupiter (Johnson *et al.*, 1995). *Galileo* data not only greatly reinforced the McEwen and Soderblom classification, but also greatly refined understanding of why there are two main plume types and the mechanisms by which they are generated. Characteristics of Prometheus- and Pele-type plumes are summarized in Table 16.2.

16.3.1 Pele-type plumes

Pele-type plumes (Table 16.3) are the result of explosive volcanism with eruption velocities (estimated for the ballistic case) in excess of 1 km s^{-1} (Strom *et al.*, 1981).

Table 16.3 *Pele-type plumes and deposits*

Name	Longitude W	Latitude	Area (km²)	Range (km)	Radius (km)	Comments
Pele	255°	18°S	1 000 000	720 (max)	540+	Persistent since *Voyager*, sometimes stealthy
Tvashtar	122°	63°N	1 571 700	780	720	Deposits 275 km wide
Dazhbog	300°	55°N	581 900	480	425	Orange, oval ring
Surt	335°	42°N	470 875	390	390	Orange ring. *Voyager* deposit, *Galileo* plume
Aten	311°	48°N	1 500 000	700	–	*Voyager* plume deposit
South of Karei	13°	12°S	932 700	730	690	Giant ring, faded by C22
North Polar Ring	100°	80°N	625 000	446	446	Orange deposit

SSI data from Geissler *et al.* (2004a).
Key:
Red deposits unless otherwise stated.
Range = maximum extent of plume deposits, if present.
Radius = radius of ring, if present.

The plumes are umbrella-shaped and symmetrical. In the final stages of magma ascent, exsolving gases accelerate the magma column to high speeds, generating a spray of gas and fine particulates. The plumes themselves are rich in gas and are difficult to see except in the ultraviolet, which was possible with *Voyager* and *Cassini*. For example, during the I31 Io fly-by (August, 2001), no plume was seen at Tvashtar by SSI, although high concentrations of SO_2 were later detected by the *Galileo* Plasma Subsystem (Frank and Paterson, 2001). *Galileo* SSI detected this type of plume only once in daylight – at Pele during Orbit E4 (November, 1996) (McEwen *et al.*, 1998a). The Pele plume was, however, detected on numerous occasions during the *Galileo* years with the Hubble Space Telescope (Spencer *et al.*, 1997b, 2000a) (Plate 9f). Non-detections within 24 hours of detections suggested considerable and rapid variability in the level of activity or composition (Spencer *et al.*, 1997b). The Pele plume was observed by *Voyager 1* but was not detected by *Voyager 2* (Strom *et al.*, 1981).

The brightness distribution in the Pele plume was unique as observed by *Voyager*. Regardless of filter, brightness within other plumes decreased from core to top monotonically with height, but Pele had a bright core and top (Strom *et al.*, 1981) and, unlike other plumes, showed similar sizes in both visible and ultraviolet images. The bright top is explained by the presence of a shock front near the top of the plume where descending gas and dust meet the rising plume (Strom *et al.*, 1981).

Generally, plumes in this class are large, with heights that can exceed 500 km. They are rich in sulphur and lay down extensive deposits typically 1000 km across (Table 16.3); and they are also associated with intense eruptions of short duration, generally at higher latitudes (Geissler, 2003). The exception to the description just given is, ironically, the plume after which the class is named: Pele. The persistence of the red, sulphur-rich deposits at Pele is a testament to the continuous volcanic activity and degassing of magma supplying the lava lake. It is a mystery as to why, of all of these eruptions, the Pele plume and resulting deposits are unique on Io. As discussed in Chapter 10, there must be a source of volatiles feeding the magma beneath the surface because the amount of sulphur and sulphur dioxide in the plume exceeds the limit of what can be dissolved in the magma.

Pele-type plumes are richer in sulphur than others, with a distinguishing characteristic of S_2/SO_2 ratios of 0.08–0.3, as opposed to a ratio of <0.005 at Prometheus-type plumes (Spencer *et al.*, 2000a; Jessup *et al.*, 2004).

Other plume deposits similar to Pele are emplaced quickly and can fade rapidly. For example, a giant ring deposit from an unnamed patera south of Karei (13°W, 12°S) was emplaced around May 31, 1998 (the time of Orbit E15), but had faded by August 14, 1999 (Orbit C22) (Geissler *et al.*, 2004a). Most of the Pele-type plume deposits are found at latitudes greater than 30°. The exceptions are Pele and the location south of Karei, at latitudes of 18°S and 12°S, respectively. Persistent hot spots are located preferentially at lower latitudes (see Chapter 17) and are indicative of a greater availability of magma to mobilize sulphurous deposits. The plume deposits at Pele and Karei would not be expected to last long without plume reactivation. To be visible, plume deposits need to be at least 10 μm thick (Geissler *et al.*, 2004a). The plume material is cold when emplaced and easily covered by later deposits. A dramatic example of this process at work is illustrated by the overlapping deposits at Pele and Pillan (Plates 9a–c). The red deposits from Pele were overlain by black, silicate-rich deposits during the 1997 Pillan eruption, which left a "black eye" ≈700 km in diameter on Io's surface (Plate 9b). Pele's plume was present at that time because interaction between the two plumes was seen in the September, 1997 image. In that image, a segment of the Pele deposits is covered by and distorted by gas and particle flow from Pillan. Ten months later (July, 1998), deposition of material from Pele and the Kaminari hot spot were well on the way to burying the Pillan deposits (Plate 9c).

16.3.2 Prometheus-type plumes

Prometheus-type plumes are the second major plume class (Table 16.4) identified by McEwen and Soderblom (1983). These plumes are smaller than Pele-type plumes. From 66 measurements, the mean height was determined to be 82 ± 15 km

Table 16.4 *Prometheus-type plumes and deposits*

Name	Longitude W	Latitude	Area (km²)	Range (km)	Radius (km)	Comments
Prometheus	155°	1°S	21 800–125 600	110–200	100–195	Variable activity during *Galileo* epoch
Reiden Patera	236°	13°S	2 800 & 10 300	≈60	≈45	Various small eruptions
Amirani	115°	23°N	123 100	220	200	Various episodes, white rings, dark deposits
Ra	323°	10°S	170 100	260	–	Yellow deposits to north, G1 to E6 (June 27, 1996, to Feb. 20, 1997)
Acala Fluctus	335° & 331°	10°N & 12°N	142 000 & 149 800	250–260	140	Various bright deposits; stealth plume
Masubi	54°	59°S & 44°S	110 100 & 252 900	215–230	180–215	Dark rings and bright deposits
West Zal	84°	41°N	70 500 & 55 400	135–170	170–135	Semi-circular to circular rings
Kanehekili	34°–37°	16°S–19°S	69 600–207 900	106–350	195	Irregular deposits, diffuse deposits; many minor changes
Marduk	208°–212°	22°S–15°S	12 500–177 300	115–270	208–250	Sulphur flow, bright and red deposits, ring deposit; stealthy phase
Zamama	171°–176°	18°N–19°N	13 500–135 800	85–250	140–185	Multiple episodes; bright deposits, red deposits, flows
Culann	161°	20°S	7 100–131 200	75–210	165	Bright and red deposits, some in rings, some minor; stealthy phase
Arinna Fluctus	149°	31°N	84 100 & 120 600	200	200	Ring, white and red deposits
Volund	177°	21°N	≈19 000	≈77	47–77	Observed by *Voyager 1.* Plume 98 km high
Maui	122°	19°N	≈17 000	74	74	Observed by *Voyager 1.* Light and dark diffuse halo
Unnamed	146°	26°S	75 000	200	125	*Voyager*-era Prometheus-like deposit centered on dark flows
Other candidates						
Thor	131°	39°N	279 000	298	298	Bright ring, dark spot
South Polar Ring	170°	70°S	190 100	245	237	Small, gray. Circular deposit

After Geissler *et al.* (2004a).

(Geissler and McMillan, 2006). The largest of the small plumes that was directly observed was 118 km high. These plumes lay down broad circular deposits, up to 100 km to 200 km from the vent (Geissler *et al.*, 2004a).

Some Prometheus-type plumes demonstrate more persistence than the large, gas-rich plumes, although the underlying volcanic activity can still be episodic or sporadic (e.g., Davies *et al.*, 2006c). Additionally, Prometheus-type plumes contain more particulates, especially SO_2 snow, which are laid down in white or yellow annular rings typically 50 km wide (Geissler, 2003). Compared to Pele-type plumes, Prometheus-type plumes are easier to detect and may have central cores that are visually opaque (Strom *et al.*, 1981). This opacity is demonstrated in *Voyager* images in Figure 1.3 and in the *Galileo* image in Plate 10c, where the shadow of the Prometheus plume can be seen. In comparison, Pele had only a faint central column.

The primary class example is the Prometheus plume (Figures 1.3a–c; Plates 6 and 7). The characteristics of the Prometheus hot spot have already been described (Chapter 12). A major *Galileo* discovery was the revelation of how this plume type is generated. The important clue came from SSI images showing that the plume deposits at Prometheus had moved ≈80 km westward since *Voyager* days (McEwen *et al.*, 1998a). The center of the plume deposits – the point of plume origin – was now located at the western end of the lava flows that had been erupted between *Voyager* and *Galileo* (McEwen *et al.*, 1998a). This plume source movement was inconsistent with a fixed plume vent, as originally required by earlier models of the Prometheus plume, where the plume originated in the heating of SO_2 at depth by contact with hot sulphur or silicates (Kieffer, 1982). There is nothing, however, to suggest that this is not happening elsewhere on Io.

Rather than explosive volcanic activity caused by volatiles exsolving from magma, these plumes are instead created on the surface by newly erupted silicate lava flows, mobilizing predominantly SO_2 surface ice beneath (Kieffer *et al.*, 2000). The single vent at Prometheus proposed by Kieffer *et al.* was not seen in high-resolution SSI images, although the vent region may not have been imaged at high enough resolution to identify it. Instead, high-resolution SSI images (Plate 6e) showed new, dark lava flows at the edge of the Prometheus lava flow field and bright jets of SO_2 emanating from those flows (Milazzo *et al.*, 2001), suggesting that multiple SO_2 sources were combining in some manner to form the plume.

Earlier modeling of the Prometheus plume by Strom *et al.* (1981) considered the single-vent and multiple-vent cases. Figure 16.4 shows a computer simulation of Prometheus-type plumes. Assuming ballistic trajectories, an ejection velocity of 0.5 km s^{-1}, and the vent as a 15-km-radius disk with multiple point sources (Strom *et al.*, 1981), the resulting profile is very similar to the observed Prometheus plume as seen by *Voyager* (Figures 1.3a–c). The primary deposition ring (mainly sulphur

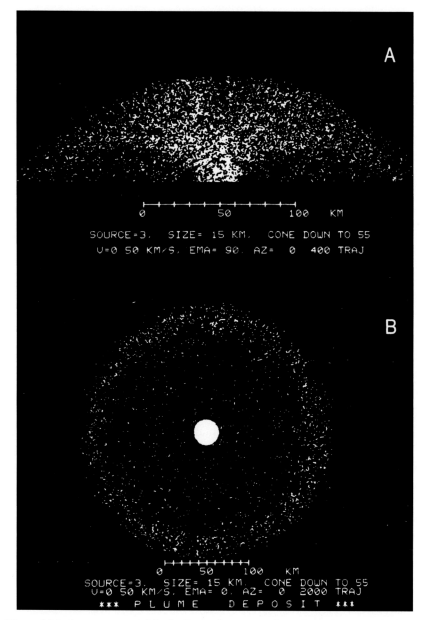

Figure 16.4 A computer model of a Prometheus-type plume from Strom *et al.* (1981). This particular model is of particles emanating from a disk containing multiple sources and following ballistic trajectories. The resulting structure and deposits are consistent with *Voyager* and *Galileo* images of Prometheus (Figure 1.3a–c) and support the plume generation mechanism of Milazzo *et al.* (2001), where SO_2 is being mobilized by multiple contacts with hot lava. Courtesy of Strom *et al.* (1981).

Table 16.5 *Other probable Prometheus-type plumes and deposits*

Name	Longitude W	Latitude	Area (km²)	Range (km)	Comments
Loki 1	305°	19°N	≈200 000	>300	Observed by *Voyager*
Loki 2	301°	17°N	>150 000	>35	Observed by *Voyager*
Pillan	242°	11°S	205 400	254	Explosive activity; deposits mostly silicate pyroclasts. SO_2 mobilized from surface

dioxide) is shown in the lower image of Figure 16.4. This simulation produces an excellent correlation with the plume deposits observed by both *Voyager* and *Galileo* as mapped by Douté *et al.* (2002). The multiple sources correspond to the observed venting of sulphur dioxide from the margins of the Prometheus flows in high-resolution SSI images.

Other instances of lava flows mobilizing surface SO_2 were seen at Zamama, Amirani, and other locations (Table 16.4). Often the plumes were accompanied by small, diffuse red deposits, probably short-chain sulphur degassing from the lava.

Thor is possibly the largest of the Prometheus-type plumes, although it is not a persistent feature like the other plumes in Table 16.4. The eruption at Thor produced a dust plume 500 km high that was observed by *Galileo* during Orbit I31 (August, 2001). Lava flows more than 40 km long were emplaced during this eruption, and its associated white circular ring, nearly 600 km in diameter (Plate 14a), partially buried the plume deposits from the Tvashtar plume (Turtle *et al.*, 2004). It is not known why the Thor plume was so large. The lava may have tapped a particularly rich source of SO_2, such as a thick surface deposit or a near-surface liquid SO_2 reservoir, resulting in a short-lived, but larger-than-usual Prometheus-type plume.

Another unusual member of this class is the South Polar Ring, which laid down a gray ring deposit nearly 500 km across (Turtle *et al.*, 2004). This ring may have contained a larger proportion of silicates than typical Prometheus-type plumes, leading to the gray coloration.

Other probable Prometheus-type plumes and deposits are listed in Table 16.5.

16.3.3 Variants

Loki and Loki Patera

Some plumes do not at first glance fit neatly into the bi-modal classification as described, although closer investigation in the wake of *Galileo* allows more robust identification. The two Loki plumes observed by *Voyager 1* and *2* were classified

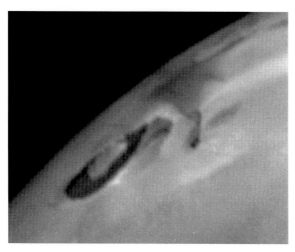

Figure 16.5 The Loki plumes as seen by *Voyager*. From *Voyager* image FDS 16377.52. Courtesy of NASA.

as a hybrid of the Prometheus and Pele types (McEwen and Soderblom, 1983). The irregular Loki plumes originated from the ends of what appeared to be a fissure (Loki) ≈80 km long and 20 km wide, located to the north of Loki Patera (Plate 11). These two plumes lacked the distinctive umbrella shape of Pele and Prometheus, nor was any Pele-like shock front seen. During the *Voyager 1* encounter, the plume from the west end of Loki (*Voyager* Plume 2) was up to 165 km high. In oblique views of Loki (Figure 16.5), Plume 2 was seen to have a dark central column, evidence of a plume rich in dust or snow. Enhancement of the image revealed an optically thin outer cloud of fine (≈1 μm) particles (Collins, 1981; Strom *et al.*, 1981). It was in this plume that IRIS detected gaseous SO_2 (Pearl *et al.*, 1979b). The plume emanating from the eastern end of Loki (*Voyager* Plume 9) was smaller, about 16 km to 20 km high during the *Voyager 1* encounter. Diffuse deposits in the region around Loki suggest that Loki Patera is also a source of volatiles, with SO_2 degassing from magma in the lava lake. No substantial plume forms, which is consistent with the quiescent style of resurfacing in the patera (Davies, 2003a).

The plumes demonstrated considerable changes between *Voyager* encounters. In *Voyager 2* clear filter images, the combined Loki plume envelope was ≈338 ± 68 km high (Strom *et al.*, 1981) and the core of Plume 9 had brightened considerably (Strom *et al.*, 1981).

No plumes at Loki were detected during the *Galileo* mission, and the Loki region appeared more like it did during the *Voyager 1* encounter than during that of *Voyager 2*. It appears that the Loki plumes were dust-rich Prometheus-class plumes created by lava flows.

Pillan

The 1997 eruption of Pillan (see Chapter 11) produced a large dark deposit rich in silicate pyroclastic material. A plume >140 km high was observed by *Galileo* SSI from June, 1997 through October, 1997 (Plate 10c), as well as by the Hubble Space Telescope. Despite a huge outpouring of lava at a volumetric rate at least two orders of magnitude larger than that estimated for Pele, there was no evidence of similar amounts of sulphur degassing from the Pillan lavas. No large red or orange deposit formed, such as those seen at sites of other large short-lived eruptions (such as at Tvashtar). Instead, the dark, silicate-rich deposits were surrounded by a thin white aureole (Plate 9b), most likely rich in SO_2, originating from volatiles exsolving from the lava and from frozen surface deposits mobilized by the advancing flows. The presence of "pseudo-craters" or "rootless cones," another sign of the interactions between hot lava and Io's volatile-rich surface, strengthens this view. The bright, dust-rich plume was best observed on the limb of Io in June, 1997 (Orbit C9) by SSI, at the height of the 1997 eruption. The plume deposit may have been so black because of the inclusion of an unusually large amount of silicate material, possibly due to vigorous lava–ice interactions and thermal exchange as a result of the enormous effusion rate.

Stealth plumes

Stealth plumes (Johnson *et al.*, 1995), a subset of the Pele class inasmuch as they originate in a reservoir at some depth, are the result of eruptions that proceed purely in the gas phase. The plumes are formed through a high-entropy interaction between hot silicates and a reservoir of SO_2, characterized by Kieffer (1982) as "SO_2 superheated vapor in contact with or degassing from a hypothetical silicate melt at 1.5 km depth. Pressure = 40 bar; Temperature = 1400 K."

   The resulting plumes are of low enough concentrations to be hard to detect at visible wavelengths (hence "stealth"). However, the plumes can be detected in ultraviolet images as well as by auroral glow. It may be that Pele is sometimes a stealth plume, which would explain the difficulty in observing a plume that is known to be present because of its interactions with other plumes and the persistence of its deposits.

   Other likely stealth plumes include those seen at Acala Fluctus. At Acala, a plume that laid down a white ring deposit was seen in eclipse by auroral glow during April to November, 1997 (*Galileo* Orbits G7 to E11) and May, 1998 (Orbit E15), but not in daylight images during June, 1996 (Orbit G1) and September, 1997 (Orbit C10) (McEwen *et al.*, 1998a; Geissler *et al.*, 1999).

   Plumes at Culann and Marduk also went through stealthy phases, when no dust load was seen; the opacity of Marduk was an order of magnitude less than that of a similar-sized plume at Zamama (Geissler and McMillan, 2006).

Table 16.6 *Seepages and sapping (selection)*

Name	Longitude W	Latitude	Area (km^2)	Spacecraft	Comments
Zal Montes	81°	40°N	40 250	*Galileo*	SO$_2$ seepage
Haemus Montes	53°	69°S	69 700	*Galileo*	SO$_2$ seepage
Dorian Montes	204°	32°S	15 325	*Galileo*	SO$_2$ seepage
N. of Pillan	234°	8°S	1 775–47 725	*Galileo*	From Kaminari, or deposit remobilized by Pillan lavas
Nemea Planum	266°	71°S	8 700	*Voyager*	Image ID 141J1+000
Iopolis Planum	330°	37°S	1 780	*Voyager*	Image ID 139J1+000

Seepages and sapping events

The final class of events are hardly plumes at all but rather are deposits of SO$_2$ that have seeped out onto the surface along the margins of mountains (Table 16.6). These minor "plumes" are more akin to terrestrial hydrothermal springs than Prometheus- and Pele-type plumes. On Io, these events take place at the edges of mountain blocks, with SO$_2$ seeping up along faults, and are perhaps triggered by proximity to intrusive or extrusive silicate activity. McCauley *et al.* (1979) and Moore *et al.* (2001) described the escape of SO$_2$ under artesian pressure. This process preferentially takes place in regions of crustal fracturing, wholly in keeping with the model of Io's mountains as tilted crustal blocks (Schenk and Bulmer, 1998; Turtle *et al.*, 2001). SO$_2$ driven toward the surface would partially crystallize and partially vaporize at the surface, erupting as a mixture of gas and ice crystals at velocities of up to 350 m s^{-1} to form high-albedo surface deposits. After condensation of the mixture, the sulphur dioxide snow would then be deposited typically within tens of kilometers from the source. After depletion of the underground reservoir, withdrawal of support from beneath the solid crust would cause collapse of the overlying terrain, thus generating irregular surfaces. The observed presence of debris down the scarps at the bases of the mountains supports this scenario (Turtle *et al.*, 2001).

16.4 Plume models

Modeling plume dynamics and the resulting emplacement of fallout is not trivial. The complexities are described by Kieffer (1982), who found a wide range of possible plume types erupting from sub-surface reservoirs, the result of different interactions among hot silicates, sulphur, and SO$_2$.

A model of plume dynamics must consider the absence of a thick atmosphere (as found on Earth, Venus, and Mars), the resulting effects of rapid pressure decrease in the vent, the change of state of both gas and liquid magma during violent decompression, the effect on plume dynamics of the formation of shocks in the vent, a 25% decrease in gravity at the top of a 300-km-high plume, and a surface curvature of 38° at the plume base, if the deposits extend 600 km from the source.

Methods for modeling plumes under these conditions are described by Zhang *et al.* (2003), who modeled Prometheus- and Pele-type plumes under both day and night conditions using Monte Carlo simulations. Zhang *et al.* found that, for Pele-type plumes, a large (≈17-km-diameter) circular conduit, with gas exiting at Mach 3, produced a 300-km-high plume, deposits 400 km to 600 km in radius, and also the canopy-shaped shock reported by Strom *et al.* (1981). In short, the model successfully reproduced the main morphological features observed by *Voyager* and *Galileo*. In the model, the bright shock is caused by the gravity field, not atmospheric back-pressure. Heating of gas and frozen particles takes place due to compression against the shock, after which material flows outward and downward toward the surface, expanding and cooling along the way, to condense on the cold surface as frost (Zhang *et al.*, 2003). The momentum of the gas flow causes further flow outward (Plates 14b–c), leading to additional deposition, possibly explaining the concentric rings seen at Pele by *Voyager* (Figures 1.3d, e).

Prometheus-type plumes were also modeled by Zhang *et al.*, producing lower ejection velocities (≈500 m s^{-1}), and a vent temperature of ≈300 K, lower than that of Pele-type plumes (≈650 K). The model produced a multiple-bounce shock structure around Prometheus-type plumes that compares well with observations of both plumes and deposits.

Additional modeling is needed to better understand vent conditions, especially at Pele. In particular, the detailed relationship between the plume formation mechanism and the postulated lava lake is not understood at all, assuming the plume originates from the lake itself and not from an adjacent vent.

16.5 Summary

In their meticulous examination of plume activity on Io, Geissler *et al.* (2004a) concluded that plumes are responsible for most observed changes on the surface of Io. They further concluded that, although most heat was lost through volcanic centers, most emplacement of lava was areally confined within paterae or took place on top of older lava flows and contributed little on a global scale to resurfacing (no more than 10%; see Davies *et al.* [2000a]). Two main classes of plumes were present: small, dusty, persistent, SO_2-rich plumes and large, gas-rich, sulphur-rich, sporadic plumes. Smaller seepages of SO_2 were also identified. Repeated eruptions of SO_2 are required to lay down deposits of sufficient thickness to be visible and represent a significant contribution to Io's resurfacing.

17

Hot spots

Using *Galileo* data, volcanic activity at individual volcanic centers was monitored and quantified on timescales of weeks to months, for a period of seven years. This monitoring was a huge step beyond the two snapshots of activity, four months apart, obtained by the *Voyager* spacecraft. *Galileo*'s most exciting discovery was the large number of active volcanoes on Io, of which at least 50 had temperatures >1000 K (identified in Appendix 1). Having identified the locations of these high-temperature lavas, similar locations (e.g., dark-floored paterae; McEwen *et al.*, 1985) were inferred also to be the result of silicate volcanism (see Chapter 9). By the end of the *Galileo* mission, more than 160 active hot spots had been identified from *Voyager*, *Galileo*, and Earth-based telescope data (see lists in Lopes *et al.* [2001, 2004]). In NIMS regional observations around Prometheus, where spatial resolution improved by a factor of 10 (Plate 7), the number of detected hot spots increased by a factor of ≈3 (Lopes *et al.*, 2004), confirming earlier predictions that these small hot spots were present (Blaney *et al.*, 1998, 2000). PPR detected many more low-temperature thermal sources not detectable by SSI or NIMS (Rathbun *et al.*, 2004). Extrapolating to a global scale suggested as many as 300 large and small active hot spots on Io (Lopes *et al.*, 2001), although the region covered by the aforementioned NIMS observation has a particularly high density of volcanic features (Schenk *et al.*, 2001). Appendix 1 contains a list of all detected hot spots, and their locations are plotted in Plate 13b.

The application of models of thermal emission and eruption processes (described in Chapters 6 and 7) quantified volumetric and thermal fluxes, and the charting of hot-spot locations has helped constrain heat supply mechanisms from the interior.

How to classify hot spots? Chapters 10 through 14 looked at individual sites and classified activity using geomorphology and effusive or explosive style, based on surface imagery, thermal emission spectra, and spectral evolution. Eruptions fell naturally into several classes: insulated lava flows, lava lakes, lava fountains and sheet flows, and activity within paterae – which could be lava lakes, ponded flows, or lava flows. The magnitude of thermal emission, the volume of material erupted, the flux at any given wavelength, and ratios of thermal emission at different wavelengths all have been used to classify eruption class.

   This chapter looks at the distribution, size, style, and number of hot spots on Io and assesses their role in transporting heat to the surface of Io.

17.1 Variability and style of activity

Quantification of thermal output, areal coverage rate, and effusion and eruption rates allows comparison of eruptions past and present on Io, Earth, and elsewhere. Io's volcanoes exhibit a broad range of activity across timescales, areal scales, style of eruption and lava emplacement, persistence of activity, and power output. Some volcanoes proved to be always active (Prometheus, Pele, and Loki Patera, for example) and thus were designated "persistent" (Lopes-Gautier *et al.*, 1999); others were "sporadic," only occasionally at detectable levels of activity (e.g., Shamash, Monan, and Gish Bar Paterae [see Appendix 1]); whereas others were seen only once (e.g., Altjirra Patera), as far as researchers could tell from uneven spatial and temporal coverage (e.g., Lopes-Gautier *et al.*, 1999; Lopes *et al.*, 2001). Such variability in activity implied different mechanisms of magma supply and eruption style in different locations, a situation very much like that on Earth.

17.1.1 Thermal signature

Visible imagery and the evolution of infrared thermal emission spectra allowed identification of eruption style (Chapters 8, 10–14). The differences in model-derived thermal emission spectra for the different styles of eruption can be seen in Figure 17.1. As described in previous chapters, Pele exhibited the thermal signature of an active, boiling lava lake (a preponderance of short-wavelength thermal emission without flow emplacement) that is (so far) unparalleled on Io; Pillan, that of a lava-fountain episode and the emplacement of flows that subsequently cooled as predicted using silicate cooling models; and Prometheus, Zamama, and Amirani, wandering hot spots, mostly within an existing flow field, with a thermal emission spectrum that did not change shape but changed intensity as individual eruption episodes waxed and waned (Ennis and Davies, 2005). Between 1988 and 2001, Loki Patera underwent *periodic* activity, with regular brightening episodes followed by a cooling-off period, on a cycle timescale of roughly 540 days (Rathbun *et al.*, 2002). This periodicity was something that had not been seen anywhere else on Io – or on Earth for that matter – and modeling of the data concluded that this was the thermal signature of a huge, quiescently overturning lava lake: a "magma sea" (Matson *et al.*, 2006b).

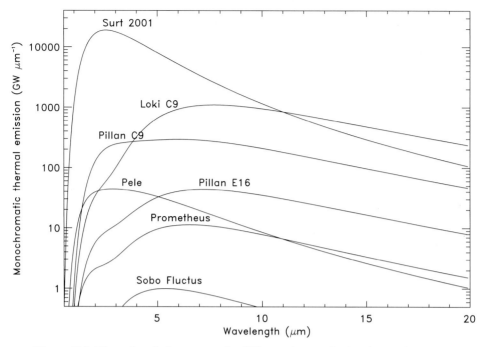

Figure 17.1 Thermal emission spectra for different classes of volcanic eruption on Io. Eruptions range from the small hot spot at Sobo Fluctus to the massive Surt thermal outburst (see Chapter 14). Thermal emission spans over four orders of magnitude. The possibility exists that even smaller hot spots below current levels of detection are present.

Pele and Loki Patera exhibit thermal and temporal characteristics consistent with their both being lava lakes but on greatly different scales (Pele of order 100 km²; Loki Patera, >21 000 km²) and with different mechanisms replacing the crust. Only by charting the temporal variability in thermal emission at multiple wavelengths could these styles of activity be constrained.

Other hot spots were more difficult to classify. Gish Bar Paterae and Tupan Patera may be lava lakes or may be resurfaced in part by lava flows. The Tvashtar Paterae eruption began with explosive activity and the formation of lava fountains that emplaced flows. Tvashtar subsequently exhibited variable levels of activity in different locations within the paterae, with temperature distributions that are difficult to explain in simple terms of a lava lake or lava flows (Milazzo *et al.*, 2005). Volcanism at Tvashtar after early 2000 was probably a mixture of styles of lava effusion, lakes or ponds, and flows.

17.1.2 Classification and comparison using 2:5-μm intensity ratio

Eruption spectral signature can be expressed by the ratio of 2-μm to 5-μm thermal emission (Davies and Keszthelyi, 2005). Figure 17.2 shows the 2-μm and 5-μm

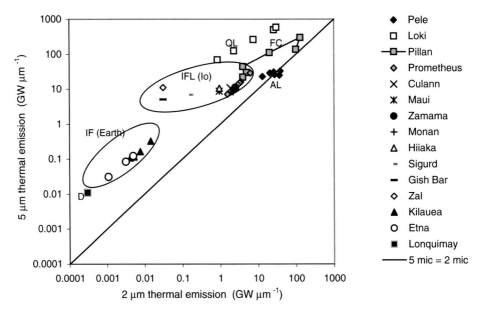

Figure 17.2 Plot of 2-μm vs. 5-μm monochromatic thermal emission (flux) for different volcanoes on Io and Earth. The terrestrial data were derived from *Landsat* Thematic Mapper (TM) data supplied by A. J. Harris. Pele generates a 2-μm flux greater than that at 5 μm, an indication of the continual exposure of incandescent lava. Only Pillan, at the height of the 1997 eruption, approaches this relative level of 2-μm emission. Such data, at only two wavelengths, can be used to constrain effusion style. Key: FC = fountains and channeled flows; AL = active lava lake; IFL = insulated flows, quiescent lakes and ponds, Io; QL = quiescent lava lake (Loki Patera); IF = insulated flows, Earth; D = lava dome (Earth only). After Davies and Keszthelyi (2005).

fluxes for a small number of different volcano types on Earth and Io, which incidentally illustrates once again the greater scale of activity on Io when compared with Earth.

Individual eruptions can also be compared using mass and flux densities. Mass density is the mass of lava being erupted every second per unit active area; flux density is the total thermal emission per unit area. Different eruption styles exhibit different mass and flux densities. Style types are shown in Table 17.1.

Table 17.2 shows eruptions ranked by flux density: note how similar the flux densities are for Pele and Kupaianaha during rapid overturn. Not surprisingly, the highest flux densities are generated by disrupted lava lakes and lava-fountain events. Despite differences in environment and possibly differences in magma composition, the mass and flux densities are similar for similar styles of activity on Io and Earth. A single square meter of an active pahoehoe flow on Io would be much like a square meter of an active pahoehoe flow on Earth. Heat loss is dominated by radiation, and the different environment temperatures (≈288 K on Earth, ≈100 K on Io) have little effect.

Table 17.1 *Thermal characteristics of different active volcanoes on Io and Earth*

Ionian volcanoes			Terrestrial volcanoes		
Eruptive Style	Volcano	Thermal emission characteristics	Eruptive Style	Volcano	Thermal emission characteristics
Lava lake	*Pele*	Steady, fixed emission 210–280 GW ($\approx$15 kW m^{-2})	Lava lake	*Mt. Erebus, Antarctica*	Steady 0.05–0.1 GW (1–6 kW m^{-2})
Fire fountains and open channel flow; transition to insulated sheet flows	*Pillan 1997*	Dropped from 3600 GW–350 GW as hottest area decreased from 30–0.5 km^2 ($\approx$32 kW m^{-2})	Open channel and tube-fed a'a flows	*Mt. Etna, Italy, 1992*	Peak at 7.3 GW, dropped to 0.5 GW (1.2–3.5 kW m^{-2}). Hot area decreased from 3.7 to 0.1 km^2
Insulated flow field	*Prometheus*	Variable thermal emission, typically $\approx$100 GW ($\approx$2 kW m^{-2}). Moved 80 km in 20 years	Tube-fed pahoehoe	*Kilauea, Hawai'i, USA, 1991*	Thermal emission typically 1.2–3.3 GW (2.9–3.1 kW m^{-2})
Insulated flow field	*Amirani*	Variable thermal emission, typically 300 GW ($\approx$2 kW m^{-2}). Source shifted $\approx$100 km	Lava curtains and open sheet flows	*Krafla, Iceland, 1984*	26–50 GW ($\approx$0.5–1.5 kW m^{-2})
Large active patera	*Loki Patera*	Periodic activity for many years: regular brightenings. Average 13 TW ($\approx$0.3 kW m^{-2}), from 2.1 × 10^4 km^2	Large active lava lake (>10 km diameter)	*No terrestrial equivalent*	–
Lava dome	*No ionian equivalent*	–	Lava dome	*Lonquimay, Chile, 1989*	0.1 GW (3.2 kW m^{-2})

From Davies and Keszthelyi (2007).

Table 17.2 *Thermal emission, mass eruption rates, and flux and mass densities*

Volcano	Date	Thermal emission F_{tot} (GW)	Mass flux[a] M (kg s^{-1})	Flux density (kW m^{-2})	Mass density (kg s^{-1} m^{-2})
Mt. Erebus	Jan., 1985	0.016–0.103	38–76	89–572	0.2–0.4
Pu'u 'O'o lava lake	July, 1991	0.508–0.674	1553–2079	128–170	0.4–0.5
Nyiragongo	Jan., 1972	5.150	8162–10350	114	0.2
Mt. Erebus	Jan., 1989	0.013–0.108	30–69	43–359	0.1–0.2
Erta 'Ale	Jan., 1973	4.728	14902–18204	50	0.2
Pillan 1 (G8)	May 7, 1997	1010	3.21×10^6	32	0.103
Kupaianaha Stage 1[b]	Oct., 1987/Jan., 1988	0.018	41–57	$\approx$22	0.018–0.025
Pele, NIMS (E16)	July 20, 1998	280	644000–889000	17	0.04–0.05
Nyiragongo	Aug., 1987	0.0002–0.0006	1–2	15–20	0.03–0.1
Erta 'Ale	Jan., 1986	0.018–0.031	44–104	6–11	0.02–0.04
Pele, IRIS[c]	Mar., 1979	1160	$2.67 \times 10^6 - 3.68 \times 10^6$	10	0.02–0.03
Pele, IRIS[d]	Mar., 1979	1333	$3.07 \times 10^6 - 4.23 \times 10^6$	6.0	0.01–0.02
Kupaianaha Stage 2[b]	Oct., 1987/Jan., 1988	0.014	32–44	5.3	0.014–0.019
Kupaianaha Stage 3[b]	Oct., 1987/Jan., 1988	0.011	25–35	4.9	0.011–0.015
Pillan 3 (C10)	Sept. 18, 1997	1130	3.58×10^6	3.0	0.011
Culann (G1)	June 28, 1996	139	270400	2.9	0.008
Kilauea	July 23, 1991	1.14	13000	3.3	0.03
Kilauea	Oct. 17, 1991	3.3	36000	2.9	0.03
Amirani (G1)	June 28, 1996	327	637000	2.6	0.019
Tupan (G1)	June 28, 1996	79	153400	2.5	0.005

Zamama (G1)	June 28, 1996	126	244 400	2.4	0.005
Monan (G1)	June 28, 1996	85	166 400	2.2	0.003
Pillan 2 (C9)	June 28, 1997	3610	1.15×10^7	2.0	0.006
Prometheus (G1)	June 28, 1996	118	228 800	2.0	0.003
Maui (G1)	June 28, 1996	96	184 600	1.9	0.001
Pillan 6 (C20)	May 2, 1999	351	1.11×10^6	2.0	0.005
Pillan 4 (E15)	May 31, 1998	817	2.59×10^6	1.0	0.004
Pillan 5 (E16)	July 20, 1998	507	1.61×10^6	1.0	0.005
Loki Patera (C9)[e]	June 25, 1997	13 084	3.23×10^7	0.62	0.0015
Loki hot spot[f]	Mar., 1979	12 600	$2.9 \times 10^7 - 4 \times 10^7$	0.42	0.0009–0.001
Loki Patera (G7)[e]	April 5, 1997	6053	1.49×10^7	0.29	0.0007
Loki Patera (G29)[e]	Dec. 28, 2000	5689	1.01×10^7	0.27	0.0005
Loki Patera (E16)[e]	July 21, 1999	3295	8.13×10^6	0.16	0.0004
Loki Patera (C22)[e]	Aug. 14, 1999	1943	4.80×10^6	0.09	0.0002

Note: Names in italics are terrestrial lava lakes.
From Davies and Keszthelyi (2007).

[a] Mass flux calculated using Harris *et al.* (1999a), Eqn. 18. See also Glaze *et al.* (1989).

[b] Flynn *et al.* (1993).

[c] Pearl and Sinton (1982). Component of 113 km^2 at 654 K used. The component of 20 106 km^2 at 175 K would not be detected by NIMS. NIMS has a low-temperature detection limit of 180 K, as long as the source fills the NIMS field of view (Smythe *et al.*, 1995). The 654 K and 175 K components yield the best fit to the IRIS Pele data.

[d] Pearl and Sinton (1982). A two-component fit to IRIS data yields 29.7 km^2 at 854 K and 181.5 km^2 at 454 K.

[e] *Galileo* NIMS data, Loki Patera low-albedo area = 2.1×10^{10} m^2 (Matson *et al.*, 2006b).

[f] *Voyager* IRIS data, using Loki Patera low-albedo area of 2.1×10^{10} m^2 (excluding "island").

Table 17.3 *Total thermal emission and 3.5-µm flux as a function of eruption class*

Dates of observations	Location	Total peak output (GW)	3.5-µm flux (GW µm⁻¹)	Eruption class
Feb. 20, 2001	Surt	74 000	15 000[a]	Outburst
Jan. 9, 1990	Loki vicinity	≈10 000	21 000[b]	Outburst
June 28, 1997 (C9)	Loki Patera	≈16 000	2 900	Peak lake overturn[c]
June 28, 1997 (C9)	Pillan	3 610	260	Lava fountains[d]
July 21, 1998 (C16)	Pillan	510	62	Cooling flows[d]
May 5, 1999 (C20)	Pillan	350	10	Cooler flows[d]
1996–2000 (G2-I27)	Pele	280	Typically 30	Active lava lake[d]
June 28, 1996 (G1)	Prometheus	118	5	Insulated flows[d]
June 28, 1996 (G1)	Average of 14 volcanoes	≈120	5 ± 2.6[e]	Insulated flows[d]
June 28, 1996 (G1)	Total of 14 volcanoes	≈1 660	70	Insulated flows[d]
Oct. 10, 1999 (I24)	Sobo Fluctus	≈8	Typically <0.1−0.5	Small hot spots[f]

[a] Marchis *et al.* (2002). The Surt 2001 eruption nearly matched Io's total thermal output. Data from Keck Observatory, Mauna Kea, Hawai'i.
[b] Veeder *et al.* (1994). Flux derived by Davies (1996), based on model fit to 4.8-µm and 8.7-µm data. Data from IRTF, Mauna Kea, Hawai'i.
[c] Davies *et al.* (2000b), from fit to *Galileo* NIMS C9INCHEMIS06 data at height of brightening of Loki Patera. See Matson *et al.* (2006b) for modeling of Loki Patera's thermal emission cycle.
[d] Davies *et al.* (2001). *Galileo* NIMS data.
[e] Davies *et al.* (2000a). *Galileo* NIMS data.
[f] Lopes *et al.* (2001). *Galileo* NIMS data.

17.2 Thermal emission comparisons

Table 17.3 shows total thermal emission and thermal emission at 3.5 µm for the volcano type examples shown in Figure 17.1. The largest eruptions by magnitude of thermal emission and implied rate of areal coverage are relatively rare "outburst" events, as seen at Pillan, Gish Bar Patera, Tvashtar Paterae, Surt, and Thor. Thermal emission quickly waxes in hours or days and then wanes as effusion rate drops and lava fountaining ceases. In the case of the Pillan 1997 eruption, the cooling flows from an eruption that started with lava fountains were detectable for years (Davies *et al.*, 2001). As with volcanic eruptions on Earth, a rule of thumb is: the larger the volumetric effusion rate, the faster the magma supply is depleted and the shorter the duration of activity. Additionally, the largest eruptions are also the least frequent. The massive Surt outburst of February, 2001 was the most powerful thermal event ever seen on Io (Marchis *et al.*, 2002) and very short-lived (about a few days). Smaller lava-fountain events on the scale of Pillan 1997 are more common, perhaps

Table 17.4 *Effusion rate* (Q_F) *comparisons: Io and Earth, assuming basalt composition*

Eruption (style)	Effusion rate ($m^3\ s^{-1}$)	References
Io		
Jan. 1990 (outburst)	10^5–10^6	Blaney *et al.* (1995); Davies (1996)
Pillan (lava fountains, open channel flow)	$\geq 10^4$ (max.)	Davies *et al.* (2001, 2006d); Williams *et al.* (2001a)
Pele (lava lake)	≈250–350	Carr (1986); Davies *et al.* (2001)
Amirani (insulated flow field)	≈58	Davies (2003b)
Prometheus (insulated flow field)	≈35 to >100	Davies (2003b); Davies *et al.* (2006c)
14 volcanoes seen by NIMS G1	≈9–65	Davies (2003b)
Earth		
Laki, Iceland (open channel flow)	$>10^4$ (max.)	Thordarson and Self (1993)
Mauna Loa, Hawai'i 1984 (channeled a'a flows)	10–1000	Malin (1980)
Kilauea Iki, Hawai'i (lava fountains)	100	Malin (1980)
Kupaianaha, Hawai'i (lava lake)	≈5	Flynn *et al.* (1993)
Kilauea, Hawai'i (pahoehoe flow field)	Typically ≈2–5	Malin (1980); Heliker *et al.* (1998); Kauahikaua *et al.* (1998)

as many as 20 per year (Spencer and Schneider, 1996). The next most common classes are the persistent hot spots (sites of gas-poor magma and effusive activity) and then the widespread smallest hot spots. Loki Patera and Pele are in a class of their own on the basis of longevity, the persistence of the Pele plume, and the magnitude of Loki Patera's thermal emission. One striking comparison in Table 17.3 is how the total thermal emission from 14 volcanoes identified in the nighttime portion of NIMS observation G1INNSPEC01 (Davies *et al.*, 2000a; Davies, 2003b), 1660 GW, is greatly exceeded by the total thermal emission from Pillan at the peak of activity in June, 1997, and how the peak of the Pillan eruption is in turn dwarfed by the thermal output of the Surt 2001 eruption.

17.3 Effusion rates

Effusion rate is a parameter that can be used to quantitatively compare eruptions on Io with past and present eruptions on Earth. Volumetric effusion rates have been calculated from the total thermal emission from each hot spot for each class of eruption, using appropriate equations based on derivation of effusion style (see Chapters 7 and 8). Effusion rates are shown in Table 17.4, assuming a mafic (e.g., basalt) composition. As already discussed, although the exact composition of ionian lavas is not known, the general morphology of lava flows seen in medium- to high-resolution observations, coupled with the observed cooling rate of flows, suggest fluid, low-viscosity, mafic lavas.

Again, outbursts dominate. The 1990 outburst in the Loki region may have yielded peak effusion rates on the order of 10^5 to 10^6 m^3 s^{-1} (Blaney *et al.*, 1995; Davies, 1996). The peak Pillan 1997 effusion rate probably exceeded 10^4 m^3 s^{-1} (Davies *et al.*, 2006d), akin to volumetric rates proposed for terrestrial flood basalts. The extensive flows at Lei-Kung Fluctus on Io may have been emplaced in a similar fashion, at similar effusion rates. The volume of magma circulating through the Pele lava lake, however, is orders of magnitude larger than that at Kupaianaha and other terrestrial lava lakes (see Table 17.2). The quiescent style of lava eruption at Kilauea, Hawai'i, is exceeded by more than an order of magnitude by activity at Prometheus and other mid-sized Io eruptions. The small eruptions identified in NIMS data are closer in effusion rate to activity at Kilauea in 2005–2006.

These comparisons reveal the true scale of activity of Io's volcanism, with activity on a much larger areal and volumetric scale than contemporary analogous eruptions on Earth.

17.4 Distribution of hot spots

On Earth, volcanoes are found in three distinct environments: at ocean ridges, along the margins of continents, and at isolated hot spots. The surface of the Earth is divided into ocean and continental plates, with most volcanic activity confined to the plate margins. At constructive margins, new basaltic ocean plate forms along mid-ocean ridges (e.g., Mid-Atlantic Ridge, East Pacific Rise). At destructive margins, denser oceanic plates descend beneath continental plates (e.g., along the west coast of South America). The descending ocean plate carries with it sediments rich in water and other volatiles, which aid in the melting of the plate and the continental material above. Plotting the locations of volcanic activity over the past 10 000 years clearly defines the margins of tectonic plates.

Volcanic activity also takes place at intra-plate hot spots, away from plate margins. Although in some locations there is evidence of magma transport along transverse faults, most hot spots are isolated above the heads of proposed mantle plumes. Canonical thought utilizes the presence of stationary plumes beneath moving plates to explain the Hawaiian–Emperor island chain and a series of large-volume eruptions across North America, from Oregon and Washington (Columbia River Flood Basalts) to Idaho (Snake River) to its current position under Wyoming (Yellowstone).

The locations of active hot spots, plume origin sites, and paterae on Io are plotted in Plate 13b. There is no distribution indicative of plate tectonics, no ionian equivalent of the Pacific "Ring of Fire." The absence of plate tectonics suggests that

Io's surface is immobile and that there are no large-scale, highly vigorous convective processes capable of initiating such large-scale lateral motion. The lithosphere may be too cold and thick for tectonics to develop, and of too low a density to sink into the mantle. As a result, any surface expression of volcanism on Io is controlled by more subtle convective processes in the aesthenosphere and mantle (McEwen *et al.*, 2004).

At first glance, even with the uneven spatial coverage of spacecraft observations, the distribution of active hot spots on Io showed no obvious correlation with longitude or latitude. Persistent hot spots were, however, more likely to be found at lower latitudes, and sporadic, high-energy, but short-lived activity was more likely at higher latitudes (Lopes-Gautier *et al.*, 1999; Geissler, 2003). The possibility that high-latitude hot spots were not being seen, perhaps because instrument line of sight was being blocked by topography during equatorial plane fly-bys, was discounted because NIMS demonstrated that hot spots at lower latitudes could be detected at high emission angles (in excess of 80°) (Lopes-Gautier *et al.*, 1999).

A different approach to hot-spot distribution involved estimating the average areal density of hot spots above and below 60° latitude, and for all of Io (Milazzo *et al.*, 2005). One hot spot per 2.5×10^5 km^2 was found for all of Io (166 hot spots), one hot spot per 5.1×10^5 km^2 for polar regions (11 hot spots), and one per 2.3×10^5 km^2 for the region below 60° latitude (Milazzo *et al.*, 2005). It is not clear whether this factor of two fewer hot spots per unit area at high latitudes is significant, but it is consistent with other studies of the distribution of all volcanic features, including all paterae and non-active (to limits of detection) volcanoes. Schaber (1982) and Radebaugh *et al.* (2001) showed that paterae at higher latitudes were less numerous and larger in size than those found at lower latitudes, suggesting that the style of volcanism was different.

A deeper understanding of the distribution of volcanic activity was obtained by considering all volcanic centers, not just active ones. Schenk *et al.* (2001) catalogued and mapped the distribution of all volcanic features on Io: 541 volcanic centers in total, including 341 paterae, 46 shield volcanoes, and 7 fissure-type vents. Two broad concentrations (of 32 to 30 centers per 10^6 km^2) were found in the distribution of volcanic centers, centered at 165°W, 5°N, and 352°W, 10°S.

Comparison of volcanic center distribution with predictions of heating models produced some additional clues about the interior of Io. Models of dissipation and interior heat transfer generally utilize a convecting silicate mantle (see summary in Tackley *et al.* [2001]). Gaskell *et al.* (1988), Segatz *et al.* (1988), and Ross *et al.* (1990) considered two end-member cases. In the first model, heat was dissipated deep in a homogeneous mantle. Heating would be at a maximum at the poles and a minimum at the equator. High heat flow and volcanism would therefore be expected at the poles. In the second model, dissipation took place in the upper mantle or an aesthenosphere some 50 km to 100 km thick. Heating was preferentially equatorial, in particular in two lobes facing away from and toward Jupiter (at longitudes of 180°W and 360°W), and at a minimum at the poles (Segatz *et al.*, 1988).

Io's volcanic center distribution showed a good agreement with the aesthenosphere/upper mantle model. Interestingly, this analysis also showed an anti-correlation of volcanic centers with the distribution of mountains, offset by roughly 90° longitude (Schenk *et al.*, 2001). This anti-correlation may suggest, generally, thicker lithosphere under mountains than under volcanoes. One detail of this analysis showed an interesting correlation between some mountain types and paterae (Jaeger *et al.*, 2003), which was discussed in Chapter 15. Plate 16a shows all active hot spots plotted on the predicted heating distribution of the aesthenosphere/upper mantle model. Persistent hot-spot locations are consistent with this heating model (Lopes-Gautier *et al.*, 1999). Sporadic eruptions, however, appear to be preferentially at higher latitudes (Geissler, 2003), more in keeping with the deep mantle heating model.

The higher-than-expected polar temperatures detected by PPR (Spencer *et al.*, 2000b) indicate either enhanced polar heat flow, possibly due to volcanic activity (Veeder *et al.*, 2004), or some unexpected surface property, such as a high thermal inertia (Rathbun *et al.*, 2004). The former case requires the release of endogenic heat from old but extensive flows preferentially emplaced at high latitudes.

It is likely that actual internal heating distribution lies somewhere between the two end-member models. The sporadic eruptions that take place at higher latitudes may result from a deep source, yielding eruptions of large volume and short duration with possibly higher temperature magmas than their smaller, persistent brethren at lower latitudes (Geissler, 2003).

17.5 Heat transport by eruption class

It is possible to estimate the total thermal contribution from ongoing eruptions to Io's total thermal budget in order to gauge the relative importance of each class of volcanic activity to Io's heat flow. The yearly thermal emission from various hot-spot classes is shown in Table 17.5. Outburst activity may be the most spectacular and easily detected type of eruption, but the total heat transport from such activity, even assuming 25 such events a year, makes up less than 2% of Io's yearly output. For example, a single Pillan-like event (>50 km³ erupted) is responsible for transporting more than 6×10^{19} J of heat energy to the surface. Heat loss is initially very high but rapidly decays, and the remaining energy from the cooling flows is lost only over many years.

Based on observations of thermal emission, steady Pele is responsible for less than 1% of Io's total annual heat flow. (This number does not include the mechanical energy within the plume.) Loki Patera makes up the largest portion of observed

Table 17.5 *Estimates of yearly heat transport by eruption class*

	Typical output (W)	Duration (days)	Number per year	Yearly output (J)	% of Io total emission[a]
Outbursts	5.0×10^{12}	5	25^b	5.4×10^{19}	1.7
Loki Patera[c]	1.5×10^{13}	Continuous	1	4.7×10^{20}	15.0
Pele[d]	2.8×10^{11}	Continuous	1	8.8×10^{18}	0.3
Mid-sized, persistent[e]	2.6×10^{11}	Continuous	40	3.2×10^{20}	10.3
Small hot spots	10^{10}	Continuous	250^f	7.9×10^{19}	2.5
Total				9.4×10^{20}	29.7
Remainder[g]				2.2×10^{21}	70.3

[a] Io total thermal emission is 10^{14} W, or 3.2×10^{21} J year^{-1}.
[b] Estimate, Spencer and Schneider (1996).
[c] Veeder *et al.* (1994); Matson *et al.* (2006b).
[d] Davies *et al.* (2001).
[e] Davies *et al.* (2000b).
[f] Lopes *et al.* (2001, 2004), extrapolated to all of Io.
[g] Other sources not seen by NIMS.

volcanic heat transport (typically 10–20%). Mid-sized, persistent hot spots – including Prometheus, Amirani, and Zamama – globally contribute another 10% (Davies *et al.*, 2000a). Small hot spots, even assuming a number as high as 250 (based on the estimate of Lopes *et al.* [2004]) and with an average thermal output of 10 GW, contribute only ≈3% to Io's total emission. A chart of detected or postulated volcanic

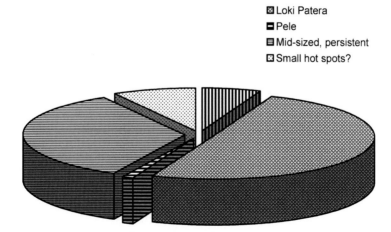

Figure 17.3 Partitioning, by eruption class, of the ≈30% of Io's total heat flow detected in or extrapolated from NIMS data.

thermal emission is shown in Figure 17.3. The remainder of the thermal emission, 70% of Io's heat flow, comes from (a) a suite of eruptions conceivably emplaced over the last century, resulting in a distribution of cooling flows down to Io's ambient temperature (Veeder *et al.*, 1994; Blaney *et al.*, 1995); and (b) additional, as yet not quantified, conductive heat transport in and around volcanically active centers. This latter hot-spot class includes heat conducted to the surface from intrusions that almost certainly exist, but which may not show any manifestation of volcanic activity other than increased surface heat flow.

The above analysis assumes little or no heat loss by conduction through the lithosphere, a factor strongly dependent on the resurfacing rate. Io is being resurfaced at a prodigious rate, great enough to erase any impact crater to the limit of detection (McEwen *et al.*, 1998a). If the resurfacing rate is 1 cm year^{-1} (e.g., Johnson and Soderblom, 1982), then at least 430 km^3 of material has to be erupted or remobilized every year to coat Io with a layer that thick (Blaney *et al.*, 1995). If the resurfacing rate is much less than 1 cm year^{-1}, then the thermal gradient in the lithosphere increases, the lithosphere becomes warm, and the heat flow by conduction to the surface plays a greater role in removing heat from the interior.

Active volcanism from mid-sized hot spots is responsible for about 10% of Io's heat flow, representing a volume of $\approx$43 km^3 of magma erupted per year (Davies *et al.*, 2000a). Much of that lava is emplaced effusively within paterae and/or on older lava flows (Prometheus and Amirani are prime examples), so there was little contribution to global resurfacing during the *Galileo* epoch from effusive activity from mid-sized hot spots.

It appears that most resurfacing comes not from lava flows but from plume deposits (Davies *et al.*, 2000a; Geissler, 2003), at least based on the *Galileo* dataset. Even so, if an outburst eruption takes place outside a patera, then a large area can be covered by lavas (e.g., Pillan, 1997). Lei-Kung Fluctus covers more than 1.25×10^5 km^2, and these flows were emplaced recently enough to still be warm when observed by *Galileo* (Rathbun *et al.*, 2004). Plume deposits have the advantage of rapidly covering much greater areas, albeit with very thin deposits (see Chapter 16).

17.5.1 Volcanic heat transport on Earth

Whereas almost all heat loss on Io is through advection – the "direct transport of heat from the interior by the melting of rock at depth and its transport upward as intrusions, lava flows, and airborne eruptions" (Head, 1990) – on Earth most heat is lost through the processes of plate recycling and conduction through the oceanic and continental crusts (see Chapter 4). Extrusive volcanic activity, producing $\approx$4 km^3 of magma per year, is responsible for less than 2% of the Earth's heat loss (Table 17.6).

Table 17.6 *Heat transport by extrusive and intrusive volcanism on Earth*

	Volume[a] (km³ yr⁻¹)	Heat content[b] (J)	% of global total[c]	% of process heat flow[d]
Extrusive volcanism				
Mid-ocean ridges	3	1.3×10^{19}	1.33	76.1
Oceanic intraplate volcanoes	0.4	1.7×10^{18}	0.18	10.1
Subduction-zone volcanoes	0.6	2.0×10^{18}	0.21	11.8
Continental intraplate volcanoes	0.1	3.4×10^{17}	0.03	2.0
Total extrusive volcanism[c]	4.1	1.7×10^{19}	1.75	100.0
Intrusive volcanism				
Mid-ocean ridges	18	7.8×10^{19}	7.99	65.8
Oceanic intraplate volcanoes	2	8.7×10^{18}	0.89	7.3
Subduction-zone volcanoes	8	2.7×10^{19}	2.75	22.6
Continental intraplate volcanoes	1.5	5.0×10^{18}	0.52	4.2
Total intrusive volcanism[c]	29.5	1.2×10^{20}	12.14	100.0
Total	**33.6**	$\mathbf{1.4 \times 10^{20}}$	**13.89**	

[a] Volumes from Crisp (1984) and Schminke (2004).
[b] Assumes average global heat flow of 31 TW, or 0.061 W m⁻² (see Hamilton, 2003), and a total heat content over a temperature range of 1450 K to 300 K of 1.55 MJ kg⁻¹ for oceanic volcanism and 1.2 MJ kg⁻¹ for continental volcanism, with magma solid density of 2800 kg m⁻³ for all cases.
[c] Ratio of extrusive to intrusive volcanism is 1 to 8, globally. Intrusive/extrusive ratios are estimated to be 5 to 1 for oceanic locations and 10 to 1 for continental locations (Crisp, 1984).
[d] For example, mid-oceanic volcanism makes up 76.1% of extrusive volcanism.

Seventy-six percent of this volume is extruded at mid-ocean ridges. The occasional large effusive or explosive eruption has little effect on global heat flow. In 1980, Mt. St. Helens erupted about 0.25 km³ of lava (dense rock equivalent). This is a contribution of ≈0.1% of the average global heat loss, assuming all heat content from the 0.25 km³ of lava was lost over one year.

The volume of magma intruded globally, ≈30 km³, is nearly eight times larger than the extruded volume (Crisp, 1984). The total heat content of all this volcanic material is ≈14% of the global heat loss total.

17.5.2 Implications for Io

Crisp (1984) concluded that the ratio of intrusive to extrusive activity on Earth was larger for continental crust because it was thick and cold. Magma ascending through continental crust loses more heat by conduction than magma rising through thinner, hotter oceanic crust. Therefore, magma rising through continental crust has a greater likelihood of stalling and forming an intrusion. The continental intrusive-to-extrusive ratio was estimated to be 10 to 1. On Io, with a thick and very cold lithosphere, with at least the upper few kilometers rich in low-density pyroclasts and sulphurous volatiles, the intrusive-to-extrusive ratio would be expected to be at least equal to and possibly much higher than that encountered on Earth.

Heat lost from intrusions, making its way slowly to the surface by conduction and via thermal exchange with volatiles, may make up a substantial part of the 70% of Io's heat flow that is not accounted for by obvious hot spots seen by NIMS.

Section 6

Io after *Galileo*

18

Volcanism on Io: a post-*Galileo* view

By the end of the *Galileo* mission, many characteristics of Io's volcanism had been identified for the first time. Processes were observed that continue to yield insight into the way volcanic activity helped shape the Earth and other terrestrial planets. This chapter summarizes the post-*Galileo* view of Io.

18.1 Volcanism and crustal structure

Galileo revealed a world dominated by silicate volcanism, putting to rest a major point of contention that dated back to *Voyager*. *Galileo* discovered many active volcanic centers on Io (Appendix 1), considerably more than were detected by *Voyager*. The global distribution of activity and volcanic features favors the heating of Io by tidal dissipation taking place, in large part, in the aesthenosphere – the partially molten upper mantle.

The upper few kilometers of Io's crust appear to be mostly mafic silicates interbedded with deposits of sulphur, sulphur dioxide, and silicate pyroclasts. The thickness of this volatile-rich layer probably varies from location to location but provides plenty of material that is easily mobilized by thermal interaction with mafic magma. Ultramafic magmas may be present, but the wide error bars ascribed to the data do not show this conclusively. If recent data analyses (Keszthelyi *et al.*, 2004a, 2005a) are confirmed, ultramafic volcanism may not be present or widespread, and the presence of a global magma ocean (e.g., Keszthelyi *et al.*, 1999) may not be required. If, however, ultra-high derived temperatures (>1800 K) are representative of true magma eruption temperatures, then Io's interior structure and composition are truly bizarre (Kargel *et al.*, 2003b).

Io is being resurfaced at such a prodigious rate that impact craters are rapidly erased, although how this is achieved while retaining Io's mostly unchanging appearance is still not known. The resurfacing rate has broad implications for temperature structure in the lithosphere. The coldest lithosphere profile is achieved with the highest resurfacing rate (O'Reilly and Davies, 1981; Jaeger and Davies, 2006), which also generates the highest compressive lithospheric pressures. These pressures contribute to the formation of Io's mountains. The resulting faults from the mountain-building process appear to create pathways to the surface for magma (Jaeger *et al.*, 2003).

There is no consensus on the average thickness of the lithosphere (see summary in McEwen *et al.* [2004]), although this thickness is probably at least 30 km in order to support mountains as high as 18 km. Lithospheric thickness estimates range up to 100 km (McKinnon *et al.*, 2001). Lithospheric thickness may be greatly variable across Io, based on the location of highest dissipation of tidal heating and mechanisms of heat transport.

A post-*Galileo* view of volcanism and lithospheric structure on Io is shown in Figure 18.1.

18.2 Magma composition

If Io has maintained the current observed level of volcanic activity over even a small part of the age of the Solar System, then the entire silicate portion of Io should have undergone hundreds of episodes of partial melting (Carr, 1986; Keszthelyi and McEwen, 1997a), resulting in a highly differentiated crust and mantle. As noted by Keszthelyi and McEwen (1997a), the resulting fractionation process should ultimately produce a wide range of silicate compositions. Repeated distillation, however, would ultimately yield high-silica-content, viscous magmas, rich in sodium and potassium, of relatively low density and low melting point. The remaining melt would be rich in magnesium, of high density and high melting point (Keszthelyi and McEwen, 1997a). In the latter case, these very dense magmas would have difficulty reaching the surface through a thick, lower-density lithosphere.

Yet, *Galileo* showed that low-viscosity magmas were the global norm with eruption temperatures and geomorphologies indicative of mafic compositions. The implication is that an extremely efficient recycling of materials back into a primitive melt is taking place, perhaps with material delaminating from a rapidly buried lithosphere and being subducted deep into Io by vigorous solid-state mantle convection (Tackley *et al.*, 2001).

The possibility also exists that, because Io's tidal heating is not steady-state (see Chapter 4), Io undergoes long periods of quiescence followed by a period of intense heat buildup, leading to the formation of a global magma ocean, widespread, intense volcanic activity, and perhaps even catastrophic crustal replacement. Volcanic activity then decreases, ending when sufficient heat has been lost to solidify the magma ocean and restart the heating via dissipation cycle once again.

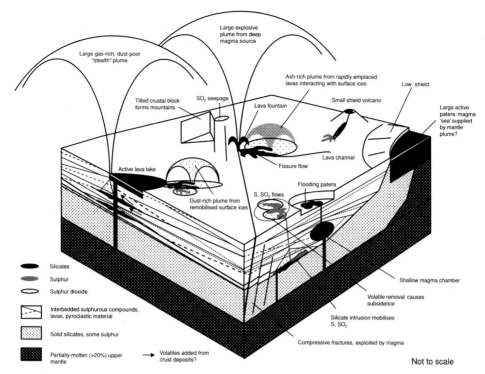

Figure 18.1 A section through the Io crust in light of what has been learned from *Galileo* data analysis. In places, the upper few kilometers of the lithosphere may be rich (>10%) in sulphur and SO_2. Beneath this layer, compressive and thermal stresses in the lithosphere contribute to mountain building. Faults in the mostly cold lithosphere act as pathways for magma to reach the surface. Silicate magma from deep sources, possibly superheated, erupts in short-lived, gas-rich, explosive events. Silicate magma from shallow reservoirs erupts effusively, mobilizing surface volatiles to form dust-rich plumes. Intrusions also mobilize sulphurous material that may erupt at the surface. Paterae form by removal of the overlying material, collapse of the magma chamber, or magma drain-back after overlying volatiles have been removed. The volume of silicate material in intrusions may exceed that extruded by an order of magnitude or more. The lithosphere is at least 30 km thick and overlies a partially molten upper mantle of silicate composition with a melt fraction of perhaps 20% or more.

18.3 Crust volatile content

Although SO_2 is a mobile, very-low-viscosity liquid throughout much of the cold lithosphere, sulphur is mostly solid and may be trapped and carried down to the base of the lithosphere, where it would be incorporated into molten magma (Keszthelyi *et al.*, 2004a).

There are indications of effusive sulphur and SO_2 volcanism, almost certainly due to the remobilization of existing near-surface and surface deposits by thermal interaction with silicate magma.

If the upper few kilometers of the lithosphere are rich in volatiles, then bulk density considerations indicate a preponderance of intrusive silicate activity, leading to removal of layers of overlying volatiles. This is one possible mechanism for forming paterae.

18.4 Hot-spot variability

From the *Galileo* dataset, augmented by observations from Earth-based telescopes and *Cassini*, the evolution of volcanic activity at many sites revealed the broad scale of volcanic activity present, with areal coverage rates, effusion and eruption rates, and thermal emission spanning many orders of magnitude. Temporal and spectral coverage was sufficient to constrain eruption style in several cases, with sequences of low-spatial-resolution infrared observations often yielding better constraints on activity than what was achievable from a single high-resolution observation of the same target. In visible imagery, surface changes caused by the emplacement of lava and plume deposits were identified and mapped. Some volcanoes showed great variations in level as well as style of activity, which is also the case with volcanoes on Earth. Internal plumbing of volcanoes and magma composition evolve, and changes can manifest themselves as different styles of eruption and different centers of activity on the same volcano.

18.5 Eruption styles

The most violent eruptions are outbursts, now known to be episodes of extreme silicate lava fountaining that may last from hours to days. In an extreme case, thermal emission from such an event has almost matched Io's global thermal emission. Yet the contribution to global heat flow from these brief events is relatively small.

The next largest events are large effusive outpourings of lava, as at Pillan in 1997. This event may have been the largest outpouring of lava ever witnessed on *any* planet, with a peak effusion rate surpassing the peak effusion of the Mauna Loa 1984 eruption by more than an order of magnitude (see Table 17.4). These extraordinary effusive events start with lava fountaining and progress via open channel flows to the emplacement of large, inflating flows in a laminar flow regime, possibly akin to inflating pahoehoe sheet flows, and probably illustrating how similar voluminous flows were erupted on Earth and Mars in the distant past (e.g., Keszthelyi *et al.*, 2004b; Davies *et al.*, 2006d).

Pahoehoe-like flows appear to be common within paterae and on the inter-paterae plains, and are indicative of a low volatile content, with a combination of rheology and effusion rate that results in a laminar flow regime. These flows are relatively thin (≈ 1 m; see Section 12.1.2), suggesting viscosities comparable to or lower than terrestrial basalts, possibly as a result of the superheating of magma during ascent. Not surprisingly, mass and flux densities are similar to Io's terrestrial counterparts for this and other styles of volcanic activity.

Galileo confirmed the importance of paterae in the context of global volcanism. Paterae appear to be where most volcanic heat is lost at near-infrared and thermal wavelengths (1–15 μm). Loki Patera alone accounts for about 10% to 20% of Io's heat loss. Paterae are probably formed by several processes, including collapse after eruption or withdrawal of lava and removal of overlying volatiles that can unroof silicate intrusions (Keszthelyi *et al.*, 2004a).

Lava lakes are important because they are open systems connecting the magma in the lake to a deep magma supply. Recycling of material through the system allows transfer of heat from the interior to the surface without additional resurfacing. Few hot spots on Io consistently exhibit the characteristic thermal signatures of overturning, active lava lakes (Chapter 8), although if quiescent lava lakes like that proposed at Loki Patera are the norm, it is not possible to identify them conclusively from thermal signature alone.

Even among the three leading lava-lake candidates, very different styles of lava lake resurfacing activity are exhibited because of the different volatile contents of the magmas involved:

- Loki Patera is episodically resurfaced through crustal foundering, a process controlled by crust bulk density. This style of resurfacing is quiescent (Stage 3 activity; see Section 8.1.3 for stage descriptions), indicating magma with a very low volatile content.
- Pele, in contrast, is a highly active lava lake, where the surface crust is constantly disrupted by volatiles degassing from the lava (Stage 1 activity).
- Janus Patera may be an intermediate case, where cycles of activity are closest to that observed at Hawaiian lava lakes (such as Kupaianaha). More or less constant thermal emission is punctuated by periods of enhanced thermal emission – as would result from rapid overturning and lava fountaining on a small scale (Stages 1, 2, and 3).

Loki Patera and Janus Patera did not produce detectable plumes during the *Galileo* epoch. The persistence of the Pele plume testifies to volatiles being added to ascending magma. Poor temporal coverage and the paucity of nighttime high-resolution NIMS data prevent conclusive identification of lava lakes at other locations.

18.6 Plumes

Galileo confirmed that two main types of plumes are found on Io and revealed the mechanisms by which these plumes form (see review by Geissler, 2003). The large gas-rich plumes result from the degassing of magma and often contain sulphur as a driving volatile. The smaller dust-rich plumes contain more sulphur dioxide and entrained silicate material and result from the interaction between relatively gas-poor lava and frozen volatiles on the surface (see Chapter 16). The presence of red deposits, rich in sulphur, is a prime indicator of current or recent effusive silicate activity. Plumes were the primary means of resurfacing of Io during the *Galileo* epoch. Resurfacing was found to be mostly by multiple coatings of relatively thin plume deposits (micrometers to millimeters), rather than by thicker lava flows (meters to tens of meters), although such flows have played a major role in the recent past (e.g., fluctūs).

18.7 Volcanism on Io and Earth

Io is wracked by dynamic processes on a much larger scale than on Earth today, allowing study of large, voluminous eruptions without having to deal with the local and regional consequences. There appear to be no global-scale tectonic controls on hot-spot distribution on Io, unlike on Earth, which raises the question: why does Io *not* have plate tectonics? Whereas transfer of internal heat on Earth to the surface is via conduction coupled with ocean plate recycling, on Io the bulk of heat is transferred by advection.

Io's volcanoes consistently emit larger thermal fluxes than their contemporary terrestrial analogues, the result of larger ionian volumetric fluxes and possibly higher-temperature magmas, although in many cases the flux and mass densities are similar for ionian and terrestrial eruptions of the same effusive style.

18.8 Questions

Many questions about Io's volcanism remain unanswered or have incomplete answers. Here is a sample.

18.8.1 The global view

- What is the current state of Io's tidal resonance?
- Where is tidal heat dissipated internally?
- What is the composition of Io's mantle?
- What is the extent of partial melting in Io's mantle?
- Is heat transported by solid-state mantle convection or by convection in a liquid layer?
- How thick is Io's lithosphere, and does thickness vary (if so, by how much)?
- What is the composition of Io's lithosphere?

18.8.2 Composition and rheology

- Are there magmas of ultramafic composition erupting on Io?
- If so, how common are they?

- What is the range of compositions of the silicate lavas?
- What are the processes that can superheat magma and by how much can magmas be so heated?
- How does the eruption of superheated basalt affect resulting geomorphology, and how does it differ from non-superheated morphology?
- Are lava flows ever emplaced in turbulent flow regimes?
- What compositional and rheological differences are there between terrestrial basaltic lavas and lavas found on Io?
- How common are sulphur and SO_2 effusive activity, and how important are they to global heat transfer?

18.8.3 Regional volcanism

- Why are paterae larger at higher latitudes?
- Is there a link between latitude and large, gas-rich eruptions?
- Are the anomalously high polar surface temperatures due to enhanced polar heat flow?
- Globally, how many active hot spots are there, and what is their total heat flow?
- Is there a causal link between mountains and paterae?
- Why are there no large volcanic shields on Io?
- What are the particle size distributions and mass loads in the different plume classes?

18.8.4 Local volcanism

- What is the relationship between the plume and the lava lake at Pele?
- How does the plume form its coherent shape from degassing magma?
- Is reservoir pumping maintaining the volatile supply at Pele?
- Is Loki Patera a "magma sea"? If so, how deep is it?
- How is Loki Patera's heat flow sustained: by tidal dissipation or a mantle plume?
- By what mechanisms do paterae form? How do they evolve?
- How many of Io's paterae contain active lava lakes?

19

The future of Io observations

To unlock the secrets of Io's volcanism and, therefore, the story of the formation and evolution of the jovian system, observations must continue to extend the time-series data that have proven to be so valuable up to now. Taking a broader view, no other body in the Solar System is subject to as much tidal heating as Io, but tidal heating does play an important dynamic role in heating other planetary satellites, such as Europa and Enceladus. To better understand the process, it is therefore logical to study Io, where tidal heating is at its most extreme. Observations of Io can be made from spacecraft and from telescopes, both on the ground and in space. It will be interesting to see how Io changes over the next 20, 50, and 100 years in observations at increasing temporal and spatial resolutions.

19.1 Spacecraft observations

At the time of this writing, the only high-spatial-resolution spacecraft observations of Io that are likely in the next decade will be in February, 2007 from the NASA New Frontiers Program *New Horizons* spacecraft, as it passes at high velocity through the jovian system on its way to a rendezvous with Pluto in 2016. The following is a description of planned Io observations as presented by John Spencer, a *New Horizons* Science Team Member, at a meeting of the *ad hoc* Io Working Group in June, 2006.

A series of global observations are planned at resolutions of up to ≈ 12 km pixel^{-1} at visible to short-infrared wavelengths (0.1 μm to 1 μm) and ≈ 140 km pixel^{-1} in the infrared (1.25 μm to 2.5 μm). These resolutions are for the closest approach of ≈ 2.7 million km, a distance somewhat greater than most *Galileo* fly-bys during its Prime Mission (Table 3.2).

These observations have been planned to

- yield global color maps to detect changes since the last SSI observations in 2001;
- detect plumes on the limb at high phase angles at selected sites (Pele, Tvashtar Paterae, Prometheus);

- allow a time-variability study of the Pele plume, if active;
- constrain magma composition from temperatures derived from eclipse images obtained at 0.4 μm to 1 μm;
- monitor thermal emission from known hot spots and discover new ones;
- detect small, faint hot spots with the sensitive camera (designed to image Pluto; many Io observations will be made by reflected Jupiter-shine to avoid saturation); and
- image auroral emissions in eclipse observations.

A campaign of Earth-based telescope observations has been planned in conjunction with this fly-by. These observations will provide valuable contemporaneous data across a wide wavelength range and wide longitude range to help constrain volcanic activity on local, regional, and global scales and will allow correlation of volcanic activity on Io with changes in the immediate environment, such as the plasma torus.

Observations at this time are also possible with the HiRise camera on the *Mars Reconnaissance Orbiter* (*MRO*), which can resolve Io from Mars orbit (Laszlo Keszthelyi, pers. comm., 2006).

19.1.1 Juno

After *New Horizons*, the next mission to the jovian system may be *Juno*, a mission concept selected for further development under the auspices of NASA's New Frontiers Program. *Juno* will investigate Jupiter's magnetic field, atmosphere, and interior. In a high-inclination polar orbit around Jupiter, distant observations of Io (including observations of the poles, which were poorly imaged by *Galileo*) will be possible. Changes due to large-scale eruptions could be easily identified. The target date for launch of *Juno* is 2011, with operations at Jupiter beginning in 2016. *Juno* should be able to measure the tidal response of Jupiter's atmosphere to the influences of the Galilean satellites, although establishing a value of Q is much more difficult (Scott Bolton, pers. comm., 2006). Nevertheless, a better understanding of the interior of Jupiter may indirectly shed light on Io's tidal response. Currently, the best way to assess Jupiter's Q would be through modeling the long-term evolution of the Galilean satellites and developing a better understanding of the dissipation mechanism in Jupiter (and other gas giants).

19.1.2 Europa mission

The next mission dedicated to the Galilean satellites will probably not have Io as its prime target. The most likely target is Europa. The National Research Council Decadal Survey of Solar System Exploration for 2003–2013 (SSES, 2003) ranked Europa as the highest-priority target for the short term, although it remains to be seen whether Europa can maintain this position in light of discoveries made in the saturnian system by *Cassini*, specifically at Titan and Enceladus. Europa-orbiting mission concepts have been examined many times by NASA, with previous concepts (e.g., *Europa Orbiter, Jupiter Icy Moons Orbiter*) failing to clear various technological and funding hurdles. The latest study for a high-priority, large NASA "flagship" mission (Johnson *et al.*, 2006) envisions a two-year mission at Jupiter, including at least three months in orbit around Europa, to carry out investigations of the structure of the Europa surface and lithosphere. The mission may include a lander with a suite of instruments for *in situ* investigations (e.g., a seismometer for geophysical measurements).

From Europa orbit, volcanism on Io can be monitored at high temporal and spatial resolutions, with the closest approach to Io (anti-jovian hemisphere) being ≈250 000 km. From Europa, Io at closest approach would appear to be about 1.5 times the size of the Moon as seen from Earth.

Any constraint on orbital parameters at Europa, a detection of variability of libration, or change within the Jupiter-satellites resonance will immediately apply constraints on the transfer of energy from planet to satellites. A particularly useful measurement would be the determination of the rate of change of mean satellite motion ($\dot{n}$).

As discussed in Chapter 4, significant acceleration of Io's mean motion should be observed if Io is losing more orbital energy from dissipation than it gains from Jupiter's orbital tidal torque (e.g., Lainey and Tobie, 2005). Europa would be similarly affected via the orbital resonance. Such a determination may be possible from precise measurement of spacecraft trajectories, although a sequence of high-resolution telescope observations over a sufficiently long period may be the most reliable way of constraining $\dot{n}$.

19.1.3 Dedicated Io mission

The Decadal Survey (SSES, 2003) that flagged Europa as the highest-priority target for exploration also supported an Io-dedicated mission in the following decade (2013–2023). The case for and objectives of such a mission are described by Spencer *et al.* (2002). A number of formidable technological hurdles would have to be overcome for such a mission, the greatest of which is the harsh radiation environment. *Galileo* demonstrated that spacecraft could not only survive but also operate successfully close to Io. There was a progressive degradation of performance, however, and success or failure of planned operations was very much a hit-or-miss proposition (see Turtle *et al.*, 2004).

As noted by Johnson *et al.* (2006), radiation-hardened technology has been developed by NASA to improve spacecraft performance and survivability. The most important areas are the flight computer, memory, electronics, and communications, all of which can be disrupted by the effects of ionizing radiation. In particular, a reliable, powerful flight processor, the RAD750, has been successfully flight-demonstrated on the *Deep Impact* and *MRO* missions.

The drive to miniaturize instrumentation and spacecraft bus systems allows more mass for shielding. Advanced autonomous fault detection, diagnosis, and system recovery software (such as used by the Autonomous Sciencecraft Experiment [ASE], described later) reduces the chances of a prolonged spacecraft shutdown during an Io fly-by.

The ideal *Io Volcano Observer* mission would orbit Io in a circular orbit. Given the inherent difficulties, however, the mission would more likely adopt a highly elliptical orbit and conduct as many as 50 to 100 close fly-bys of Io, minimizing time in the depths of the radiation belts and using lead time before encounters to plan the most useful observations and instrument settings.

The placing of a seismic station with a transponder (ideally, three seismic stations some distance apart) on the surface would allow precise measurement of the distortion of Io. Such instruments would have to survive not only a landing on an airless moon but also radiation at 4 MRads per day for about four days (two Io orbits of Jupiter, one orbit of Europa). This level of survivability is beyond current technological capabilities, but NASA has shown over many decades that it is up to such daunting challenges.

An upgrade of the NASA Deep Space Network, used to receive data from a fleet of spacecraft spread across the Solar System, and the development of advanced optical communications would allow a data return in excess of 10 Gb per day – five times greater than currently possible with *Cassini* (already exceeding by many orders of magnitude what was possible with *Galileo*'s crippled communications system). A future mission could return hundreds of gigabits per encounter. Advanced data-mining tools would be used to identify the most relevant areas for closer investigation on future encounters.

The missions described above are primarily NASA missions, but other space agencies may decide to visit the giant planets. The European Space Agency (ESA) has already done so, having built the successful *Huygens* Titan lander that was delivered by *Cassini*. While currently focused on the terrestrial planets (with missions to Mars, Venus, and Mercury), in future ESA may turn its attention to missions to the outer planets. The Russian Federal Space Agency is currently confining itself to operations in Earth orbit but has a heritage of planetary exploration and notable "firsts" on which to construct a deep-space exploration strategy. The list of nations with space exploration programs is growing. The Japanese Aerospace Exploration Agency has launched probes to Mars and inserted the *Hayabusa* spacecraft into orbit around the near-Earth asteroid Itokawa. *Hayabusa* then touched down on it. The Indian Space Research Organization is building *Chandrayaan*, which will orbit the Moon. The China National Space Administration has an emerging space program with goals that include placing a man on the Moon. Development of lift vehicles capable of boosting a spacecraft out of Earth orbit would produce a deep-space exploration capability. It is likely that a future flagship mission to Jupiter will include extensive collaboration between space agencies.

19.2 Imaging Io

To gain a deeper understanding of Io's volcanic activity, what observations would be most desirable on a return to Io? Given advances in design over the past 30 years, the spatial and spectral resolutions of a new generation of imagers, perhaps based on the *EO-1* Hyperion hyperspectral imager (Pearlman *et al.*, 2003), the *MRO* Compact Reconnaissance Imaging Spectrometers for Mars (CRISM) (Murchie *et al.*, 2003), and the *MRO* HiRise camera (Eliason *et al.*, 2003; McEwen, 2003), would be at least an order of magnitude better than *Galileo*-era instrumentation.

An ultraviolet imager would be used to compile a full global inventory of gas plumes, charting variations in location, appearance, and level of activity with time, to be correlated with eruption data obtained by other instruments.

Some important visible-wavelength imaging goals include global coverage at $200 \, \text{m pixel}^{-1}$ and a relatively high incidence angle, conditions found by the *Galileo* SSI Team to be particularly useful for interpreting regional geologic relationships (Turtle *et al.*, 2004). Low-phase-angle imaging accentuates color and albedo variations and global coverage would again be desirable. A gap exists between $320°\,\text{W}$ and $30°\,\text{W}$, a region not imaged at $<10 \, \text{km pixel}^{-1}$ by *Voyager* or *Galileo*. High-resolution stereo imaging would reveal topographic relationships and yield slopes that can be used to constrain lava flow rheology (Turtle *et al.*, 2004). Color global mapping at the same illumination and phase angles would remove difficulties, encountered with *Galileo* imagery, in comparing observations of different areas obtained under different lighting conditions.

Galileo images at very high resolutions were often obtained without context and of only a tiny portion of the surface, including a handful of volcanic targets. Many planned observations were lost during the final orbits (Turtle *et al.*, 2004). Some high-priority targets that remain after *Galileo* include high-resolution daytime images of (a) the lava lake and plume origin site at Pele; (b) the vent region of the Pillan flows (if it still exists!); and (c) Loki Patera, to conclusively determine the resurfacing mechanism and the nature of volcanic activity at the active southwest margin and patera edges, and to take a close look at the mysterious "island" that makes up about 25% of the patera area. Images obtained at Loki over a suitable time period would chart the resurfacing cycle.

Perhaps the most desirable observations of all would be unsaturated multi-filter and hyperspectral observations of all of Io's active volcanoes. These observations would set minimum eruption temperatures, which would immediately set compositional constraints.

The evolution of volcanic activity can best be monitored with a near-infrared mapping instrument with an additional thermal infrared capability. A hyperspectral imager would also identify compounds and minerals on the surface. The single greatest loss of instrument functionality induced by Jupiter's hostile radiation environment was the great reduction of NIMS spectral resolution just before the Io close fly-by campaign began. High-resolution spectra at high spatial resolutions of active or recently active surfaces, hot enough to be free from sulphurous deposits, would reveal the mineralogical makeup of silicate lavas on Io's surface. These results would immediately apply constraints on interior composition and evolution. A full global inventory of active and recently active hot spots would be used to map global heat flow patterns and quantify volcanic heat transport and effusion rate.

The most valuable observations obtained by NIMS were time-series data of volcanic activity, allowing the thermal evolution of different eruptions to be charted (e.g., Davies *et al.*, 2001). Modeling of these data revealed the physical processes taking place, allowing quantification of the eruption process. These infrared data were of greater science value and content than single observations of a target obtained at higher resolution. High-temporal-resolution infrared data (including movies at multiple frames per second) of eruptions as they wax and wane would be highly desirable, even at the expense of stand-alone high-spatial-resolution observations.

Dual observations by visible and infrared imagers proved to be of the highest value during the *Galileo* mission. Few near-contemporaneous SSI and NIMS observations were obtained during the period of full NIMS operation, although many "ride-along" contemporaneous observations were obtained during the later *Galileo* Io encounters.

Timing of dual-instrument observations becomes an issue because great variability in thermal emission was observed at visible wavelengths (e.g., at Pele) on short timescales (Radebaugh *et al.*, 2004). Dual-instrument observations of areas of the most vigorous volcanic activity would have to be contemporaneous to provide useful, mutually supporting datasets.

Observations also have to be precisely timed to quantify thermal emission with a precision that guarantees accurate derivation of temperature from multi-wavelength data. Many subsequent data analyses use these temperatures as input for further modeling.

Any derived temperature sets a minimum temperature for the magma because lava erupted at liquidus temperatures of 1500 K (mafic lava) and 1800 K (ultramafic lava) cools over the distinguishing temperature difference extremely quickly (see Figure 7.8). The areas at temperatures unambiguously diagnostic of ultramafic composition are therefore very small and hard to detect, often swamped by larger areas at lower – but still very high – temperatures.

The best chance of detecting the highest surface temperatures, closest to eruption (conduit) temperature, comes from observations at visible and very short infrared wavelengths. A re-examination of temperature derivations from *Galileo* data (Chapter 9) revealed the importance of knowing, to a high degree of accuracy, the timing of instrument functions and how detector responses had changed over the course of the mission.

The best opportunity for detecting these high temperatures therefore comes during outburst eruptions, when the largest areas at very high temperatures are produced. To understand the observations, further modeling is necessary to understand distribution and cooling of pyroclasts from such events on an airless world.

At the other end of the thermal imaging range, a thermal infrared imager/radiometer covering 8 μm to 50 μm would yield passive, non-volcanic background temperatures and would be used to investigate the strange polar nighttime surface temperature distribution on Io, as well as to identify and chart the cooling of lava surfaces not warm enough to be detected by other instruments.

19.3 Artificial intelligence, autonomy, and spacecraft operations

A dedicated mission to Io – or those phases of missions to the jovian system that are dedicated to monitoring Io – would have to be operated very differently from previous missions. Io is so much more dynamic than any other moon that it requires a new strategy in order to meet science requirements and maximize mission science return. For much of the era of spaceflight, robotic spacecraft have collected data according to pre-planned operations sequences. After acquisition, data were returned to Earth, analyzed on the ground, and the results used to adjust the next observation sequence, in the event that another fly-by of the same target was possible. This was the case with successive fly-bys of Mars by *Mariner* spacecraft, the *Voyager* Jupiter encounters, and the multiple fly-bys of *Galileo*. In the case of planetary fly-by missions like *Mariner*, *Pioneer*, and *Voyager*, the observations obtained were the first high-resolution look at the surfaces, revealing strange new worlds. This was the discovery phase of planetary exploration. All the data collected were of high value due to their uniqueness and allowed the planning of the next stage of exploration, reconnaissance missions such as *Viking* and *Galileo*, to further investigate planets and satellites. Again, regardless of usefulness, all data collected were returned within the limits of downlink. In the case of *Galileo*, downlink was severely restricted, and a significant amount of collected data was not returned.

Even today, *Cassini* observations, including instrument settings, are pre-planned, firmly established, and unalterable, weeks before encounter.

Many missions today have moved beyond the reconnaissance and mapping phases into a deeper investigative mission role, driven primarily by high-level science goals. Prime examples are the fleet of assets in orbit around and on the surface of Mars, currently consisting of the *Mars Exploration Rovers*, *Mars Odyssey*, *Mars Express*, and *MRO*. *Mars Global Surveyor* ceased operations in November 2006.

For Io observations, rather than focus on global mapping (the traditional first step of planetary missions), investigations directed toward identification and monitoring of specific processes should be given higher priority when opportunities appear. This scenario requires more flexibility than has been traditionally available, requiring the ability to rapidly detect dynamic, short-lived events and quickly react to them, changing spacecraft operations as needed. This process would need to be independent of communications with the ground and would be attained by placing data analysis and mission planning software on the spacecraft.

Such autonomous spacecraft operations were successfully demonstrated by NASA's New Millennium Program Autonomous Sciencecraft Experiment (ASE). ASE is software that was uploaded to the Earth-orbiting *Earth Observing 1* spacecraft (*EO-1*) in 2004 (Chien *et al.*, 2005). ASE enabled operational decisions to be made autonomously onboard the spacecraft, including resource allocation and fault detection and mitigation. This latter ability would be highly desirable on a mission to the harsh Io environment, considering the number of times *Galileo* instruments entered "safe" mode or stopped operating during fly-bys. As demonstrated on *EO-1*, sequenced operations commands can be altered as necessary in response to results from onboard data analysis, with *mission science goals driving subsequent operations*.

The benefits of such a spacecraft autonomy capability on an Io mission can be illustrated by considering how best to detect and monitor an outburst event from a spacecraft during a fly-by encounter (Davies *et al.*, 2006e). For the purpose of the example, the science goal would be to determine the temperature of Io's lavas and constrain possible lava compositions.

For *Galileo*, instrument settings and exposure times were pre-ordained as part of the operation sequence. Although lava fountains were observed, in one case at high spatial resolution (at Tvashtar Paterae; see Chapter 11), exposures were planned to image the non-thermally active background, and the intense thermal emission saturated both the visible imaging system (SSI) and the infrared imager (NIMS). There was no opportunity to quantify the intensity of the thermal emission and to change observation sequencing and instrument settings. By the time data had been returned to Earth and analyzed, the spacecraft had moved on and the science event was over.

An onboard artificial intelligence (AI) would do things very differently. By using classifiers akin to those used by ASE to detect volcanic thermal emission in Hyperion hyperspectral observations (Davies *et al.*, 2006a), onboard data processing could quickly identify an intense thermal source at a great distance, calculate the opportune moment to make observations (selecting visible and infrared imagers in the 0.4-μm to 10-μm range to capture the full thermal emission spectrum), and set the appropriate instrument gain state or exposure time to obtain unsaturated data. Additional instrumentation would be brought to bear on the eruption: an ultraviolet spectrometer would be used to study erupting gas, and a thermal imager would map the temperatures in the plume. On subsequent orbits, this location would be flagged for in-depth investigation, for example, by passing a "high-priority" goal of obtaining high-resolution spectra to the onboard planner for insertion into the future operations sequence.

The value of the returned data is therefore increased from that obtained from an acquisition queue using pre-set observation sequencing: the need for communications (data transfer, analysis, new commands, and accompanying time lag) between spacecraft and Earth for spacecraft re-tasking is eliminated, and the use of bandwidth is optimized through data editing and product prioritization. Using onboard autonomy, important science questions can be answered by making decisions on the spot.

19.4 Telescope observations

Once more, the long-term monitoring of volcanism on Io rests in the hands of ground-based astronomers. Some astonishing technological advances over the past 15 years (multi-segmented, deformable mirrors; faster computer processors; actuators capable of 1000 movements per second; and adaptive optics) have resulted in large (8–10 m) ground-based telescopes that can image Io at multiple infrared wavelengths at resolutions better than $\approx$160 km pixel^{-1}, equal to or better than many observations collected by NIMS during the *Galileo* Prime Mission and GEM (see Table 3.2). Earth-based telescopes are currently examining Io with increasing spectral resolution to detect compounds being erupted into Io's vicinity, delivered into space by erupting plumes. In this way, composition can be studied and magma chemistry constrained. Examples of such telescopes are found on Mauna Kea, Hawai'i (Subaru, Keck I and II, Gemini North and South), and at the European Southern Observatory (ESO) in Chile (Very Large Telescope [VLT], consisting of a cluster of four 8.2-m telescopes). Examples of recent Io observations obtained with AO can be found in de Pater *et al.* (2004) and Marchis *et al.* (2005).

The development of Next Generation Adaptive Optics (NGAO), invoking data-processing tools such as the Adaptive Image Deconvolution Algorithm (AIDA)

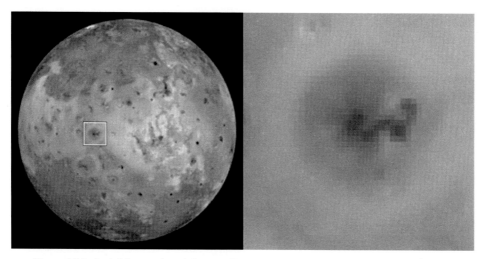

Figure 19.1 A visible-wavelength image of Io at a spatial resolution of 10 km pixel^{-1} (left). The enlargement (right) is of the volcano Prometheus, shown at the same resolution. Such resolutions may be attained with a new generation of very large Earth-based optical telescopes that utilize adaptive optics.

(Hom *et al.*, 2007), would increase resolution in the infrared to ≈90 km pixel^{-1}. Combining new AO systems with larger telescopes, such as the proposed Thirty-Meter Telescope (TMT), would result in resolutions in the infrared as high as 45 km pixel^{-1} (Franck Marchis, pers. comm., 2006), perhaps by 2015.

In the concept and design stages are even larger telescopes, such as the ESO Extremely Large Telescope (ELT), which has a 50-m-diameter mirror and OWL, the OverWhelmingly Large telescope: a 100-m-diameter mirror behemoth with a resolution of 1 milli-arcsecond, even without AO. With advanced AO and instrumentation, these huge light collectors could image Io at resolutions of only a few kilometers in the near infrared (e.g., Figure 19.1). OWL will cost one billion Euros, equivalent to two New Frontiers Missions. The OWL proponents promise "a planetary fly-by every night" (ESO website).

Observations of Io can also be made from orbiting platforms, building on the work already performed by the Hubble Space Telescope. The Spitzer Space Telescope can detect thermal emission from Io in the short-wavelength to thermal infrared. The final "Great Observatory" – the James Webb Space Telescope, with a 6.5-m mirror – is due to be launched in 2013.

Appendix 1

Io hot-spot locations

This table of Io's active hot spots has been compiled from identifications of hot spots in data from *Voyager* IRIS (Pearl and Sinton, 1982; McEwen *et al.*, 1992, 1996); SSI (McEwen *et al.*, 1997, 1998a; Keszthelyi *et al.*, 2001a; Geissler, 2003; Turtle *et al.*, 2004); NIMS (Lopes-Gautier *et al.*, 1997, 1999, 2000; Lopes *et al.*, 2001, 2004); PPR (Spencer *et al.*, 2000b; Rathbun *et al.*, 2004); and terrestrial telescopes (Goguen *et al.*, 1998; Marchis *et al.*, 2001; de Pater *et al.*, 2004; Marchis *et al.*, 2005). Hot spots in bold type have temperatures in excess of 1000 K as determined from SSI data (summarized in Geissler [2003]), NIMS data (Davies *et al.*, 1997, 2000b; Lopes-Gautier *et al.*, 2000; Davies *et al.*, 2001, 2003b), and ground-based instruments (Marchis *et al.*, 2002). This list will no doubt expand as more data are obtained and analyzed. Provisional names for features are in italics. Detections are indexed as follows: I = *Voyager* IRIS; S = *Galileo* SSI; N = *Galileo* NIMS; P = *Galileo* PPR; E = Earth-based or Earth-orbiting telescopes; C = *Cassini* ISS. All of the hot spots are plotted in Plates 13b and 16a.

The positions of hot spots are subject to uncertainties caused by different observation resolutions and movement of thermal sources, as well as by reference to nearest feature rather than the hot spot itself. For example, the position of the thermal source at Donar Fluctus in a SSI observation during Orbit E11 was reported as 189.3°W, 27.8°N (Keszthelyi *et al.*, 2001a). The location of the same hot spot was subsequently reported in terms of the nearest feature as 187.3°W, 21.8°N (Geissler, 2003) and 186.2°W, 24.3°N (Lopes *et al.*, 2004). McEwen *et al.* (1998a) reported a hot spot at 189.4 ($\pm$ 0.6) °W, 28.5 ($\pm$ 0.3) °N, probably the same hot spot detected at Donar Fluctus by SSI during Orbit E11, whereas Lopes *et al.* (2001) reported another possible hot spot at "Donar" at 185.5°W, 24°N.

Appendix 1

Io hot-spot locations

Volcanic center or nearest feature to hot spot	Hot-spot longitude (° W)	Hot-spot latitude (°)	Detected by
Acala Fluctus	337.0	11.0 N	S
Ah Peku Patera	105 ± 1	9 ± 1 N	N
Aidne Patera	178 ± 3	2 ± 3 S	N
Altjirra Patera	108	35 S	N
Amaterasu Patera	309 ± 4	40 ± 4 N	I, N, P
Amirani	112–113	27–30 N	I, S, N, E
Argos Planum	317.8	48 S	P
Arinna Fluctus	147 ± 1	30 ± 1 N	N
Arusha Patera	100 ± 2	39 ± 2 S	N, E
Atar Patera	278.6	31.1 N	P
Aten Patera	310.5	48.2 S	I, P, E
Babbar Patera	271.8	39.4 S	I, N, P
Camaxtli Patera	136.4	15 N	S, N
Catha Patera	105 ± 1	53 ± 1 S	N
Chaac Patera	157	10 N	N
Chors Patera	251	69 N	P
Creidne/Nusku Paterae	355	53 S	I
Cuchi Patera	144 ± 1	2 ± 1 S	N
Culann Patera	161.5	19.9 S	S, N
Daedalus Patera	275	19 N	I, N, P, E
Dazhbog Patera	301.5	55.1 N	P, E
Donar Fluctus	185.5	24.3N	N
Donar Fluctus 2	189.3	27.8 N	S
Dusurra Patera	125 ± 7	39 ± 7 N	N
East Creidne/Aten	335	58 S	I
Ekhi Patera	87.6	28.3 S	S
Emakong Patera	119 ± 1	3 ± 1 S	N
Estan Patera	87 ± 2	21 ± 2 N	N
Estan Patera vicinity	81 ± 1	20 ± 1 N	N
Fjorgynn Fluctus	358.0	11.5 N	S
Fo Patera	192	40.5 N	S, N
Fuchi Patera	327.9	28.4 N	S
Gabija Patera	204 ± 3	52 ± 3 S	N, P
Gibil/Kibero Paterae	294.6	15.0 S	I?, P
Girru Patera	238 ± 3	22 ± 3 N	S, N, P
Gish Bar Patera	89.1	15.6 N	S, N, E
Haokah Patera	185 ± 3	19 ± 3 S	N
Heno Patera	311.5	57.1 S	P
Hephaestus Patera	289.5	2.0 N	P, E
Hi'iaka Patera	79	3 S	N, E
Isum Patera North	204.7	32.9 N	I, S, N, P, E
Isum Patera South	206.8	30.3 N	I, S, N, P, E
Itzamna Patera	97 ± 3	15 ± 3 S	N, E
Janus Patera	39	3.9 S	S, N, E
Kanehekili Fluctus	33.6	18 S	S, N, E

(continued)

Io hot-spot locations (continued)

Volcanic center or nearest feature to hot spot	Hot-spot longitude (° W)	Hot-spot latitude (°)	Detected by
Karei Patera	15.3	4.7 N	S
Kinich Ahau Patera	311	49 N	E
Kurdalagon Patera	219 ± 3	47 ± 3 S	P
Lei-Kung Fluctus	204	38 N	S, N, P
Lerna Regio	291.9	62.3 S	P
Llew Patera	240 ± 2	10 ± 2 N	N, P
Loki Patera	308.8	12.7 N	I, S, N, P, E, C
Malik Patera	128 ± 2	34 ± 2 S	I, N
Manua Patera	321.6	35.2 N	S, E
Marduk	209.9	28.4 S	I, S, N, P
Masubi	60	48 S	S, N, E
Maui Patera	122	17 N	I, S, N
Mazda Paterae	314.9	9.4 S	I, P
Mbali Patera	6.8	31.4 S	I, E
Michabo Patera	169 ± 2	2 ± 2 S	N
Mihr Patera	305.5	16.5 S	I?, S, P, E
Mithra Patera	266.5	58.8 S	P
Monan Patera	104	20 N	S, N
Mulungu Patera	217.5	17.2 N	S, N, P, E
Nemea Planum	320	78 S	I, P
Nusku Patera	4.9	64.4 S	I, E
Ot Patera	218 ± 3	2 ± 3 S	N, P
Pele	255.7	18.4 S	I, S, N, P, E, C
Pillan Patera	244	12 S	S, N, P, E, C
Poliahu	82	19 S	E
Prometheus	153.9	1.5 S	S, N, P, E
Pyerun Patera	251.1	55.4 S	I
Radegast Patera	160 ± 0.5	27 ± 0.5 S	N
Rarog Patera	304.4	41.7 S	P, E
Rata Patera	199.5	35.5 S	S, N, P
Reiden Patera	236	13 S	S, N, P
Ruaumoko Patera	139 ± 1	15 ± 1 N	N
Ruwa Patera	2.7	0.5 N	S, E
Sengen Patera	304	32.5 S	P, E
Sethlaus Patera	195 ± 3	50 ± 3 S	N
Shakuru Patera	265.7	24.1 N	P
Shamash Patera North	153 ± 1	34 ± 1 S	I, N
Shamash Patera South	151 ± 1	36 ± 1 S	I, N
Shamshu	67 ± 4	10 ± 4 S	N
Sigurd Patera	100 ± 4	5 ± 4 S	N, E
Sobo Fluctus	150 ± 1	14 ± 1 N	N
Surt	337.1	44.9 N	E
Surya Patera	152 ± 1	22 ± 1 N	N
Susanoo Patera	222 ± 3	21 ± 3 N	N, P, E
Svarog Patera	267.5	48.3 S	I, S, N, P, E
Tawhaki Patera	75.1	3.1 N	S, N, E

(continued)

Appendix 1

Io hot-spot locations (continued)

Volcanic center or nearest feature to hot spot	Hot-spot longitude (° W)	Hot-spot latitude (°)	Detected by
Thor 1	131 ± 1	38 ± 1 N	N, E
Thor 2	135 ± 1	39 ± 1 N	N
Tien Mu Patera	134 ± 1	12 ± 1 N	N
Tsui Goab Fluctus	164	0 N	N
Tupan Patera	141	19 S	N
Tvashtar Paterae 1	120.2	61.5 N	S, N, E
Tvashtar Paterae 2	123	62 N	S, N, E
Tvashtar Paterae 3	120 ± 1	61 ± 1N	N
Tvashtar Paterae 4	126	64.8 N	N, E
Ulgen Patera	288	40.4 S	I, N, P, E
unnamed	4.7	15.3 N	S
unnamed	11	11 S	E
unnamed	13.3	2.8 S	S
unnamed	13.3	11.5 S	S, E
unnamed	14.2	10.7 N	S
unnamed	14 ± 1	31 ± 1 N	E
unnamed	19	6 S	S
unnamed	21	1 N	S
unnamed	22.8	13.8 S	S
unnamed	23	1 S	S
unnamed	23	5 N	S
unnamed	27.9	16.5 S	S
unnamed	30 ± 15	70 ± 15 N	E
unnamed	41 ± 3	46 ± 1 N	E
unnamed	51 ± 4	34 ± 1 N	E
unnamed	59	11 N	S, E
unnamed	60 ± 15	60 ± 15 N	E
unnamed	68 ± 1	49 ± 1 S	N, E?
unnamed	79 ± 3	37 ± 3 S	N
unnamed	87 ± 1	19 ± 1 S	N
unnamed	91 ± 2	44 ± 2 N	S, N
unnamed	92.4	46.1 N	S, E
unnamed	95 ± 1	7 ± 1 N	N
unnamed	106 ± 2	20 ± 2 N	N
unnamed	108.3	10.8 N	S
unnamed	109 ± 2	48 ± 2 S	N
unnamed	109 ± 1	24 ± 1 N	N
unnamed	118 ± 2	37 ± 2N	N
unnamed	117 ± 0.5	31 ± 0.5 N	N
unnamed	117 ± 1	59 ± 1 N	N
unnamed	125 ± 1	67 ± 1 N	N
unnamed	127 ± 1	11 ± 1 S	N
unnamed	130 ± 1	20 ± 1 N	N
unnamed	132 ± 1	5 ± 1 S	N
unnamed	132.0	65.2 N	S

(continued)

Io hot-spot locations (continued)

Volcanic center or nearest feature to hot spot	Hot-spot longitude (° W)	Hot-spot latitude (°)	Detected by
unnamed	137 ± 1	35 ± 1 N	N
unnamed	139 ± 1	45 ± 1 S	N
unnamed	144	66 N	S
unnamed	145 ± 1	22 ± 1 N	N
unnamed	147 ± 1	26 ± 1 S	N
unnamed	149 ± 2	37 ± 2 S	N
unnamed	166 ± 1	68 ± 1 S	N
unnamed	172 ± 2	45 ± 2 S	N
unnamed	199	32 N	N
unnamed	206	55 S	P
unnamed	215.6	10.2 S	S
unnamed	227 ± 2	28 ± 2 N	N
unnamed	224	24 S	P
unnamed	233	4 S	P
unnamed	233	28 S	P
unnamed	236	49 S	P
unnamed	242.5	35.6 S	S
unnamed	261	37 N	P
unnamed	264	53 N	P
unnamed	277	7 S	P
unnamed	278	13 S	P
unnamed	279	50 N	E
unnamed	305 ± 2	61 ± 2 S	E
unnamed	332.9	17.0 N	S
unnamed	355.1	4.4 N	S
unnamed	356.1	4.8 N	S
unnamed	35	7 S	S, N
unnamed	109 ± 2	69 ± 2 S	N
Uta Patera	24.5	35.3 S	S, E
Viracocha Patera	281	61.4 S	I
Vivasvant Patera	294	75.1N	P
Volund	174	25 N	I, S, N
Wayland Patera	226.0	32.2 S	N, P, C
Yaw Patera	132 ± 1	9.5 ± 1 N	N
Zal Patera	74.9	40.5 N	S, N
Zamama	174	18 N	S, N, E

Detected-by key: C = ISS (*Cassini*); I = IRIS (*Voyager*); N = NIMS (*Galileo*); S = SSI (*Galileo*); P = PPR (*Galileo*); E = Earth-based; ? = possible detection.

Appendix 2

Io maps

The maps in this appendix were produced by the U.S. Geological Survey Astro-geology Branch, Flagstaff, Arizona. Features designated with an asterisk have provisional names, pending approval by the International Astronomical Union.

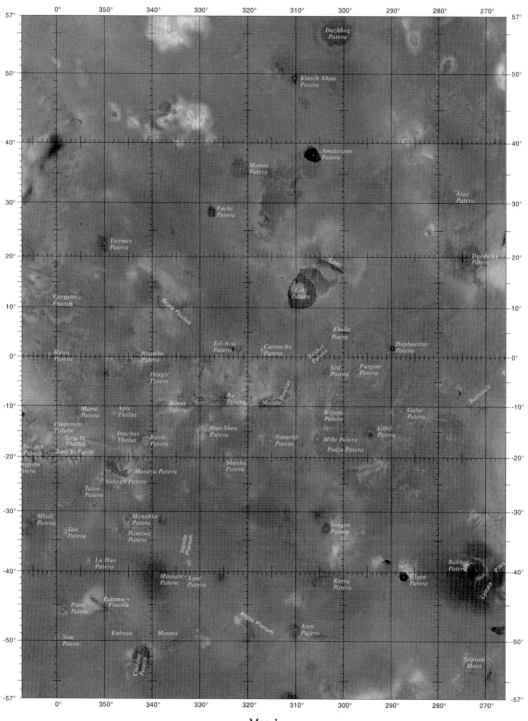

Map 1

Appendix 2

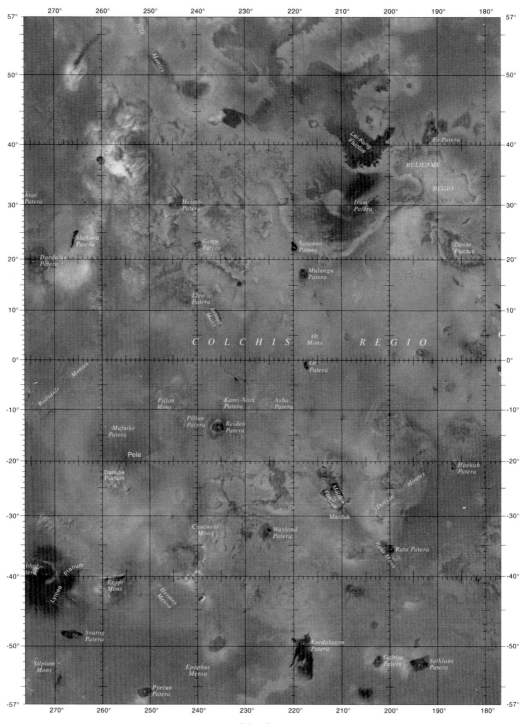

Map 2

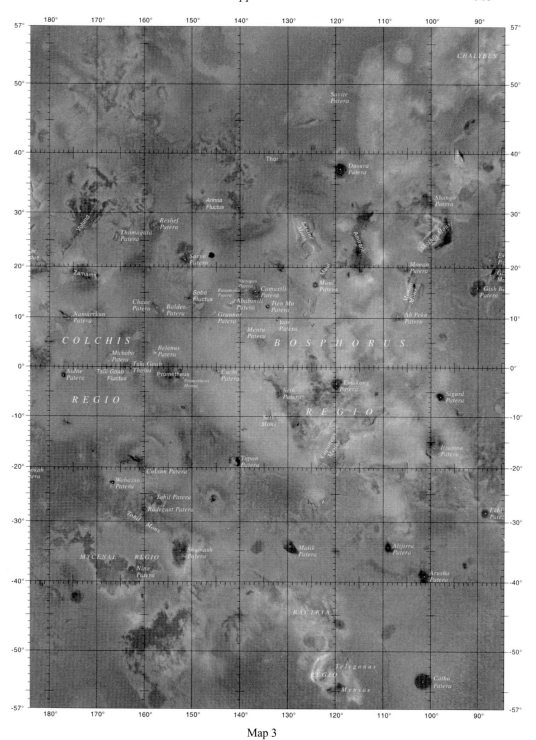

CHALYBES

Savitr
Patera

Thor

Dusara
Patera

Shango
Patera

Arinna
Fluctus

Volund

Reshef
Patera

Thomagata
Patera

Enlil
Mons

Amarani

Sibylus Mons

Surya
Patera

Mihan
Patera

Zamama

Maui

Masubi
Mons

Gish Ba
Patera

Chaac
Patera

Sobo
Fluctus

Rata'umaka
Patera

Steropes
Patera

Camaxtli
Patera

Maui
Patera

Balder
Patera

Ababinili
Patera

Tien Mu
Patera

Ah Peku
Patera

Namarrkun
Patera

Granno
Patera

Mentu
Patera

Yaw
Patera

COLCHIS

BOSPHORUS

Michabo
Patera

Belenus
Patera

Tsũi Goab
Thelus

Aidne
Patera

Tsũi Goab
Fluctus

Prometheus

Cuchi
Patera

Seth
Patera

Enakong
Patera

Sigurd
Patera

Prometheus
Mensa

REGIO

REGIO

Seth
Mons

Copernicus
Mesa

Itzanma
Patera

okah
era

Tupan
Patera

Culann Patera

Wabasso
Patera

Tohil Patera

Radegast Patera

Ekhi
Pate?

Tohil Mons

Shamash
Patera

Malik
Patera

Altjirra
Patera

MYCENAE REGIO

Arusha
Patera

Nina
Patera

BACTRIA

Telegonus

Catha
Patera

REGIO

Mensae

Map 3

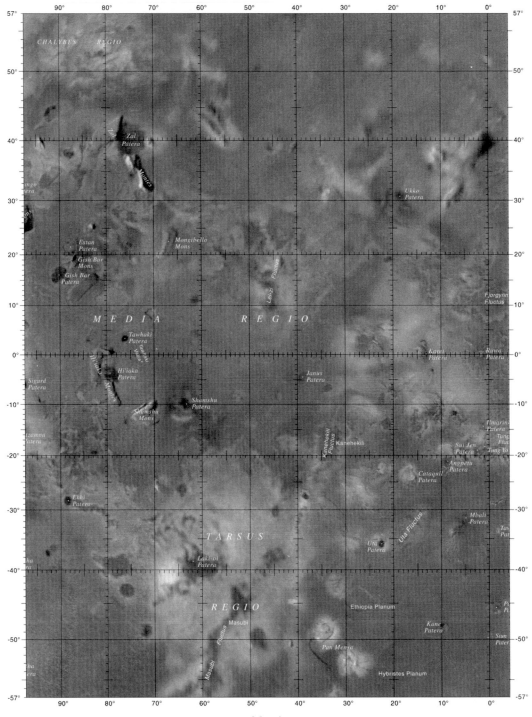

Map 4

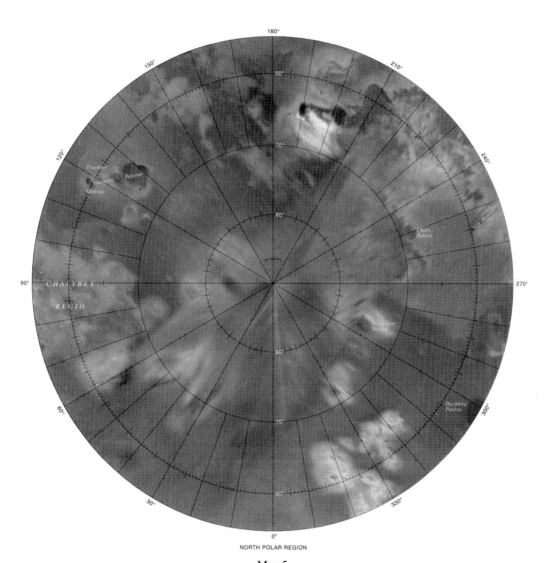

NORTH POLAR REGION

Map 5

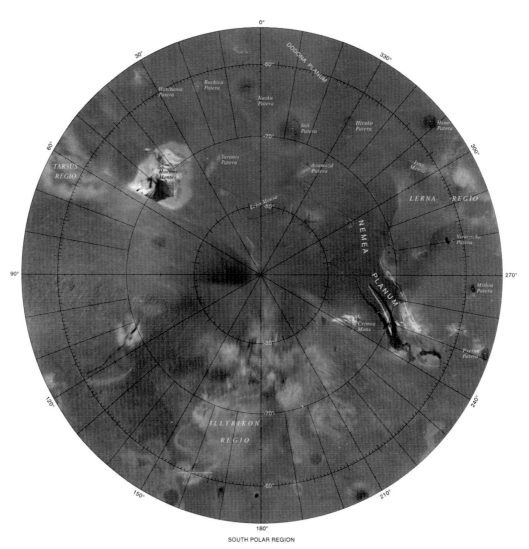

SOUTH POLAR REGION

Map 6

References

Abrams, M. (2000). The Advanced Spaceborne Thermal Emission and Reflection Radiometer (ASTER): data products for the high-spatial resolution imager on NASA's *Terra* platform. *International Journal of Remote Sensing*, **21**, no. 5, 847–59.

Abtahi, A. A, A. B. Kahle, E. A. Abbott, *et al.* (2002). Emissivity changes in basalt cooling after eruption from Pu'u 'O'o, Kilauea, Hawai'i, *Eos Transactions, AGU,* **83**(47), Fall Meeting. Supplement, Abstract V71A-1263.

Aksnes, K., and F. A. Franklin. (2001). Secular acceleration of Io derived from mutual satellite events. *Astronomical Journal*, **122**, 2734–9.

Amelin, Y., A. N. Krot, I. D. Hutcheon, *et al.* (2002). Lead isotope ages of chondrules and calcium-aluminum-rich inclusions. *Science*, **297**, 1678–83.

Anderson, D. L. (1999). Planet Earth. In *The New Solar System*, 4th edn., ed. J. K. Beatty *et al*. Cambridge, UK: Cambridge University Press, pp. 111–24.

Anderson, G., J. Chetwynd, and J. Theriault. (1993). MODTRAN2: suitability for remote sensing. *Proceedings of the SPIE*, 1968, 514–25.

Anderson, J. D., G. W. Null, and S. K. Wong. (1974). Gravity results from *Pioneer 10* Doppler data. *Journal of Geophysical Research*, **79**, 3661–4.

Anderson, J. D., W. L. Sjogren, and G. Schubert. (1996). *Galileo* gravity results and the internal structure of Io. *Science*, **272**, 709–12.

Anderson, J. D., R. A. Jacobson, E. L. Lau, *et al.* (2001). Io's gravity field and interior structure. *Journal of Geophysical Research*, **106**, 32963–70.

Anderson, J. D., G. Schubert, A. Anabtawi, *et al.* (2002). Recent results on Io's gravity field and interior structure. AGU Spring Meeting Abstracts, P21B-01.

Arndt, N. T., and E. G. Nisbet. (1982). *Komatiites*. London: George Allen and Unwin.

Arndt, N. T., D. M. Francis, and A. J. Hynes. (1979). The field characteristics and petrology of Archean and Proterozoic komatiites. *The Canadian Mineralogist*, **17**, 147–63.

Arthur, D. W. (1981). Vertical dimensions of the Galilean satellites. NASA TM-81776. Washington, DC: NASA.

Baloga, S., and D. Pieri. (1986). Time-dependent profiles of lava flows. *Journal of Geophysical Research*, **91**, 9543–52.

Barberi, F., and J. Varet. (1970). The Erta 'Ale volcanic range. *Bulletin Volcanologique*, **34**, 848–917.

Bartha, G. (1981). Earth tides in geodynamics. *Acta Geology Geophysics et Montanist*, **16**, 265–8.

Becker, T., and P. E. Geissler. (2005). *Galileo* global color mosaics of Io. *Lunar and Planetary Science Conference, XXXVI, Abstract 1862*.

317

Belton, M. J. S., R. A. West, and J. Rahe (Eds.). (1989). Time-variable phenomena in the jovian system. NASA-SP-494. Washington, DC: NASA.

Belton, M. J. S., K. P. Klaasen, M. C. Clary, *et al.* (1992). The *Galileo* Solid-State Imaging experiment. *Space Science Reviews*, **60**, 413–55.

Binder, A. B., and D. P. Cruikshank. (1964). Evidence for an atmosphere on Io. *Icarus*, **3**, 299.

Blake, S. (1981). Volcanism and the dynamics of open magma chambers. *Nature*, **289**, 783–5.

Blaney, D. L., T. V. Johnson, D. L. Matson, *et al.* (1995). Volcanic eruptions on Io: heat flow, resurfacing, and lava composition. *Icarus*, **113**, 220–5.

Blaney, D. L., D. L. Matson, T. V. Johnson, *et al.* (1998). The role of thermal emission in determining SO_2 frost band depths in the *Galileo* NIMS G2 data. *Bulletin of the American Astronomical Society*, **30**, 1121.

Blaney, D. L., D. L. Matson, T. V. Johnson, *et al.* (2000). Myriads of small, hot eruptions on Io. *Lunar and Planetary Science Conference XXXI, Abstract 1617*.

Boehler, R. (1986). The phase diagram of iron to 430 kbar. *Geophysical Research Letters*, **13**, 1153–6.

Boehler, R. (1992). Melting of the Fe-FeO and Fe-FeS systems at high pressure: constraints on core temperatures. *Earth and Planetary Science Letters*, **111**, 217–27.

Brett, R. (1973). Lunar core of Fe-Ni-S. *Geochimica et Cosmochimica Acta*, **37**, 165–70.

Brown, R. A. (1974). Optical line emission from Io. *IAU Symposium 65: Exploration of the Planetary System*, **65**, 527–31.

Burns, J., and M. S. Matthews (Eds.). (1986). *Satellites*. Tucson: University of Arizona Press.

BVSP. (1981). *Basaltic Volcanism on the Terrestrial Planets – Basaltic Volcanism Study Project*. New York: Pergamon Press.

Calvari, S., M. Cotelli, M. Neri, *et al.* (1994). The 1991–1993 Etna eruption: chronology and lava flow-field evolution. *Acta Volcanology*, **4**, 1–14.

Carlson, R. W., P. R. Weissman, W. D. Smythe, *et al.* (1992). Near-Infrared Mapping Spectrometer experiment on *Galileo*. *Space Science Reviews*, **60**, 457–502.

Carlson, R. W., W. D. Smythe, R. M. C. Lopes-Gautier, *et al.* (1997). Distribution of sulfur dioxide and other infrared absorbers on the surface of Io. *Geophysical Research Letters*, **24**, 2479.

Carr, M. H. (1985). Volcanic sulphur flows on Io. *Nature*, **313**, 735–6.

Carr, M. H. (1986). Silicate volcanism on Io. *Journal of Geophysical Research*, **91**, 3521–32.

Carr, M. H., H. Masursky, R. G. Strom, *et al.* (1979). Volcanic features of Io. *Nature*, **280**, 729–33.

Carr, M. H., A. S. McEwen, K. A. Howard, *et al.* (1998). Mountains and calderas on Io: possible implications for lithosphere structure and magma generation. *Icarus*, **135**, 146–65.

Casadevall, T. J., J. B. Stokes, L. P. Greenland, *et al.* (1987). SO_2 and CO_2 emission rates at Kilauea volcano, 1979–1984. In *Volcanism in Hawaii*, ed. R. W. Decker *et al.* U.S. Geological Survey Professional Paper 1350, pp. 771–80.

Cassen, P., and R. T. Reynolds. (1974). Convection in the Moon: effect of variable viscosity. *Journal of Geophysical Research*, **79**, 2937–44.

Cassen, P., S. J. Peale, and R. Reynolds. (1982). Structure and thermal evolution of the Galilean satellites. In *Satellites of Jupiter*, ed. D. Morrison. Tucson: University of Arizona Press, pp. 93–128.

Castillo, J. C., D. L. Matson, T. V. Johnson, *et al.* (2005). [26]Al in the saturnian system – new interior models for the saturnian satellites. AGU Fall Meeting Abstracts. P32A-05.

Cattermole, P. J. (1996). *Planetary Volcanism – A Study of Volcanic Activity in the Solar System*, 2nd edn. New York: Wiley.

Cervelli, P. F., and A. Miklius. (2003). The shallow magmatic system of Kilauea volcano. In *The Pu'u 'O'o-Kupaianaha Eruption of Kilauea Volcano, Hawai'i: The First 20 Years*, ed. C. Heliker *et al*. U.S. Geological Survey Professional Paper 1676, pp. 149–64.

Chien, S., R. Sherwood, D. Tran, *et al*. (2005). Using autonomy flight software to improve science return on *Earth Observing One*. *Journal of Aerospace Computing, Information, & Communication*, **2**, 196–216.

Clark, R. N., and T. B. McCord. (1980). The Galilean satellites – new near-infrared spectral reflectance measurements (0.65–2.5 microns) and a 0.325–5 micron summary. *Icarus*, **41**, 323–39.

Clark, S. P. (1966). Viscosity. In *Handbook of Physical Constants*, ed. S. P. Clark. U.S. Geological Survey Memoir 97, pp. 291–300.

Clow, G. D., and M. H. Carr. (1980). Stability of sulfur slopes on Io. *Icarus*, **44**, 268–79.

Cohen, B. A., and R. F. Coker. (2000). Modelling of liquid water on CM meteorite parent bodies and implications for amino acid racemization. *Icarus*, **145**, 369–81.

Collins, S. A. (1981). Spatial color variations in the volcanic plume at Loki, on Io. *Journal of Geophysical Research*, **86**, 8621–6.

Consolmagno, G. J. (1981). Io – thermal models and chemical evolution. *Icarus*, **47**, 36–45.

Crisp, J., and S. Baloga. (1990a). A model for lava flows with two thermal components. *Journal of Geophysical Research*, **95**, 1255–70.

Crisp, J., and S. Baloga. (1990b). A method for estimating eruption rates of planetary lava flows. *Icarus*, **85**, 512–15.

Crisp, J. A. (1984). Rates of magma emplacement and volcanic output. *Journal of Volcanology and Geothermal Research*, **20**, 177–211.

Cruikshank, D. P., T. J. Jones, and C. B. Pilcher. (1977). Absorptions in the spectrum of Io, 3.0–4.2 microns. *Bulletin of the American Astronomical Society*, **9**, 465.

Cruikshank, D. P., T. J. Jones, and C. B. Pilcher. (1978). Absorption bands in the spectrum of Io. *Astrophysical Journal*, **225**, L89–L92.

Darwin, G. H. (1880). On the secular change of the orbit of a satellite revolving about a tidally-distorted planet. *Philosophical Transactions of the Royal Society of London*, **171**, 713–891.

Davies, A. G. (1988). Sulphur-silicate interactions on the jovian satellite Io. Unpublished Ph.D. thesis, Lancaster University, Lancaster, UK.

Davies, A. G. (1996). Io's volcanism: thermo-physical models of silicate lava compared with observations of thermal emission. *Icarus*, **124**, 45–61.

Davies, A. G. (2001). Volcanism on Io: the view from *Galileo*. *Astronomy and Geophysics*, **42**, 10–12.

Davies, A. G. (2002). A tale of two hot spots: charting thermal output variations at Prometheus and Amirani from *Galileo* NIMS data. AGU Fall Meeting Abstract P71B-0461.

Davies, A. G. (2003a). Temperature, age and crust thickness distributions of Loki Patera on Io from *Galileo* NIMS data: implications for resurfacing mechanism. *Geophysical Research Letters*, **30**, 2133–6.

Davies, A. G. (2003b). Volcanism on Io: estimation of eruption parameters from *Galileo* NIMS data. *Journal of Geophysical Research (Planets)*, **108**, 5106–20.

Davies, A. G., and L. P. Keszthelyi. (2005). Classification of volcanic eruptions on Io and Earth using low-resolution remote sensing data. *Lunar and Planetary Science Conference XXXVI, Abstract 1963*.

Davies, A. G., and L. P. Keszthelyi. (2007). The thermal signature of volcanic eruptions on Io and Earth, manuscript in prep.

Davies, A. G., and P. Kyle. (2006). Spacecraft and in-situ observations of the Mt. Erebus, Antarctica, lava lake: a terrestrial analogue for Pele on Io. *Lunar and Planetary Science Conference XXXVII, Abstract 2284.*

Davies, A. G., and L. Wilson. (1987). Photoclinometric determination of surface topography and albedo variations on Io. *Lunar and Planetary Science Conference XVIII, Abstract,* **18**, 221–2.

Davies, A. G., and L. Wilson. (1988). Silicate-sulphur interactions on Io – implications for Pele type plumes. *Lunar and Planetary Science Conference XIX, Abstract,* **19**, 247–8.

Davies, A. G., A. S. McEwen, R. M. C. Lopes-Gautier, *et al.* (1997). Temperature and area constraints of the South Volund volcano on Io from the NIMS and SSI instruments during the *Galileo* G1 orbit. *Geophysical Research Letters,* **24**, 2447.

Davies, A. G., R. Lopes-Gautier, W. D. Smythe, *et al.* (2000a). Silicate cooling model fits to *Galileo* NIMS data of volcanism on Io. *Icarus,* **148**, 211–25.

Davies, A. G., L. P. Keszthelyi, R. M. C. Lopes-Gautier, *et al.* (2000b). Eruption evolution of major volcanoes on Io: *Galileo* takes a close look. *Lunar and Planetary Science Conference XXXI, Abstract 1754.*

Davies, A. G., L. P. Keszthelyi, D. A. Williams, *et al.* (2001). Thermal signature, eruption style, and eruption evolution at Pele and Pillan on Io. *Journal of Geophysical Research,* **106**, 33079–104.

Davies, A. G., J. Radebaugh, L. W. Kamp, *et al.* (2002). The lava lake at Pele: an analysis of high-resolution, multi-wavelength *Galileo* data. *Lunar and Planetary Science Conference XXXIII, Abstract 1162.*

Davies, A. G., D. L. Matson, G. J. Veeder, *et al.* (2005). Post-solidification cooling and the age of Io's lava flows. *Icarus,* **176**, 123–37.

Davies, A. G., S. Chien, V. Baker, *et al.* (2006a). Monitoring active volcanism with the Autonomous Sciencecraft Experiment on *EO-1. Remote Sensing of Environment,* **101**, 427–46.

Davies, A. G., S. Chien, R. Wright, *et al.* (2006b). Sensor web enables rapid response to volcanic activity. *Eos,* **87**, 1, 5.

Davies, A. G., L. Wilson, D. L. Matson, *et al.* (2006c). The heartbeat of the volcano: the discovery of episodic activity at Prometheus on Io. *Icarus,* **184**, 460–77.

Davies, A. G., L. P. Keszthelyi, and L. Wilson. (2006d). Estimation of maximum effusion rate for the Pillan 1997 eruption on Io: implications for massive basaltic flow emplacement on Earth and Mars. *Lunar and Planetary Science Conference XXXVII, Abstract 1155.*

Davies, A. G., S. Chien, T. Doggett, *et al.* (2006e). Improving mission survivability and science return with onboard autonomy. Paper presented at the International Planetary Probes Workshop 4, June 27–30, 2006, Pasadena, CA.

Davies, M. E. (1982). Cartography and nomenclature for the Galilean satellites. Proceedings of the *Satellites of Jupiter Conference,* January, 1982, pp. 911–33.

Davies, M. E., T. R. Colvin, J. Oberst, *et al.* (1998). The control networks of the Galilean satellites and implications for global shape. *Icarus,* **135**, 372–6.

de Pater, I., F. Marchis, B. A. Macintosh, *et al.* (2004). Keck AO observations of Io in and out of eclipse. *Icarus,* **169**, 250–63.

de Sitter, W. (1931). Jupiter's Galilean satellites (George Darwin Lecture). *Monthly Notices of the Royal Astronomical Society,* **91**, 706–38.

Desch, M. D. (1980). Io control of Jovian radio emission. *Nature,* **287**, 815–17.

Deschamps, P., J. E. Arlot, W. Thuillot, *et al.* (1992). Observations of the volcanoes of Io, Loki and Pele, made in 1991 at the ESO during an occultation by Europa. *Icarus,* **100**, 235–44.

Dessler, A. J., and T. W. Hill. (1979). Jovian longitudinal control of Io-related radio emissions. *Astrophysical Journal,* **227**, 664–75.

Dollfus, A. (1975). Optical polarimetry of the Galilean satellites of Jupiter. *Icarus,* **25**, 416–31.

Douté, S., B. Schmitt, R. Lopes-Gautier, *et al.* (2001). Mapping SO_2 frost on Io by the modeling of NIMS hyperspectral images. *Icarus,* **149**, 107–32.

Douté, S., R. Lopes, L. W. Kamp, *et al.* (2002). Dynamics and evolution of SO_2 gas condensation around Prometheus-like volcanic plumes on Io as seen by the Near Infrared Mapping Spectrometer. *Icarus,* **158**, 460–82.

Douté, S., R. Lopes, L. W. Kamp, *et al.* (2004). Geology and activity around volcanoes on Io from the analysis of NIMS spectral images. *Icarus,* **169**, 175–96.

Dragoni, M. (1989). A dynamical model of lava flows cooling by radiation. *Bulletin of Volcanology,* **51**, 88–95.

Dunbar, N. W., K. V. Cashman, and R. Dupré. (1994). Crystallization processes of anorthoclase phenocrysts in the Mount Erebus magmatic system: evidence from crystal composition, crystal size distributions, and volatile contents of melt inclusions. In *Volcanological and Environmental Studies of Mount Erebus, Antarctica, Antarctic Research Series,* vol. 66, ed. P. Kyle. Washington, DC: AGU, pp. 129–46.

Dzurisin, D., L. A. Anderson, G. P. Eaton, *et al.* (1980). Geophysical observations of Kilauea volcano, Hawaii, 2: constraints on the magma supply during November 1975–September 1977. *Journal of Volcanology and Geothermal Research,* **7**, 241–69.

Eliason, E. M., C. J. Hansen, A. S. McEwen, *et al.* (2003). Operation of *MRO*'s High Resolution Imaging Science Experiment (HiRise): maximizing science participation. Sixth International Conference on Mars, July 20–25, 2003, Pasadena, CA, Abstract 3122.

Ennis, M. E., and A. G. Davies. (2005). Thermal emission variability of Zamama, Culann and Tupan on Io using *Galileo* Near-Infrared Mapping Spectrometer (NIMS) data. *Lunar and Planetary Science Conference XXXVI, Abstract 1474.*

Fanale, F. P., R. H. Brown, D. P. Cruikshank, *et al.* (1979). Significance of absorption features in Io's IR reflectance spectrum. *Nature,* **280**, 761–3.

Fink, J. H., S. O. Park, and R. Greeley. (1983). Cooling and deformation of sulfur flows, *Icarus,* **56**, 38–50.

Fink, U., N. H. Dekkers, and H. P. Larson. (1973). Infrared spectra of the Galilean satellites of Jupiter. *Astrophysical Journal,* **179**, L155.

Fink, U., H. P. Larson, L. A. Lebofsky, *et al.* (1978). The 2–4 micron spectrum of Io. *Bulletin of the American Astronomical Society,* **10**, 580.

Fischer, D. (2001). *Mission Jupiter: The Spectacular Journey of the* Galileo *spacecraft.* New York: Springer-Verlag.

Fischer, H. J., and T. Spohn. (1990). Thermal-orbital histories of viscoelastic models of Io (J1). *Icarus,* **83**, 39–65.

Flynn, L. P., and P. J. Mouginis-Mark. (1992). Cooling rate of an active Hawaiian lava flow from nighttime spectroradiometer measurements. *Geophysical Research Letters,* **19**, 1783–6.

Flynn, L. P., P. J. Mouginis-Mark, J. C. Gradie, *et al.* (1993). Radiative temperature measurements at Kupaianaha lava lake, Kilauea volcano, Hawai'i. *Journal of Geophysical Research,* **98**, 6461–76.

Flynn, L. P., P. Mouginis-Mark, and K. A. Horton. (1994). Distribution of thermal areas on an active lava flow field. *Bulletin of Volcanology*, **56**, 284–96.

Flynn, L. P., A. J. L. Harris, D. A. Rothery, *et al.* (2000). High-spatial-resolution thermal remote sensing of active volcanic features using *Landsat* and hyperspectral data. In *Remote Sensing of Active Volcanism, Geophysical Monograph 116*, ed. P. Mouginis-Mark *et al.* Washington, DC: American Geophysical Union, pp. 161–77.

Francis, P., and C. Oppenheimer. (2004). *Volcanoes – A Planetary Perspective*, 2nd edn. Oxford, UK: Oxford University Press.

Francis, P. W. (1993). *Volcanoes – A Planetary Perspective*. Oxford, UK: Oxford University Press.

Francis, P. W., and D. A. Rothery. (1987). Using *Landsat* Thematic Mapper to detect and monitor volcanic activity: an example from Lascar volcano, north Chile. *Geology*, **15**, 614–17.

Frank, L. A., and W. R. Paterson. (2001). Passage through Io's ionospheric plasmas by the *Galileo* spacecraft. *Journal of Geophysical Research*, **106**, 26209–24.

Fujii, N., and S. Uyeda. (1974). Thermal instabilities during flow of magma in volcanic conduits. *Journal of Geophysical Research*, **79**, 3367–9.

Gaskell, R. W., S. P. Synnott, A. S. McEwen, *et al.* (1988). Large-scale topography of Io – implications for internal structure and heat transfer. *Geophysical Research Letters*, **15**, 581–4.

Gawarecki, S. J., R. J. P. Lyon, and W. Nordberg. (1965). Infrared spectral returns and imagery of the Earth from space and their applications to geological problems. *Science and Technology Series*, vol. 4, American Astronautical Society, pp. 13–33.

Geissler, P. E. (2003). Volcanic activity on Io during the *Galileo* era. *Annual Review of Earth and Planetary Sciences*, **31**, 175–211.

Geissler, P. E., and M. McMillan. (2006). *Galileo* observations of volcanic plumes on Io. *Lunar and Planetary Science Conference XXXVII, Abstract 1913*.

Geissler, P. E., A. S. McEwen, L. Keszthelyi, *et al.* (1999). Global color variations on Io. *Icarus*, **140**, 265–82.

Geissler, P. E., A. McEwen, C. Phillips, *et al.* (2004a). Surface changes on Io during the *Galileo* mission. *Icarus*, **169**, 29–64.

Geissler, P. E., A. McEwen, C. Porco, *et al.* (2004b). *Cassini* observations of Io's visible aurorae. *Icarus*, **172**, 127–40.

Giggenbach, W. F., P. R. Kyle, and G. L. Lyon. (1973). Present volcanic activity on Mount Erebus, Ross Island, Antarctica. *Geology*, **1**, 135–6.

Glaze, L. S., P. W. Francis, and D. A. Rothery. (1989). Measuring thermal budgets of active volcanoes by satellite remote sensing. *Nature*, **338**, 144–6.

Goguen, J. D., and W. M. Sinton. (1985). Characterization of Io's volcanic activity by infrared polarimetry. *Science*, **230**, 65–9.

Goguen, J. D., D. L. Matson, W. M. Sinton, *et al.* (1988). Io hot spots – infrared photometry of satellite occultations. *Icarus*, **76**, 465–84.

Goguen, J. D., A. Lubenow, and A. Storrs. (1998). HST NICMOS images of Io in Jupiter's shadow. *Bulletin of the American Astronomical Society*, **30**, 1120.

Goldreich, P., and S. Soter. (1966). Q in the Solar System. *Icarus*, **5**, 375–89.

Goldstein, S. J., Jr., and K. C. Jacobs. (1986). The contraction of Io's orbit. *Astronomical Journal*, **92**, 199–202.

Goldstein, S. J., Jr., and K. C. Jacobs. (1995). A recalculation of the secular acceleration of Io. *Astronomical Journal*, **110**, 3054.

Gounelle, M., and S. S. Russell. (2005). Spatial heterogeneity of short-lived isotopes in the solar accretion disk and early Solar System chronology. In *ASP Conference Series*, vol. 341, ed. A. N. Krot, *et al.*, pp. 588–601.

Gounelle, M., F. H. Shu, H. Shang, *et al.* (2006). The irradiation of beryllium radioisotopes and other short-lived radionuclides. *Astronomical Journal*, **640**, 1163–70.

Graham, F., and B. Hapke. (1986). Observational evidence for red polar caps on Io. *NASA TM 88383: Reports of Planetary Geology and Geophysics Program 1985*, p. 73.

Grattan, J. (2005). Pollution and paradigms: lessons from Icelandic volcanism for continental flood basalt studies. *Lithos*, **79**, 343–53.

Greeley, R., and J. Iverson. (1987). *Wind as a Geological Process on Earth, Mars, Venus and Titan*, 2nd edn. Cambridge, UK: Cambridge University Press.

Greeley, R., E. Theilig, and P. Christensen. (1984). The Mauna Loa sulfur flow as an analog to secondary sulfur flows on Io. *Icarus*, **60**, 189–99.

Greeley, R., S. W. Lee, D. A. Crown, *et al.* (1990). Observations of industrial sulfur flows – implications for Io. *Icarus*, **84**, 374–402.

Greenberg, R. (1987). Galilean satellites – evolutionary paths in deep resonance. *Icarus*, **70**, 334–47.

Greenberg, R. (1989). Time-varying orbits and tidal heating of the Galilean satellites. In *NASA Special Publication Series*, NASA-SP-494, ed. J. S. M. Belton *et al.* Washington, DC: NASA, pp. 100–15.

Gregg, T. K. P., and R. M. Lopes. (2004). Lava lakes on Io: new perspectives from modeling. *Lunar and Planetary Science Conference XXXV, Abstract 1558*.

Hamilton, W. B. (2003). An alternative Earth. *GSA Today*, **13**, 4–12.

Hanel, R., B. Conrath, M. Flasar, *et al.* (1979). Infrared observations of the jovian system from *Voyager 1*. *Science*, **204**, 972–6.

Hansen, O. L. (1973). Ten-micron eclipse observations of Io, Europa, and Ganymede. *Icarus*, **18**, 237.

Hardee, H. C., and D. W. Larson. (1977). Viscous dissipation effects in magma conduits, *Journal of Volcanology and Geothermal Research*, **2**, 299–308.

Harland, D. M. (2000). *Jupiter Odyssey: The Story of NASA's Galileo Mission*. Chichester, UK: Springer-Verlag UK.

Harris, A. J. L., and M. Neri. (2002). Volumetric observations during paroxysmal eruptions at Mount Etna: pressurized drainage of a shallow chamber or pulsed supply? *Journal of Volcanology and Geothermal Research*, **116**, 79–95.

Harris, A. J. L., S. Blake, D. A. Rothery, *et al.* (1997a). A chronology of the 1991 to 1993 Etna eruption using AVHRR data: implications for real time thermal volcano monitoring. *Journal of Geophysical Research*, **102**, 7985–8003.

Harris, A. J. L., A. L. Butterworth, R. W. Carlton, *et al.* (1997b). Low cost volcano surveillance from space: case studies from Etna, Krafla, Cerro Negro, Fogo, Lascar and Erebus. *Bulletin of Volcanology*, **59**, 59–64.

Harris, A. J. L., L. P. Flynn, L. Keszthelyi, *et al.* (1998). Calculation of lava effusion rates from *Landsat* TM data. *Bulletin of Volcanology*, **60**, 52–71.

Harris, A. J. L., L. P. Flynn, D. A. Rothery, *et al.* (1999a). Mass flux measurements at active lava lakes: implications for magma recycling. *Journal of Geophysical Research (Solid Earth)*, **104**, 7117–36.

Harris, A. J. L., R. Wright, and L. P. Flynn. (1999b). Remote sensing of Mount Erebus volcano, Antarctica, using polar orbiters: progress and prospects. *International Journal of Remote Sensing*, **20**, 3051–71.

Harris, A. J. L., L. P. Flynn, K. Dean, *et al.* (2000a). Real-time satellite monitoring of volcanic hot spots. In *Remote Sensing of Active Volcanism, AGU Geophysical Monograph 116*, ed. P. Mouginis-Mark, *et al.*, pp. 139–60.

Harris, A. J. L., J. B. Murray, S. E. Aries, *et al.* (2000b). Effusion rate trends at Etna and Krafla and their implications for eruptive mechanisms. *Journal of Volcanology and Geothermal Research*, **102**, 237–70.

Harris, D. L. (1961). Photometry and colorimetry of planets and satellites. In *Planets and Satellites*, ed. G. P. Kuiper and B. M. Middlehurst. Chicago: University of Chicago Press, p. 305.

Head, J. W. (1990). Surfaces of terrestrial planets. In *The New Solar System*, 3rd edn., ed. J. K. Beatty and A. Chaikin. Cambridge, UK: Cambridge University Press, pp. 77–90.

Head, J. W. (1999). Surfaces and interiors of the terrestrial planets. In *The New Solar System*, 4th edn., ed. J. K. Beatty *et al*. Cambridge, UK: Cambridge University Press, pp. 157–73.

Head, J. W., and L. Wilson. (1981). Lunar sinuous rille formation by thermal erosion: conditions, rates and durations. *Lunar and Planetary Science Conference XII, Abstract*, **12**, 427–9.

Head, J. W., and L. Wilson. (1986). Volcanic processes and landforms on Venus – theory, predictions, and observations. *Journal of Geophysical Research*, **91**, 9407–46.

Heliker, C., and T. N. Mattox. (2003). The first two decades of the Pu'u 'O'o-Kupaianaha eruption: chronology and selected bibliography. In *The Pu'u 'O'o-Kupaianaha Eruption of Kilauea Volcano, Hawai'i: The First 20 Years*, ed. C. Heliker *et al*. U.S. Geological Survey Professional Paper 1676, pp. 1–28.

Heliker, C., M. T. Mangan, T. N. Mattox, *et al*. (1998). The character of long-term eruptions; inferences from episodes 50–53 of the Pu'u 'O'o-Kupaianaha eruption of Kilauea volcano. *Bulletin of Volcanology*, **59**, 381–93.

Hevey, P. J., and I. S. Sanders. (2006). A model for planetesimal meltdown by ^{26}Al and its implications for meteorite parent bodies. *Meteoritics and Planetary Science*, **41**, 95–106.

Hildreth, W., and J. Fierstein. (2000). Katmai volcanic cluster and the great eruption of 1912. *Geological Society of America Bulletin*, **112**, 1594–620.

Hill, R. E. T., S. J. Barnes, M. J. Gole, *et al*. (1995). The volcanology of komatiites as deduced from field relationships in the Norseman-Wiluna Greenstone-Belt, Western Australia. *Lithos*, **34**, 159–88.

Hill, R. E. T., S. J. Barnes, S. E. Dowling, *et al*. (2002). Emplacement of komatiite flow fields: an inflationary model based on field evidence and modern mafic analogues. *Geochimica et Cosmochimica Acta*, **66**, A328.

Hom, E. F. Y., F. Marchis, T. K. Lee, *et al*. (2007). AIDA: an adaptive image deconvolution algorithm application to multi-frame and three-dimensional data. *Journal of the Optical Society of America*, in press.

Hon, K., J. Kauahikaua, R. Denlinger, *et al*. (1994). Emplacement and inflation of pahoehoe sheet flows: observations and measurements of active lava flows on Kilauea volcano, Hawai'i. *Geological Society of America Bulletin*, **106**, 351–70.

Hon, K., C. Gansecki, and J. Kauahikaua. (2003). The transition from a'a to pahoehoe crust on flows emplaced during the Pu'u 'O'o-Kupaianaha eruption. In *The Pu'u 'O'o-Kupaianaha Eruption of Kilauea Volcano, Hawai'i: The First 20 Years*, ed. C. Heliker *et al*. U.S. Geological Survey Professional Paper 1676, pp. 89–104.

Hooper, P. R. (2000). Flood basalt provinces. In *Encyclopedia of Volcanoes*, ed. H. Sigurdsson. San Diego: Academic Press, pp. 345–59.

Hord, C. W., W. E. McClintock, A. I. F. Stewart, *et al*. (1992). *Galileo* Ultraviolet Spectrometer experiment. *Space Science Reviews*, **60**, 503–30.

Howell, R. R. (1997). Thermal emission from lava flows on Io. *Icarus*, **127**, 394–407.

Howell, R. R. (2006). Corrigendum to "Thermal emission from lava flows on Io, Icarus, 127, 394–407," *Icarus*, **182**, 299.

Howell, R. R., and R. M. C. Lopes. (2007). The nature of volcanic activity at Loki: insights from *Galileo* NIMS and PPR data. *Icarus*, **86**, 448–61.

Howell, R. R., and M. T. McGinn. (1985). Infrared speckle observations of Io – an eruption in the Loki region. *Science*, **230**, 63–5.

Howell, R. R., and W. M. Sinton. (1989). Io and Europa: the observational evidence for variability. In *Time-Variable Phenomena in the Jovian System*, NASA SP-494, ed. J. S. M. Belton, *et al.* Washington, DC: NASA, pp. 47–62.

Howell, R. R., J. R. Spencer, J. D. Goguen, *et al.* (2001). Ground-based observations of volcanism on Io in 1999 and early 2000. *Journal of Geophysical Research*, **106**, 33,129–40.

Hulme, G. (1974). The interpretation of lava flow morphology. *Geophysical Journal of the Royal Astronomical Society*, **39**, 361–83.

Huppert, H. E., and R. S. J. Sparks. (1985). Komatiites 1: eruption and flow. *Journal of Petrology*, **26**, 694–725.

Huppert, H. E., R. S. J. Sparks, J. S. Turner, *et al.* (1984). Emplacement and cooling of komatiite lavas. *Nature*, **309**, 19–22.

Hussmann, H., and T. Spohn. (2004). Thermal-orbital evolution of Io and Europa. *Icarus*, **171**, 391–410.

HVO. (2005). USGS – Hawaiian Volcano Observatory online eruption summary, available at http://hvo.wr.usgs.gov/Kilauea/summary/Current_table.html.

ICT. (1929). *International Critical Tables of Numerical Data, Physics, Chemistry and Technology*, vol. V. Published for the National Research Council by McGraw-Hill (1926–1933). New York.

Jacobsen, S. B. (2005). The Hf-W isotopic system and the origin of the Earth and Moon. *Annual Review of Earth and Planetary Science*, **33**, 531–70.

Jacobsen, S. B., and Q. Yin. (2003). Hf-W, accretion of the Earth, core formation and the origin of the Moon. *Lunar and Planetary Science Conference XXXIV, Abstract 1913*.

Jaeger, W. L., and A. G. Davies. (2006). Models for the crustal structure of Io: implications for magma dynamics. *Lunar and Planetary Science Conference XXXVII, Abstract 2274*.

Jaeger, W. L., E. P. Turtle, L. P. Keszthelyi, *et al.* (2003). Orogenic tectonism on Io. *Journal of Geophysical Research (Planets)*, **108**, 5093–109.

Jarvis, R. A. (1995). On the cross-sectional geometry of thermal erosion channels formed by turbulent lava flows. *Journal of Geophysical Research (Solid Earth)*, **100**, 10,127–40.

Jessup, K. L., J. Spencer, G. E. Ballester, *et al.* (2002). Spatially resolved UV spectra of Io's Prometheus plume and anti-jovian hemisphere. *Bulletin of the American Astronomical Society*, **34**, 913.

Jessup, K. L., J. R. Spencer, G. E. Ballester, *et al.* (2004). The atmospheric signature of Io's Prometheus plume and anti-jovian hemisphere: evidence for a sublimation atmosphere. *Icarus*, **169**, 197–215.

Johnson, T. V., and C. B. Pilcher. (1977). Review of satellite spectro-photometry and composition. In *Planetary Satellites*, ed. J. Burns. Tucson: University of Arizona Press, pp. 232–68.

Johnson, T. V., and L. Soderblom. (1982). Volcanic eruptions on Io: implications for surface evolution and mass loss. In *Satellites of Jupiter*, ed. D. Morrison. Tucson: University of Arizona Press, pp. 634–46.

Johnson, T. V., A. F. Cook, II, C. Sagan, *et al.* (1979). Volcanic resurfacing rates and implications for volatiles on Io. *Nature*, **280**, 746–50.

Johnson, T. V., D. Morrison, D. L. Matson, *et al.* (1984). Volcanic hotspots on Io – stability and longitudinal distribution. *Science*, **226**, 134–7.

Johnson, T. V., G. J. Veeder, D. L. Matson, *et al.* (1988). Io – evidence for silicate volcanism in 1986. *Science*, **242**, 1280–3.

Johnson, T. V., D. L. Matson, D. L. Blaney, *et al.* (1995). Stealth plumes on Io. *Geophysical Research Letters*, **22**, 3293–6.

Johnson, T. V., K. B. Clark, R. Greeley, *et al.* (2006). Europa exploration: challenges and solutions. *Lunar and Planetary Science Conference XXXVII, Abstract 1549.*

Judge, D. L., and R. W. Carlson. (1974). *Pioneer 10* observations of the ultraviolet glow in the vicinity of Jupiter. *Science*, **183**, 317–18.

Kargel, J. S., P. Delmelle, and D. B. Nash. (1999). Volcanogenic sulfur on Earth and Io: composition and spectroscopy. *Icarus*, **142**, 249–80.

Kargel, J., R. Carlson, A. Davies, *et al.* (2003a). Extreme volcanism on Io: latest insights at the end of the *Galileo* era. *Eos*, **84**, 313–18.

Kargel, J. S., B. Fegley, Jr., and L. Schaefer. (2003b). Ceramic volcanism on refractory worlds: the cases of Io and chondrite CAIs. *Lunar and Planetary Science Conference XXXIV, Abstract 1964.*

Kauahikaua, J. P., K. V. Cashman, T. N. Mattox, *et al.* (1998). Observations of basaltic lava streams in tubes from Kilauea volcano, island of Hawai'i. *Journal of Geophysical Research*, **103**, 27303–23.

Keszthelyi, L. (1995). A preliminary thermal budget for lava tubes on the Earth and planets. *Journal of Geophysical Research*, **100**, 20411–20.

Keszthelyi, L., and R. Denlinger. (1996). The initial cooling of pahoehoe flow lobes. *Bulletin of Volcanology*, **58**, 5–18.

Keszthelyi, L., and A. McEwen. (1997a). Magmatic differentiation of Io. *Icarus*, **130**, 437–48.

Keszthelyi, L., and A. McEwen. (1997b). Thermal models for basaltic volcanism on Io. *Geophysical Research Letters*, **24**, 2463.

Keszthelyi, L., A. S. McEwen, and G. J. Taylor. (1999). Note: revisiting the hypothesis of a global magma ocean in Io. *Icarus*, **141**, 415–19.

Keszthelyi, L., A. S. McEwen, and T. Thordarson. (2000). Terrestrial analogs and thermal models for martian flood lavas. *Journal of Geophysical Research*, **105**, 15027–49.

Keszthelyi, L., A. S. McEwen, C. B. Phillips, *et al.* (2001a). Imaging of volcanic activity on Jupiter's moon Io by *Galileo* during the *Galileo* Europa Mission and the *Galileo* Millennium Mission. *Journal of Geophysical Research*, **106**, 33025–52.

Keszthelyi, L., A. J. L. Harris, L. Flynn, *et al.* (2001b). Interpreting low spatial resolution thermal data from active volcanoes on Io and the Earth. *Lunar and Planetary Science Conference XXXII, Abstract 1523.*

Keszthelyi, L., W. L. Jaeger, E. P. Turtle, *et al.* (2004a). A post-*Galileo* view of Io's interior. *Icarus*, **169**, 271–86.

Keszthelyi, L., M. Milazzo, A. G. Davies, *et al.* (2006a). A simple thermal model for lava fountains: application to Io. *Lunar and Planetary Science Conference XXXVII, Abstract 2216.*

Keszthelyi, L., S. Self, and T. Thordarson. (2006b). Flood lavas on Earth, Io and Mars. *Journal of the Geological Society of London*, **163**, 253–64.

Keszthelyi, L. P., and S. Self. (1998). Some physical requirements for the emplacement of long basaltic lava flows. *Journal of Geophysical Research*, **103**, 27447–64.

Keszthelyi, L. P., T. Thordarson, A. S. McEwen, *et al.* (2004b). Icelandic analogs to martian flood lavas. *Geochemistry Geophysics Geosystems*, **5**.

Keszthelyi, L. P., M. P. Milazzo, W. L. Jaeger, *et al.* (2005a). Reconciling lava temperatures and interior models for Io. *Lunar and Planetary Science Conference XXXVI, Abstract 1902*.

Keszthelyi, L. P., W. Jaeger, M. Milazzo, *et al.* (2005b). Improved estimates for Io eruption temperatures: implications for the interior. Geological Society of America Annual Meeting, Salt Lake City, October 13–16, 2005, Abstracts with Programs, **37**, no. 7, 92.

Kieffer, S. W. (1982). Dynamics and thermodynamics of volcanic eruptions: implications for the plumes on Io. In *Satellites of Jupiter*, ed. D. Morrison. Tucson: University of Arizona Press, pp. 647–723.

Kieffer, S. W., R. Lopes-Gautier, A. McEwen, *et al.* (2000). Prometheus: Io's wandering plume. *Science*, **288**, 1204–8.

Kilburn, C. R. J. (2000). Lava flows and flow fields. In *Encyclopedia of Volcanoes*, ed. H. Sigurdsson. San Diego: Academic Press, pp. 291–305.

Kirk, R. L., L. A. Soderblom, R. H. Brown, *et al.* (1995). Triton's plumes: discovery, characteristics, and models. In *Neptune and Triton*, ed. D. P. Cruikshank. Tucson: University of Arizona Press, pp. 949–89.

Kivelson, M. G., K. K. Khurana, C. T. Russell, *et al.* (2001). Magnetized or unmagnetized: ambiguity persists following *Galileo's* encounters with Io in 1999 and 2000. *Journal of Geophysical Research*, **106**, 26121–36.

Klassen, K. P., M. J. S. Belton, H. H. Breneman, *et al.* (1997). Inflight performance characteristics, calibration, and utilization of the *Galileo* solid-state imaging camera. *Optical Engineering*, **36**, 3001–27.

Klassen, K. P., H. H. Breneman, A. A. Simon-Miller, *et al.* (2003). Operations and calibration of the solid-state imaging system during the *Galileo* extended mission at Jupiter. *Optical Engineering*, **42**, 494–509.

Kliore, A., D. L. Cain, G. Fjeldbo, *et al.* (1974). Preliminary results on the atmospheres of Io and Jupiter from the *Pioneer 10* S-Band Occultation Experiment. *Science*, **183**, 323–4.

Kliore, A. J., G. Fjeldbo, B. L. Seidel, *et al.* (1975). The atmosphere of Io from *Pioneer 10* radio occultation measurements. *Icarus*, **24**, 407–10.

Knox, K. T., and B. J. Thompson. (1974). Recovery of images from atmospherically degraded short-exposure photographs. *Astrophysical Journal*, **193**, L45–L48.

Knudson, K., and D. L. Katz. (1979). *Fluid Dynamics and Heat Transfer*. Huntingdon, NY: Robert E. Krieger.

Kumar, S. (1979). The stability of an SO_2 atmosphere on Io. *Nature*, **280**, 758–60.

Kupo, I., Y. Mekler, and A. Eviatar. (1976). Detection of ionized sulfur in the jovian magnetosphere. *Astronomical Journal*. **205**, L51–L53.

Küppers, M., and N. M. Schneider. (2000). Discovery of chlorine in the Io torus. *Geophysical Research Letters*, **27**, 513.

Kuskov, O. L., and V. A. Kronrod. (2001a). Core sizes and internal structure of Earth's and Jupiter's satellites. *Icarus*, **151**, 204–27.

Kuskov, O. L., and V. A. Kronrod. (2001b). L- and LL-chondritic models of the chemical composition of Io. *Astronomicheskii Vestnik*, **35**, 198.

Kyle, P. R., K. Meeker, and D. Finnegan. (1990). Emission rates of sulfur dioxide, trace gases and metals from Mount Erebus, Antarctica. *Geophysical Research Letters*, **17**, 2125–8.

Kyle, P. R., L. M. Sybeldon, W. C. McIntosh, *et al.* (1994). Sulfur dioxide emission rates from Mount Erebus, Antarctica. In *Volcanological and Environmental Studies of*

Mount Erebus, Antarctica, Antarctic Research Series, vol. 66, ed. P. Kyle. Washington, DC: AGU, pp. 69–82.

Lainey, V., and G. Tobie. (2005). New constraints on Io's and Jupiter's tidal dissipation. *Icarus*, **179**, 485–9.

Lawrence, T. W., D. M. Goodman, E. M. Johansson, *et al.* (1992). Speckle imaging of satellites at the Air Force Maui optical station. *Applied Optics*, **31**, 6307–21.

Le Guern, F., J. Carbonnelle, and H. Tazieff. (1979). Erta'Ale lava lake: heat and gas transfer to the atmosphere. *Journal of Volcanology and Geothermal Research*, **31**, 17–31.

Lellouch, E., G. Paubert, J. I. Moses, *et al.* (2003). Volcanically emitted sodium chloride as a source for Io's neutral clouds and plasma torus. *Nature*, **421**, 45–7.

Leone, G., and L. Wilson. (2001). Density structure of Io and the migration of magma through its lithosphere. *Journal of Geophysical Research*, **106**, 32983–96.

Lesher, C. M., N. T. Arndt, and D. I. Groves. (1984). Genesis of komatiite-associated nickel sulphide deposits at Kambalda, Western Australia: a distal volcanic model. In *Sulphide Deposits in Mafic and Ultramafic Rocks*, ed. D. L. Buchanan and M. J. Jones. London: Institute of Mineralogy and Metallogy, pp. 70–80.

Lewis, J. S. (1982). Io – geochemistry of sulfur. *Icarus*, **50**, 103–14.

Lopes-Gautier, R., A. G. Davies, R. Carlson, *et al.* (1997). Hot spots on Io: initial results from *Galileo*'s Near Infrared Mapping Spectrometer. *Geophysical Research Letters*, **24**, 2439.

Lopes-Gautier, R., A. S. McEwen, W. B. Smythe, *et al.* (1999). Active volcanism on Io: global distribution and variations in activity. *Icarus*, **140**, 243–64.

Lopes-Gautier, R., S. Douté, W. D. Smythe, *et al.* (2000). A close-up look at Io from *Galileo*'s Near-Infrared Mapping Spectrometer. *Science*, **288**, 1201–4.

Lopes, R. M. C., L. W. Kamp, S. Douté, *et al.* (2001). Io in the near infrared: Near-Infrared Mapping Spectrometer (NIMS) results from the *Galileo* flybys in 1999 and 2000. *Journal of Geophysical Research*, **106**, 33053–78.

Lopes, R. M. C., L. W. Kamp, W. D. Smythe, *et al.* (2004). Lava lakes on Io: observations of Io's volcanic activity from *Galileo* NIMS during the 2001 fly-bys. *Icarus*, **169**, 140–74.

Lunine, J. I., and D. J. Stevenson. (1985). Physics and chemistry of sulfur lakes on Io. *Icarus*, **64**, 345–67.

Lydersen, A. L. (1979). *Fluid Flow and Heat Transfer*. Hoboken, NJ: Wiley.

Macintosh, B., D. Gavel, S. Gibbard, *et al.* (1997). Volcanoes on Io: high-resolution infrared images using speckle interferometry with the Keck telescope. *Bulletin of the American Astronomical Society*, **29**, 745.

Macintosh, B. A., D. Gavel, S. G. Gibbard, *et al.* (2003). Speckle imaging of volcanic hotspots on Io with the Keck telescope. *Icarus*, **165**, 137–43.

MacKnight, W. J., and A. V. Tobolsky. (1965). Properties of polymeric sulphur. In *Elemental Sulphur: Chemistry and Physics*, ed. B. Meyer. New York: Interscience, pp. 174–212.

Mahoney, J. J., and M. F. Coffin (Eds.). (1997). *Large Igneous Provinces: Continental, Oceanic and Planetary Flood Volcanism. AGU Geophysical Monograph*, **100**. Washington, DC: AGU.

Malin, M. C. (1980). The length of Hawaiian lava flows. *Geology*, **8**, 306–8.

Marchis, F., R. Prangé, and T. Fusco. (2001). A survey of Io's volcanism by adaptive optics observations in the 3.8-micron thermal band (1996–1999). *Journal of Geophysical Research*, **106**, 33141–60.

Marchis, F., I. de Pater, A. G. Davies, *et al.* (2002). High-resolution Keck adaptive optics imaging of violent volcanic activity on Io. *Icarus*, **160**, 124–31.

Marchis, F., D. Le Mignant, F. H. Chaffee, *et al.* (2005). Keck AO survey of Io global volcanic activity between 2 and 5 microns. *Icarus*, **176**, 96–122.

Marsh, B. D. (1981). On the crystallinity, probability of occurrence, and rheology of lava and magma. *Contributions to Mineralogy and Petrology*, **78**, 85–98.

Masursky, H., G. G. Schaber, L. A. Soderblom, *et al.* (1979). Preliminary geological mapping of Io. *Nature*, **280**, 725–9.

Matson, D. L., T. V. Johnson, and F. P. Fanale. (1974). Sodium D-line emission from Io: sputtering and resonant scattering hypothesis. *Astrophysical Journal*, **192**, L43–L46.

Matson, D. L., G. A. Ransford, and T. V. Johnson. (1981). Heat flow from Io /JI. *Journal of Geophysical Research*, **86**, 1664–72.

Matson, D. L., D. L. Blaney, T. V. Johnson, *et al.* (1998). Io and the early Earth. *Lunar and Planetary Science Conference XXIX, Abstract 1650.*

Matson, D. L., T. V. Johnson, G. J. Veeder, *et al.* (2001). Upper bound on Io's heat flow. *Journal of Geophysical Research*, **106**, 33021–4.

Matson, D. L., J. C. Castillo, C. Sotin, *et al.* (2006a). Enceladus' interior and geysers – possibility for hydrothermal geochemistry and N_2 production. *Lunar and Planetary Science Conference XXXVII, Abstract 2219.*

Matson, D. L., A. G. Davies, G. J. Veeder, *et al.* (2006b). Io: Loki Patera as a magma sea. *Journal of Geophysical Research (Planets)*, 111, E09002, doi:10. 1029/ 2006JE002703.

Matson, M., and J. Dozier. (1981). Identification of sub-resolution high temperature sources using a thermal IR sensor. *Photogrammetric Engineering and Remote Sensing*, **47**, 1311–18.

Mattox, T. N., C. Heliker, J. Kauahikaua, *et al.* (1993). Development of the 1990 Kalapana flow field, Kilauea volcano, Hawaii. *Bulletin of Volcanology*, **55**, 407–13.

McAdams, W. H. (1954). *Heat Transmission*. New York: McGraw-Hill.

McBirney, A. R., and T. Murase. (1984). Rheological properties of magmas. *Annual Review of Earth and Planetary Science*, **12**, 337–57.

McCauley, J. F., L. A. Soderblom, and B. A. Smith. (1979). Erosional scarps on Io. *Nature*, **280**, 736–8.

McEwen, A. S. (1988). Global color and albedo variations on Io, *Icarus*, **73**, 385–426.

McEwen, A. S. (2003). High-resolution imaging and topography from JIMO: the HiRise model. In Forum on *Jupiter Icy Moons Orbiter*, LPI, June 12–14, 2003, Abstract 9007.

McEwen, A. S., and L. A. Soderblom. (1983). Two classes of volcanic plumes on Io. *Icarus*, **55**, 191–217.

McEwen, A. S., L. A. Soderblom, D. L. Matson, *et al.* (1985). Volcanic hot spots on Io – correlation with low-albedo calderas. *Journal of Geophysical Research*, **90**, 12345–77.

McEwen, A. S., L. A. Soderblom, T. V. Johnson, *et al.* (1988). The global distribution, abundance, and stability of SO_2 on Io. *Icarus*, **75**, 450–78.

McEwen, A. S., J. I. Lunine, and M. H. Carr. (1989). Dynamic geophysics of Io. In *NASA Special Publication Series*, NASA-SP-494, ed. J. S. M. Belton and J. Rahe. Washington, DC: NASA, pp. 11–46.

McEwen, A. S., N. R. Isbell, and J. C. Pearl. (1992). Io thermophysics: new models with *Voyager I* thermal IR spectra. *Lunar and Planetary Science Conference XXIII, Abstract*, 881–2.

McEwen, A. S., N. R. Isbell, K. E. Edwards, *et al.* (1996). Temperatures on Io: implications to geophysics, volcanology, and volatile transport. *Lunar and Planetary Science Conference XXVII, Abstract*, 843–4.

McEwen, A. S., D. P. Simonelli, D. R. Senske, *et al.* (1997). High-temperature hot spots on Io as seen by the *Galileo* Solid State Imaging (SSI) experiment. *Geophysical Research Letters*, **24**, 2443.

McEwen, A. S., L. Keszthelyi, P. Geissler, *et al.* (1998a). Active volcanism on Io as seen by *Galileo* SSI. *Icarus*, **135**, 181–219.

McEwen, A. S., L. Keszthelyi, J. R. Spencer, *et al.* (1998b). High-temperature silicate volcanism on Jupiter's moon Io. *Science*, **281**, 87–90.

McEwen, A. S., R. Lopes-Gautier, L. Keszthelyi, *et al.* (2000a). Extreme volcanism on Jupiter's moon Io. In *Environmental Effects on Volcanic Eruptions: From Deep Oceans to Deep Space*, ed. J. Zimbelman and T. K. Gregg. New York: Springer, pp. 179–204.

McEwen, A. S., M. J. S. Belton, H. H. Breneman, *et al.* (2000b). *Galileo* at Io: results from high-resolution imaging. *Science*, **288**, 1193–8.

McEwen, A. S., L. P. Keszthelyi, R. Lopes, *et al.* (2004). The lithosphere and surface of Io. In *Jupiter: The Planet, Satellites and Magnetosphere*, ed. F. Bagenal *et al.* Cambridge, UK: Cambridge University Press, pp. 307–28.

McKinnon, W. B. (2006). Formation time of the Galilean satellites from Callisto's state of partial differentiation. *Lunar and Planetary Science Conference XXXVII, Abstract 2444*.

McKinnon, W. B., P. M. Schenk, and A. J. Dombard. (2001). Chaos on Io: a model of formation of mountain blocks by crustal heating, melting, and tilting. *Geology*, **29**, 103–6.

McLeod, B. A., D. W. McCarthy, Jr., and J. Freeman. (1991). Global high-resolution imaging of hotspots on Io. *Astronomical Journal*, **102**, 1485–9.

Melchior, P. (1983). *The Tides of Planet Earth*, 2nd edn. Oxford, UK: Pergamon.

Meyer, B. (1976). Elemental sulfur. *Chemical Reviews*, **76**, 367–88.

Meyer, B. (1977). *Sulfur, Energy and Environment*. Amsterdam: Elsevier Scientific.

Milazzo, M. P., L. P. Keszthelyi, and A. S. McEwen. (2001). Observations and initial modeling of lava-SO_2 interactions at Prometheus, Io. *Journal of Geophysical Research*, **106**, 33121–8.

Milazzo, M. P., L. P. Keszthelyi, J. Radebaugh, *et al.* (2005). Volcanic activity at Tvashtar Catena, Io. *Icarus*, **179**, 235–51.

Monnereau, M., and F. Dubuffet. (2002). Is Io's mantle really molten? *Icarus*, **158**, 450–9.

Moore, H. J. (1987). Preliminary estimates of the rheological properties of 1984 Mauna Loa lava. In *Volcanism in Hawaii*, U.S. Geological Survey Professional Paper 1350, ed. R. W. Decker *et al.*, pp. 1569–88.

Moore, J. M., A. S. McEwen, E. F. Albin, *et al.* (1986). Topographic evidence for shield volcanism on Io. *Icarus*, **67**, 181–3.

Moore, J. M., A. S. McEwen, M. P. Milazzo, *et al.* (2001). Landform degradation and slope processes on Io: the *Galileo* view. *Journal of Geophysical Research*, **106**, 33223–40.

Moore, W. B. (2001). Note: the thermal state of Io. *Icarus*, **154**, 548–50.

Morabito, L. A., S. P. Synnott, P. N. Kupferman, *et al.* (1979). Discovery of currently active extraterrestrial volcanism. *Science*, **204**, 972.

Morrison, D. (Ed.). (1982). *Satellites of Jupiter*. Tucson: University of Arizona Press.

Morrison, D., and D. P. Cruikshank. (1973). Thermal properties of the Galilean satellites. *Icarus*, **18**, 224.

Morrison, D., D. P. Cruikshank, and R. E. Murphy. (1972). Temperatures of Titan and the Galilean satellites at 20 microns. *Bulletin of the American Astronomical Society*, **4**, 367.

Morrison, D., D. Pieri, T. V. Johnson, *et al.* (1979). Photometric evidence of long-term stability of albedo and colour markings on Io. *Nature*, **280**, 753–5.

Mouginis-Mark, P., and N. Domergue-Schmidt. (2000). Acquisition of satellite data for volcano studies. In *Remote Sensing of Volcanic Activity*, ed. P. Mouginis-Mark *et al.* Washington, DC: American Geophysical Union, pp. 9–24.

Mouginis-Mark, P., J. Crisp, and J. Fink (Eds.). (2000). *Remote Sensing of Active Volcanism*. AGU Monograph Series, 116. Washington, DC: American Geophysical Union.

Murase, T., and A. R. McBirney. (1970). Viscosity of lunar lavas. *Science*, **167**, 1491–3.

Murchie, S., R. W. Arvidson, K. Beisser, *et al.* (2003). CRISM: Compact Reconnaissance Imaging Spectrometer for Mars on the *Mars Reconnaissance Orbiter*. *Sixth International Conference on Mars*, 20–25 July 2003, Pasadena, CA, Abstract 3062.

Murray, J. B. (1975). New observations of surface markings on Jupiter's satellites. *Icarus*, **25**, 397–404.

Mysen, B. O. (1977). Solubility of volatiles in silicate melts under the pressure and temperature conditions of partial melting in the upper mantle. *Proceedings of the AGU Chapman Conference on Partial Melting in the Upper Mantle*, 1–12.

Nash, D., M. Carr, J. Gradie, *et al.* (1986). Io. In *Satellites*, ed. J. Burns and M. S. Matthews. Tucson: University of Arizona Press, pp. 629–88.

Nash, D. B., and F. P. Fanale. (1977). Io's surface composition based on reflectance spectra of sulfur/salt mixtures and proton-irradiation experiments. *Icarus*, **31**, 40–80.

Nash, D. B., and R. R. Howell. (1989). Hydrogen sulfide on Io – evidence from telescopic and laboratory infrared spectra. *Science*, **244**, 454–7.

Nelson, R. M., and W. Hapke. (1978). Spectral reflectivities of the Galilean satellites and Titan, 0.32 to 0.86 micrometers. *Icarus*, **36**, 304–29.

Nelson, R. M., and W. D. Smythe. (1986). Spectral reflectance of solid sulfur trioxide (0.25–5.2 micron) – implications for Jupiter's satellite Io. *Icarus*, **66**, 181–7.

Nelson, R. M., D. C. Pieri, S. M. Baloga, *et al.* (1983). The reflection spectrum of liquid sulfur – implications for Io. *Icarus*, **56**, 409–13.

Newhall, C. G., and D. Dzurisin. (1988). Historical unrest at large calderas of the world, *U.S. Geological Survey Bulletin 1855*.

O'Neil, W. J., N. E. Ausman, J. A. Gleason, *et al.* (1997). Project *Galileo* at Jupiter. Paper presented at the *47th International Astronautical Congress*, Beijing, China, October 7–11, 1996, published by International Astronautical Federation, Paris.

O'Reilly, T. C., and G. F. Davies. (1981). Magma transport of heat on Io – a mechanism allowing a thick lithosphere. *Geophysical Research Letters*, **8**, 313–16.

Ojakangas, G. W., and D. J. Stevenson. (1986). Episodic volcanism of tidally heated satellites with application to Io. *Icarus*, **66**, 341–58.

Oppenheimer, C. (1991). Lava flow cooling estimated from *Landsat* thematic mapper infrared data – the Lonquimay eruption (Chile, 1989). *Journal of Geophysical Research (Solid Earth)*, **96**, 21865–78.

Oppenheimer, C., and P. Francis. (1997). Remote sensing of heat, lava and fumarole emissions from Erta'Ale volcano, Ethiopia. *International Journal of Remote Sensing*, **18**, 1661–92.

332 *References*

Oppenheimer, C. M. M., and D. A. Rothery. (1991). Infrared monitoring of volcanoes by satellite. *Journal of the Geological Society*, **148**, 563–9.

Oppenheimer, C., and D. Stevenson. (1989). Liquid sulphur lakes at Poas volcano. *Nature*, **342**, 790–3.

Parfitt, E. A. (1991). The role of rift zone storage in controlling the site and timing of eruptions and intrusions of Kilauea volcano, Hawaii. *Journal of Geophysical Research*, **96**, 10101–12.

Parman, S. W., J. C. Dann, T. L. Grove, *et al.* (1997). Emplacement conditions of komatiite magmas from the 3.49 Ga Komati formation, Barberton Greenstone Belt, South Africa. *Earth and Planetary Science Letters*, **150**, 303–23.

Peale, S. J. (1986). Orbital resonances, unusual configurations and exotic rotation states among the planetary satellites. In *Satellites*, ed. J. A. Burns and M. S. Matthews. Tucson: University of Arizona Press, pp. 159–223.

Peale, S. J. (1989). Some unsolved problems in evolutionary dynamics in the Solar System. *Celestial Mechanics and Dynamical Astronomy*, **46**, 253–75.

Peale, S. J. (2003). Tidally induced volcanism. *Celestial Mechanics and Dynamical Astronomy*, **87**, 129–55.

Peale, S. J., P. Cassen, and R. T. Reynolds. (1979). Melting of Io by tidal dissipation. *Science*, **203**, 892–4.

Pearl, J., and W. M. Sinton. (1982). Hot spots of Io. In *Satellites of Jupiter*, ed. D. Morrison. Tucson: University of Arizona Press, pp. 724–55.

Pearl, J., R. Hanel, L. Horn, *et al.* (1979a). The jovian satellites as seen from *Voyager* IRIS. *Bulletin of the American Astronomical Society*, **11**, 585.

Pearl, J., R. Hanel, V. Kunde, *et al.* (1979b). Identification of gaseous SO_2 and new upper limits for other gases on Io. *Nature*, **280**, 755–8.

Pearl, J. C. (1985). Io: Amaterasu Patera is hot. *Bulletin of the American Astronomical Society*, **17**, 691.

Pearlman, J. S., P. S. Barry, C. C. Segal, *et al.* (2003). Hyperion, a space-based imaging spectrometer. *IEEE Transactions on Geoscience and Remote Sensing*, **41**, 1160–72.

Peck, D., T. L. Wright, and J. G. Moore. (1966). Crystallization of tholeiitic basalt in Alae lava lake, Hawaii. *Bulletin of Volcanology*, **29**, 629–56.

Peck, D. L. (1978). Cooling and vesiculation of Alae lava lake, Hawai'i. U.S. Geological Survey Professional Paper 935-B, p. 59.

Peck, D. L., M. S. Hamilton, and H. R. Shaw. (1977). Numerical-analysis of lava lake cooling models 2: application to Alae lava lake, Hawaii. *American Journal of Science*, **277**, 415–37.

Peck, D. L., T. L. Wright, and R. W. Decker. (1979). Lava lakes of Kilauea. *Scientific American*, **241**, 114–22.

Peleg, M., and C. B. Alcock. (1974). Mechanism of vaporization and morphological changes of single-crystals of alumina and magnesia at high-temperatures. *High Temperature Science*, **6**, 52–63.

Pieri, D. C., and S. M. Baloga. (1986). Eruption rate, area and length relationships for some Hawaiian lava flows. *Journal of Volcanology and Geothermal Research*, **30**, 29–45.

Pieri, D. C., R. M. Nelson, S. M. Baloga, *et al.* (1984). Sulfur flows of Ra Patera, Io. *Icarus*, **60**, 685–700.

Pieri, D. C., L. S. Glaze, and M. J. Abrams. (1990). Thermal radiance observations of an active lava flow during the June 1984 eruption of Mount Etna. *Geology*, **18**, 1018–22.

Pike, R. J., and G. D. Clow. (1981). Revised classification of terrestrial volcanoes and catalog of topographic dimensions, with new results on edifice volume. U.S. Geological Survey Open File Report 81–1038, p. 40.

Pilcher, C. B., S. T. Ridgeway, and T. B. McCord. (1972). Galilean satellites: identification of water frost. *Science*, **178**, 1087–9.

Pinkerton, H., and R. S. J. Sparks. (1976). The 1975 sub-terminal lavas, Mt. Etna: a case history of the formation of a compound lava field. *Journal of Volcanology and Geothermal Research*, **1**, 167–82.

Pinkerton, H., and L. Wilson. (1994). Factors controlling the lengths of channel-fed lava flows. *Bulletin of Volcanology*, **56**, 108–20.

Pollack, J. B., F. C. Witteborn, E. F. Erickson, *et al.* (1978). Near-infrared spectra of the Galilean satellites – observations and compositional implications. *Icarus*, **36**, 271–303.

Porco, C. C., R. A. West, A. McEwen, *et al.* (2003). *Cassini* imaging of Jupiter's atmosphere, satellites, and rings. *Science*, **299**, 1541–7.

Porco, C. C., P. Helfenstein, P. C. Thomas, *et al.* (2006). *Cassini* observes the active south pole of Enceladus. *Science*, **311**, 1393–401.

Press, W. H., B. P. Flannery, S. Teukolsky, *et al.* (1992). *Numerical Recipes: The Art of Scientific Computing*. Cambridge, UK: Cambridge University Press.

Radebaugh, J., and A. S. McEwen. (2005). Correlating hotspots on Io with surface features using *Galileo* eclipse images. *AAS/Division for Planetary Sciences Meeting Abstracts*, **37**, 58.14.

Radebaugh, J., L. P. Keszthelyi, A. S. McEwen, *et al.* (2001). Paterae on Io: a new type of volcanic caldera? *Journal of Geophysical Research*, **106**, 33005–20.

Radebaugh, J., A. S. McEwen, L. P. Keszthelyi, *et al.* (2002). Lava lakes in Io's paterae: surface expressions of subsurface processes. *AGU Fall Meeting Abstracts*, **12**, 12.

Radebaugh, J., A. S. McEwen, M. P. Milazzo, *et al.* (2004). Observations and temperatures of Io's Pele Patera from *Cassini* and *Galileo* spacecraft images. *Icarus*, **169**, 65–79.

Rampino, M. R., and R. B. Stothers. (1988). Flood basalt volcanism during the past 250 million years. *Science*, **241**, 663–8.

Ramsey, M., and J. Dehn. (2004). Spaceborne observations of the 2000 Bezymianny, Kamchatka eruption: the integration of high-resolution ASTER data into near real-time monitoring using AVHRR. *Journal of Volcanology and Geothermal Research*, **135**, 127–46.

Ramsey, M. S., and L. P. Flynn. (2004). Strategies, insights, and the recent advances in volcanic monitoring and mapping with data from NASA's Earth Observing System. *Journal of Volcanology and Geothermal Research*, **135**, 1–11.

Rathbun, J., and J. R. Spencer. (2006). Loki, Io: groundbased observations and a model for periodic overturn. *Lunar and Planetary Science Conference XXXVII, Abstract 2365*.

Rathbun, J. A., and J. R. Spencer. (2005). Loki, Io: A model for the change from periodic behavior. *AAS/Division for Planetary Sciences Meeting Abstracts*, **37**, 58.15.

Rathbun, J. A., J. R. Spencer, A. G. Davies, *et al.* (2002). Loki, Io: a periodic volcano. *Geophysical Research Letters*, **29**, no. 10, 84–8, doi:I0.1029/2002GL014747.

Rathbun, J. A., J. R. Spencer, L. K. Tamppari, *et al.* (2004). Mapping of Io's thermal radiation by the *Galileo* photopolarimeter-radiometer (PPR) instrument. *Icarus*, **169**, 127–39.

Rau, H., T. R. N. Kutty, and J. R. F. Guedes de Caravalho. (1973). Thermodynamics of sulfur vapor. *Journal of Chemical Thermodynamics*, **5**, 833–44.

Reynolds, R. T., P. Cassen, and S. J. Peale. (1980). Io – energy constraints and plume volcanism, *Icarus*, **44**, 234–9.

Richter, D. H., J. P. Eaton, K. J. Murata, *et al.* (1970). The 1959–1960 eruption of Kilauea Volcano, Hawai'i. *U.S. Geological Survey Professional Paper 537-E*, p. 73.

Ross, M. N., and G. Schubert. (1985). Tidally forced viscous heating in a partially molten Io. *Icarus*, **64**, 391–400.

Ross, M. N., G. Schubert, T. Spohn, *et al.* (1990). Internal structure of Io and the global distribution of its topography. *Icarus*, **85**, 309–25.

Rothery, D. A., and C. Oppenheimer. (1994). Monitoring Mount Erebus by satellite remote sensing. In *Volcanological and Environmental Studies of Mount Erebus, Antarctica, Antarctic Research Series*, vol. 66, ed. P. Kyle. Washington, DC: AGU, pp. 51–6.

Rothery, D. A., P. W. Francis, and C. A. Wood. (1988). Volcano monitoring using short wavelength infrared data from satellites. *Journal of Geophysical Research*, **93**, 7993–8008.

Rothery, D. A., T. L. Babbs, A. J. L. Harris, *et al.* (1996). Colored lava flows on the Earth: a warning to Io volcanologists. *Journal of Geophysical Research*, **101**, 26131–6.

Rowan, L. R., and R. W. Clayton. (1993). The three-dimensional structure of Kilauea volcano, Hawai'i, from travel time tomography. *Journal of Geophysical Research*, **98**, 4355–75.

Rowland, S. K., A. J. L. Harris, and H. Garbeil. (2004). Effects of martian conditions on numerically modeled, cooling-limited, channelized lava flows. *Journal of Geophysical Research (Planets)*, **109**, E10010, doi:10.1029/2004JE002288.

Rubero, P. A. (1964). The effect of hydrogen sulfide on the viscosity of sulfur. *Journal of Chemical and Engineering Data*, **9**, 481–606.

Rubin, A. M. (1993). Dikes vs. diapirs in viscoelastic rock. *Earth and Planetary Science Letters*, **19**, 641–59.

Russell, E. E., F. G. Brown, R. A. Chandos, *et al.* (1992). *Galileo* photopolarimeter/radiometer experiment. *Space Science Reviews*, **60**, 531–63.

Russell, S. S., M. Gounelle, and R. Hutchinson. (2001). Origin of short-lived radionuclides. *Philosophical Transactions of the Royal Society of London*, **359**, 1994–2004.

Sagan, C. (1979). Sulphur flows on Io. *Nature*, **280**, 750–3.

Sanloup, C., F. Guyot, P. Gillet, *et al.* (2000). Density measurements of liquid Fe-S alloys at high pressure. *Geophysical Research Letters*, **27**, 811–14.

Sanloup, C., F. Guyot, P. Gillet, *et al.* (2001). Physical properties of liquid Fe alloys at high pressure and their bearings on the nature of metallic planetary cores: implications for the Earth, Mars and the Galilean satellites. *Lunar and Planetary Science Conference XXXII, Abstract 1877*.

Sartoretti, P., M. A. McGrath, and F. Paresce. (1994). Disk-resolved imaging of Io with the Hubble Space Telescope. *Icarus*, **108**, 272–84.

Sartoretti, P., M. A. McGrath, A. S. McEwen, *et al.* (1995). Post-*Voyager* brightness variations on Io. *Journal of Geophysical Research*, **100**, 7523–30.

Schaber, G. G. (1980). The surface of Io – geologic units, morphology, and tectonics. *Icarus*, **43**, 302–33.

Schaber, G. G. (1982). The geology of Io. In *Satellites of Jupiter*, ed. D. Morrison. Tucson: University of Arizona Press, pp. 556–97.

Schaefer, L., and B. Fegley. (2004). A thermodynamic model of high temperature lava vaporization on Io. *Icarus*, **169**, 216–41.

Schenk, P., H. Hargitai, R. Wilson, *et al.* (2001). The mountains of Io: global and geological perspectives from *Voyager* and *Galileo*. *Journal of Geophysical Research*, **106**, 33201–22.

Schenk, P. M., and M. H. Bulmer. (1998). Origin of mountains on Io by thrust faulting and large-scale mass movements. *Science*, **279**, 1514.

Schenk, P. M., and D. A. Williams. (2004). A potential thermal erosion lava channel on Io. *Geophysical Research Letters*, **31**, 23702.

Schenk, P. M., and R. R. Wilson. (2003). Tectonic and regional topography of Io: a new high. *Lunar and Planetary Science Conference XXIV, Abstract 2097*.

Schenk, P. M., A. McEwen, A. G. Davies, *et al.* (1997). Geology and topography of Ra Patera, Io, in the *Voyager* era: prelude to eruption. *Geophysical Research Letters*, **24** 2467.

Schenk, P. M., R. R. Wilson, and A. G. Davies. (2004). Shield volcano topography and the rheology of lava flows on Io. *Icarus*, **169**, 98–110.

Schminke, H.-U. (2004). *Volcanism*. Berlin: Springer.

Schmitt, B., and S. Rodriguez. (2003). Possible identification of local deposits of Cl_2SO_3 on Io from NIMS/*Galileo* spectra. *Journal of Geophysical Research (Planets)*, **108**, (E9), 5104, doi:10.1029/2002JE001988.

Schubert, G., T. Spohn, and R. T. Reynolds. (1986). Thermal histories, compositions and internal structures of the moons of the Solar System. In *Satellites*, ed. J. Burns and M. S. Matthews. Tucson: University of Arizona Press, pp. 224–292.

Schubert, G., D. L. Turcotte, and P. Olsen. (2001). *Mantle Convection in the Earth and Planets*. Cambridge, UK: Cambridge University Press.

Schubert, G., J. D. Anderson, T. Spohn, *et al.* (2004). Interior composition, structure and dynamics of the Galilean satellites. In *Jupiter. The Planet, Satellites and Magnetosphere*, ed. F. Bagenal *et al.* Cambridge, UK: Cambridge University Press, pp. 281–306.

Segatz, M., T. Spohn, M. N. Ross, *et al.* (1988). Tidal dissipation, surface heat flow, and figure of viscoelastic models of Io. *Icarus*, **75**, 187–206.

Self, S., R. Gertisser, T. Thordarson, *et al.* (2004). Magma volume, volatile emissions, and stratospheric aerosols from the 1815 eruption of Tambora. *Geophysical Research Letters*, **31**, L20608, doi:10.1029/2004GL020925.

Settle, M. (1979). Thermal buffering by the latent heat of crystallization. *Proceedings of the Lunar and Planetary Science Conference X, Abstract*, **10**, 1107–9.

Shaw, H. R., and D. A. Swanson. (1970). Eruption and flow rates of flood basalts. In *Proceedings of the Second Conference Columbia River Basalt Symposium*, ed. E. H. Gilmour and D. A. Swanson. Cheney: East Washington State College Press, pp. 271–99.

Shornikov, S. I., I. Y. Archakov, and M. M. Shults. (1999). Mass-spectrometric study of evaporation and thermodynamic properties of silicon dioxide – II. determination of partial coefficients of silicon dioxide evaporation. *Zhurnal Obshchei Khimii*, **69**, 197–206.

Sinton, W. M. (1980a). Io's 5 micron variability. *Astrophysical Journal*, **235**, L49–L51.

Sinton, W. M. (1980b). Io – are vapor explosions responsible for the 5-micron outbursts. *Icarus*, **43**, 56–64.

Sinton, W. M. (1981). The thermal emission spectrum of Io and a determination of the heat flux from its hot spots. *Journal of Geophysical Research*, **86**, 3122–8.

Sinton, W. M., and C. Kaminski. (1988). Infrared observations of eclipses of Io, its thermophysical parameters, and the thermal radiation of the Loki volcano and environs. *Icarus*, **75**, 207–32.

Sinton, W. M., A. T. Tokunaga, E. E. Becklin, *et al.* (1980). Io – ground-based observations of hot spots. *Science*, **210**, 1015–17.

Sinton, W. M., D. Lindwall, F. Cheigh, *et al.* (1983). Io – the near-infrared monitoring program, 1979–1981. *Icarus*, **54**, 133–57.

Sinton, W. M., J. D. Goguen, T. Nagata, *et al*. (1988). Infrared polarization measurements of Io in 1986. *Astronomical Journal*, **96**, 1095–105.

Skinner, B. J. (1970). A sulfur lava flow on Mauna Loa. *Pacific Science*, **24**, 144–5.

Smith, B. A., E. M. Shoemaker, S. W. Kieffer, *et al*. (1979a). The role of SO_2 in volcanism on Io. *Nature*, **280**, 738–43.

Smith, B. A., L. A. Soderblom, R. Beebe, *et al*. (1979b). The Galilean satellites and Jupiter – *Voyager 2* imaging science results. *Science*, **206**, 927–50.

Smith, B. A., L. A. Soderblom, T. V. Johnson, *et al*. (1979c). The Jupiter system through the eyes of *Voyager 1*. *Science*, **204**, 951–7.

Smith, B. A., L. A. Soderblom, D. Banfield, *et al*. (1989). *Voyager 2* at Neptune: imaging science results. *Science*, **246**, 1422–49.

Smythe, W. D., and R. M. Nelson. (1985). Spectral reflectance of quenched sulfur glasses: implications for Io. *Bulletin of the American Astronomical Society*, **17**, 920.

Smythe, W. D., R. M. Nelson, and D. B. Nash. (1979). Spectral evidence for SO_2 frost or adsorbate on Io's surface. *Nature*, **280**, 766.

Smythe, W. D., R. Lopes-Gautier, A. Ocampo, *et al*. (1995). Galilean satellite observation plans for the Near-Infrared Mapping Spectrometer experiment on the *Galileo* spacecraft. *Journal of Geophysical Research*, **100**, 18957–72.

Smythe, W. D., R. Lopes-Gautier, S. Doute, *et al*. (2000). Evidence for massive sulfur dioxide deposit on Io. *Bulletin of the American Astronomical Society*, **32**, 1047.

Soderblom, L. A., K. J. Becker, T. L. Becker, *et al*. (1999). Deconvolution of *Galileo* NIMS day-side spectra of Io into thermal, SO_2, and non-SO_2 components. *Lunar and Planetary Science Conference XXX, Abstract 1901*.

Sohl, F., T. Spohn, D. Breuer, *et al*. (2002). Implications from *Galileo* observations on the interior structure and chemistry of the Galilean satellites. *Icarus*, **157**, 104–19.

Solomon, S. C. (1981). Thermal histories of the terrestrial planets. In *Basaltic Volcanism Study Project: Basaltic Volcanism on the Terrestrial Planets*. New York: Pergamon, pp. 1129–234.

Sotin, C., R. Jaumann, B. J. Buratti, *et al*. (2005). Release of volatiles from a possible cryovolcano from near-infrared imaging of Titan. *Nature*, **435**, 786–9.

Sparks, R. S., M. I. Bursik, S. N. Carey, *et al*. (1997). *Volcanic Plumes*. Chichester, UK: Wiley.

Sparks, R. S. J. (1978). The dynamics of bubble formation and growth in magmas: a review and analysis. *Journal of Volcanology and Geothermal Research*, **3**, 1–37.

Spencer, J. R., and N. M. Schneider. (1996). Io on the eve of the *Galileo* mission. *Annual Review of Earth and Planetary Science*, **24**, 125–90.

Spencer, J. R., M. A. Shure, M. E. Ressler, *et al*. (1990). Discovery of hotspots on Io using disk-resolved infrared imaging. *Nature*, **348**, 618–21.

Spencer, J. R., B. E. Clark, D. Toomey, *et al*. (1994). Io hot spots in 1991 – results from Europa occultation photometry and infrared imaging. *Icarus*, **107**, 195.

Spencer, J. R., W. M. Calvin, and M. J. Person. (1995a). CCD spectra of the Galilean satellites: molecular oxygen on Ganymede. *Journal of Geophysical Research*, **100**, 19049–56.

Spencer, J. R., A. S. McEwen, D. B. Nash, *et al*. (1995b). A major albedo change on Io in 1994–1995. *Bulletin of the American Astronomical Society*, **27**, 1160.

Spencer, J. R., A. S. McEwen, M. A. McGrath, *et al*. (1997a). Volcanic resurfacing of Io: post-repair HST imaging. *Icarus*, **127**, 221–37.

Spencer, J. R., P. Sartoretti, G. E. Ballester, *et al*. (1997b). Pele plume (Io): observations with the Hubble Space Telescope. *Geophysical Research Letters*, **24**, 2471.

Spencer, J. R., J. A. Stansberry, C. Dumas, *et al.* (1997c). History of high-temperature Io volcanism: February 1995 to May 1997. *Geophysical Research Letters*, **24**, 2451.

Spencer, J. R., K. L. Jessup, M. A. McGrath, *et al.* (2000a). Discovery of gaseous S_2 in Io's Pele plume. *Science*, **288**, 1208–10.

Spencer, J. R., J. A. Rathbun, L. D. Travis, *et al.* (2000b). Io's thermal emission from the Galileo Photopolarimeter-Radiometer. *Science*, **288**, 1198–201.

Spencer, J. R., F. Bagenal, A. G. Davies, *et al.* (2002). The future of Io exploration. In *The Future of Solar System Exploration (2003–2013) – First Decadal Study Contributions*, 201–16.

Spencer, J. R., J. C. Pearl, M. Segura, *et al.* (2006). *Cassini* encounters Enceladus: background and the discovery of a south polar hot spot. *Science*, **311**, 1401–5.

SSES. (2003). Solar System Exploration Survey-Space Studies Board, National Research Council, *New Frontiers in the Solar System: An Integrated Exploration Strategy*. Washington, DC: National Academy Press, 248 pp.

Stansberry, J. A., J. R. Spencer, R. R. Howell, *et al.* (1997). Violent silicate volcanism on Io in 1996. *Geophysical Research Letters*, **24**, 2455.

Stebbins, J. (1927). The light variations of the satellites of Jupiter and their applications to measures of the solar constant. *Lick Observatory Bulletin*, **13**, 1–11.

Stebbins, J., and T. S. Jacobson. (1928). Further photometric measures of Jupiter's satellites and Uranus, with tests for the solar constant. *Lick Observatory Bulletin*, **13**, 180–95.

Stevenson, D. J., A. W. Harris, and J. I. Lunine. (1986). Origins of satellites. In *Satellites*, ed. J. Burns and M. S. Matthews. Tucson: University of Arizona Press, pp. 39–88.

Stone, E. C., and A. L. Lane. (1979a). *Voyager 1* encounter with the jovian system. *Science*, **204**, 945–48.

Stone, E. C., and A. L. Lane. (1979b). *Voyager 2* encounter with the jovian system. *Science*, **206**, 925–27.

Strom, R. G., N. M. Schneider, R. J. Terrile, *et al.* (1981). Volcanic eruptions on Io. *Journal of Geophysical Research*, **86**, 8593–620.

Swanson, D. A., W. A. Duffield, D. B. Jackson, *et al.* (1979). Chronological narrative of the 1969–1971 Mauna Ulu eruption of Kilauea volcano, Hawai'i. U.S. Geological Survey Professional Paper, **1056**, p. 55.

Symonds, R. B., W. I. Rose, G. Bluth, *et al.* (1994). Volcanic gas studies: methods, results, and applications. In *Volatiles in Magmas: Reviews in Mineralogy*, vol. 30, ed. M. R. Carroll and J. R. Holloway. Mineralogical Society of America, pp. 1–60.

Tackley, P. J., G. Schubert, G. A. Glatzmaier, *et al.* (2001). Three-dimensional simulations of mantle convection in Io. *Icarus*, **149**, 79–93.

Takahashi, E. (1986). Melting of a dry komatiite KLB-1 up to 14 GPa: implications on the origin of peridotitic upper mantle. *Journal of Geophysical Research*, **91**, 9367–82.

Takahashi, E. (1990). Speculations on the archean mantle – missing link between komatiite and depleted garnet peridotite. *Journal of Geophysical Research – Solid Earth and Planets*, **95**, 15941–54.

Taylor, G., B. O'Leary, T. V. Flandern, *et al.* (1971). Occultation of Beta Scorpii C by Io on May 14, 1971. *Nature*, **234**, 405–6.

Tazieff, H. (1994). Permanent lava lakes: observed facts and induced mechanisms. *Journal of Volcanology and Geothermal Research*, **63**, 3–11.

Thompson, D. (2000). *Volcano Cowboys*. New York: Thomas Dunne Books.

Thorarinsson, S. (1969). The Lakagigar eruption of 1783. *Bulletin of Volcanology*, **33**, 910–27.

338   *References*

Thordarson, T., and S. Self. (1993). The Laki (Skaftár Fires) and Grimsvötn eruptions in 1783–1784. *Bulletin of Volcanology*, **55**, 233–62.

Thordarson, T., and S. Self. (1998). The Roza Member, Columbia River Basalt Group: a gigantic pahoehoe lava flow field formed by endogenous processes? *Journal of Geophysical Research*, **103**, 27411–45.

Thordarson, T., S. Self, N. Óskarsson, *et al.* (1996). Sulfur, chlorine, and fluorine degassing and atmospheric loading by the 1783–1784 AD Laki (Skaftár Fires) eruption in Iceland. *Bulletin of Volcanology*, **58**, 205–25.

Thornber, C. (2003). Magma-reservoir processes revealed by geochemistry of the Pu'u 'O'o-Kupaianaha eruption. In *The Pu'u 'O'o-Kupaianaha Eruption of Kilauea Volcano, Hawai'i: The First 20 Years*, ed. C. Heliker *et al.* U.S. Geological Survey Professional Paper 1676, pp. 121–36.

Tilling, R. I. (1987). Fluctuations in surface height of active lava lakes during the 1972–1974 Mauna Ulu eruption, Kilauea volcano, Hawai'i. *Journal of Geophysical Research*, **92**, 13721–30.

Tobolsky, A. V., and A. Eisenberg. (1959). Equilibrium polymerization of sulfur. *Journal of the American Chemical Society*, **81**, 780–2.

Touloukian, Y. S., and C. Y. Ho (Eds.). (1970). *Thermophysical Properties of Matter*. New York and Washington, DC: IFI/Plenum.

Touro, F. J., and T. K. Wiewiorowski. (1966a). Molten sulfur chemistry 2. Solubility of sulfur dioxide in molten sulfur. *Journal of Physical Chemistry*, **70**, 3531–5.

Touro, F. J., and T. K. Wiewiorowski. (1966b). Viscosity-chain length relationship in molten sulfur systems. *Journal of Physical Chemistry*, **70**, 239–41.

Trafton, L. (1975a). High-resolution spectra of Io's sodium emission. *Astrophysical Journal*, **202**, L107–L112.

Trafton, L. (1975b). Detection of a potassium cloud near Io. *Nature*, **258**, 690–2.

Trafton, L., T. Parkinson, and W. Macy, Jr. (1974). The spatial extent of sodium emission around Io. *Astrophysical Journal*, **190**, L85.

Tuller, W. N. (1954). *The Sulphur Data Book*. New York: McGraw-Hill.

Turcotte, D., and G. Schubert. (1986). *Geodynamics*. Cambridge, UK: Cambridge University Press.

Turtle, E. P., W. L. Jaeger, L. P. Keszthelyi, *et al.* (2001). Mountains on Io: high-resolution *Galileo* observations, initial interpretations, and formation models. *Journal of Geophysical Research*, **106**, 33175–200.

Turtle, E. P., L. P. Keszthelyi, A. S. McEwen, *et al.* (2004). The final *Galileo* SSI observations of Io: orbits G28–I33, *Icarus*, **169**, 3–28.

Ungar, S. G., J. S. Pearlman, J. A. Mendenhall, *et al.* (2003). Overview of the Earth Observing One (*EO-1*) Mission. *IEEE Transactions in Geoscience and Remote Sensing*, **41**, 1149–59.

Urey, H. C. (1955). The cosmic abundance of potassium, uranium and thorium and the heat balance of the Earth, the Moon, and Mars. *Proceedings of the National Academy of Sciences*, **41**, 127–44.

Usselmann, T. M. (1975a). Experimental approach to the state of the core. Part 1. The liquidus relations of the Fe-rich portion of the Fe-Ni-S system from 30 to 100 kb. *American Journal of Science*, **275**, 278–90.

Usselmann, T. M. (1975b). Experimental approach to the state of the core. Part 2. Composition and thermal regime. *American Journal of Science*, **275**, 291–303.

Veeder, G. J., D. L. Matson, T. V. Johnson, *et al.* (1994). Io's heat flow from infrared radiometry: 1983–1993. *Journal of Geophysical Research*, **99**, 17095–162.

Veeder, G. J., D. L. Matson, T. V. Johnson, *et al.* (2004). The polar contribution to the heat flow of Io. *Icarus*, **169**, 264–70.

Voegele, A. F., T. Loerting, C. S. Tautermann, *et al.* (2004). Sulfurous acid (H_2SO_3) on Io? *Icarus*, **169**, 242–9.

Wadge, G. (1981). The variation of magma discharge during basaltic eruption. *Journal of Volcanology and Geothermal Research*, **11**, 139–68.

Wamsteker, W., R. L. Kroes, and J. A. Fountain. (1974). On the surface composition of Io. *Icarus*, **23**, 417–24.

Watanabe, T. (1940). Eruption of molten sulphur from the Siretoko-Iosan volcano, Hokkaido, Japan. *Japanese Journal of Geology and Geography*, **17**, 289–310.

White, R. S., and D. P. McKenzie. (1995). Mantle plumes and flood basalts. *Journal of Geophysical Research*, **100**, 17543–86.

Wignall, P. B. (2001). Large igneous provinces and mass extinctions. *Earth Science Reviews*, **53**, 1–33.

Williams, D. A., A. G. Davies, L. P. Keszthelyi, *et al.* (2001a). The summer 1997 eruption at Pillan Patera on Io: implications for ultrabasic lava flow emplacement. *Journal of Geophysical Research (Planets)*, **106**, 33105–20.

Williams, D. A., R. Greeley, R. M. C. Lopes, *et al.* (2001b). Evaluation of sulfur flow emplacement on Io from *Galileo* data and numerical modeling. *Journal of Geophysical Research*, **106**, 33161–74.

Williams, D. A., R. C. Kerr, C. M. Lesher, *et al.* (2001c). Analytical/numerical modeling of komatiite lava emplacement and thermal erosion at Perseverance, Western Australia. *Journal of Volcanology and Geothermal Research*, **110**, 27–55.

Williams, D. A., J. Radebaugh, L. P. Keszthelyi, *et al.* (2002). Geologic mapping of the Chaac-Camaxtli region of Io from *Galileo* imaging data. *Journal of Geophysical Research (Planets)*, **107** (E9), 5068, doi:10.1029/2001JE001821.

Williams, D. A., P. M. Schenk, J. M. Moore, *et al.* (2004). Mapping of the Culann-Tohil region of Io from *Galileo* imaging data. *Icarus*, **169**, 80–97.

Williams, D. A., L. P. Keszthelyi, P. M. Schenk, *et al.* (2005). The Zamama-Thor region of Io: insights from a synthesis of mapping, topography, and *Galileo* spacecraft data. *Icarus*, **177**, 69–88.

Wilson, L., and J. W. Head. (1981). Ascent and eruption of basaltic magma on the Earth and Moon. *Journal of Geophysical Research*, **86**, 2971–3001.

Wilson, L., and J. W. Head. (1983). A comparison of volcanic eruption processes on Earth, Moon, Mars, Io and Venus. *Nature*, **302**, 663–9.

Wilson, L., and J. W. Head. (2001). Lava fountains from the 1999 Tvashtar Catena fissure eruption on Io: implications for dike emplacement mechanisms, eruption rates, and crustal structure. *Journal of Geophysical Research*, **106**, 32997–3004.

Wilson, L., and J. W. Head. (2003). Deep generation of magmatic gas on the Moon and implications for pyroclastic eruptions. *Geophysical Research Letters*, **30**(12), 1605, doi:10.1029/2002GL016082.

Wilson, L., and J. W. Head, III. (1994). Mars: review and analysis of volcanic eruption theory and relationships to observed landforms. *Reviews of Geophysics*, **32**, 221–63.

Wilson, L., H. Pinkerton, J. W. Head, *et al.* (1993). A classification scheme for the morphology of lava flow fields. *Lunar and Planetary Science Conference XXIV, Abstract*, 1527–8.

Witteborn, F. E., J. D. Bregman, and J. B. Pollack. (1979). Io – an intense brightening near 5 micrometers. *Science*, **203**, 643–6.

Wolfe, E. W., M. O. Garcia, D. B. Jackson, *et al.* (1987). The Pu'u 'O'o eruption of Kilauea volcano, episodes 1–20, January 3, 1983, to June 8, 1984. In *Volcanism in Hawai'i*, USGS Professional Paper 1350, vol. 1, ed. R. W. Decker *et al.*, pp. 471–508.

Wood, C. A. (1984). Calderas: a planetary perspective. *Journal of Geophysical Research*, **89**, 8391–406.

Wright, R., and L. P. Flynn. (2003). On the retrieval of lava flow surface temperatures from infrared satellite data. *Geology*, **31**, 893–6.

Wright, R., S. Blake, A. J. L. Harris, *et al.* (2001). A simple explanation for the space-based calculation of lava eruption rates. *Earth and Planetary Science Letters*, **192**, 223–33.

Wright, R., L. P. Flynn, and A. G. Davies. (2004a). The detailed surface thermal structure of active lava flows, domes, and lakes revealed by the *Earth Observing-1* Hyperion. Abstract, IAVCEI General Assembly 2004, Pucon, Chile.

Wright, R., L. P. Flynn, H. Garbeil, *et al.* (2004b). MODVOLC: near-real-time thermal monitoring of global volcanism. *Journal of Volcanology and Geothermal Research*, **135**, 29–49.

Wright, T. L., and R. T. Okamura. (1977). Cooling and crystallization of tholeiitic basalt, 1965 Makaopuhi lava lake, Hawaii. U.S. Geological Survey Professional Paper 1004.

Wright, T. L., W. T. Kinoshita, and D. L. Peck. (1968). March 1965 eruption of Kilauea volcano and the formation of Makaopuhi lava lake. *Journal of Geophysical Research*, **73**, 3181–205.

Wylie, P. J. (1988). Magma genesis, plate tectonics, and chemical differentiation of the Earth. *Reviews of Geophysics*, **26**, 370–404.

Yamaguchi, Y., A. Kahle, H. Tsu, *et al.* (1998). Overview of Advanced Spaceborne Thermal Emission and Reflection Radiometer (ASTER). *IEEE Transactions on Geoscience and Remote Sensing*, **36**, 1062–71.

Yoder, C. F. (1979). How tidal heating in Io drives the Galilean orbital resonance locks. *Nature*, **279**, 767–70.

Young, A. T. (1984). No sulfur flows on Io. *Icarus*, **58**, 197–226.

Zhang, J., D. B. Goldstein, P. L. Varghese, *et al.* (2003). Simulation of gas dynamics and radiation in volcanic plumes on Io. *Icarus*, **163**, 182–97.

Zinner, E., and C. Gopel. (2002). Aluminium-26 in H4 chondrites: implications for its production and its usefulness as a fine-scale chronometer for early Solar System events. *Meteoritics and Planetary Science*, **37**, 1001–13.

Zolotov, M. Y., and B. Fegley. (2000). Eruption conditions of Pele volcano on Io inferred from chemistry of its volcanic plume. *Geophysical Research Letters*, **27**, 2789–92.

Index